Advances in Emissions Measurement and Testing

SP-1678

GLOBAL MOBILITY *DATABASE*

All SAE papers, standards, and selected books are abstracted and indexed in the Global Mobility Database

Published by:
Society of Automotive Engineers, Inc.
400 Commonwealth Drive
Warrendale, PA 15096-0001
USA
Phone: (724) 776-4841
Fax: (724) 776-5760
March 2002

ISBN 0-7680-0946-4
SAE/SP-02/1678
Library of Congress Catalog Card Number: N98-42939

Printed in USA

PREFACE

This SAE Special Publication, <u>Advances in Emissions Measurement and Testing</u> (SP-1678) is a collection of papers from the "Emissions Measurement and Testing" sessions held at the SAE 2002 World Congress.

Special thanks to the organizers, chairpersons, and authors who made this publication possible.

TABLE OF CONTENTS

2001-01-3535 **Evaluation of Fluorocarbon Polymer Bag Material for Near Zero Exhaust Emission Measurement** 1
Edward I. Sun and Wayne N. McMahon
California Air Resources Board

2002-01-0046 **Advanced Emissions Test Site for Confident PZEV Measurements** 21
Mark Guenther, Michael T. Sherman and Mike Vaillancourt
Correlation Engineering, Ford Motor Company
Dan Carpenter
Product Analysis and Verification, Ford Motor Company
Rick Rooney and Scott Porter
Horiba Instruments, Inc.

2002-01-0047 **Evaluation of Improved Bag Mini-Diluter System for Low Level Emissions Measurements** 37
David J. Luzenski, Kyle T. Bedsole, Jim Hill,
Donald B. Nagy and Steven S. Decarteret
General Motors Corporation

2002-01-0048 **Studies on Enhanced CVS Technology to Achieve SULEV Certification** 55
H. Behrendt and O. Mörsch
DaimlerChrysler AG
C. T. Seiferth
BMW AG
G. E. Seifert
Porsche AG
J. W. Wiebrecht
Audi AG

2002-01-0049 **Improved Low-Emission Vehicle Simulator for Evaluation of Sampling and Analytical Systems** 73
Daniel Whelan, Gary Lewis and Scott Kiyabu
Horiba Instruments, Inc.
Travis Henney, Darius Harrison,
Jennifer Farrugia and Michael Gabbert
Ford Motor Co.

2002-01-0051 **Evaluation of New Bag Sampling Materials for Low Level Emissions Measurements** 89
Jim Hill, Jeffrey F. Loo and Stephen J. Swarin
General Motors Corporation

2002-01-0053 **Gas Supply and Treatment Options for use in Emissions Measurement and Test** 97
Larry Rabellino
SAES Pure Gas, Inc

2002-01-0054 **Diesel Particulate Measurement with Partial Flow Sampling Systems: A New Probe and Tunnel Design that Correlates with Full Flow Tunnels** 107
William M. Silvis, Gerald Marek and Norbert Kreft
AVL North America, Inc.
Wolfgang Schindler
AVL List GmbH

2002-01-0055 **Electrical Filter Stage for the ELPI** 123
Marko Marjamäki, Leonidas Ntziachristos,
Annele Virtanen, Jyrki Ristimäki and Jorma Keskinen
Tampere University of Technology, Institute of Physics
Mikko Moisio and Marko Palonen
Dekati Ltd.
Maija Lappi
VTT Energy, Engine Technology

2002-01-0056 **Effective Density of Diesel Exhaust Particles as a Function of Size** 129
Annele Virtanen, Jyrki Ristimäki, Marko Marjamäki,
Kati Vaaraslahti and Jorma Keskinen
Tampere Univeristy of Technology, Institute of Physics
Maija Lappi
VTT Energy

2002-01-0612 **Development of a Wet-Based NDIR and Its Application to On-Board Emission Measurement System** 135
Hiroshi Nakamura, Nobutaka Kihara,
Masayuki Adachi and Kozo Ishida
Horiba, Ltd.

2002-01-0613 **Development of On-Board System to Measure Running Condition and Actual NOx Emissions from Freight Vehicle** 143
Tomio Miyazaki, Yutaka Takada and Norimasa Iida
Keio University

2002-01-0614 **Inference of Torque and Power from Heavy-Duty Diesel Engines for On-Road Emissions Monitoring** 155
Gregory J. Thompson, Nigel N. Clark, Mridul Gautam,
Daniel K. Carder, and Donald W. Lyons
West Virginia University

2002-01-0615 **Investigation of 'Sweep' Mapping Approach on Engine Testbed** 165
M. C. Ward, C. J. Brace and N. D. Vaughan
University of Bath
R. Ceen
CP Engineering
T. Hale and G. Kennedy
Cosworth Technology

2002-01-0616 **A Transient Hydrostatic Dynamometer for Testing Single-Cylinder Prototypes of Multi-Cylinder Engines** 179
John L. Lahti and John J. Moskwa
University of Wisconsin – Madison

2002-01-1300 **An Investigation of SF6 Gas for Testing Instrumental Integrity of the Emerging SULEV/PZEV Measurement Technology** 189
Scott Porter, Richard T. Rooney and Karl Oestergaard
Horiba Instruments, Inc.
Zlatko Rauker
Ford Motor Company, Ford Research Laboratory
Satoshi Ohtsuki
Horiba, Ltd.

2001-01-3535

Evaluation of Fluorocarbon Polymer Bag Material for Near Zero Exhaust Emission Measurement

Edward I. Sun and Wayne N. McMahon
California Air Resources Board

ABSTRACT

When the California Air Resources Board (ARB) adopted automotive exhaust emission standards for Super Ultra-Low-Emission Vehicles (SULEV), new challenges were encountered for accurately measuring exhaust emissions. This is especially true for measuring NMOG emissions (NMHC and carbonyls) where the SULEV standard is 0.010 g/mi. One of the challenges in accurately measuring NMHC emissions is to find a clean sample bag material that has no or very low outgassing of hydrocarbons. Tedlar, the bag material commonly used for exhaust emission sampling, has been found to emit N,N- dimethylacetamide (DMAc), which interferes with hydrocarbon measurements and can contribute to significant error in SULEV hydrocarbon emission measurements.

Several fluorocarbon materials were tested for hydrocarbon (HC) outgassing and carbon dioxide (CO_2) permeation. The materials include Tedlar, Baked Tedlar, KynarFlex 2750, Baked KynarFlex 2800, Teflon FEP, TFM TFE, Tefzel, and Halar. The test methods included measuring HC outgassing of sample bags filled in dry zero air at 60 minutes and measuring carbon dioxide permeation of sample bags filled with 2% CO_2 at 60 minutes. Since moisture content of the source gas may affect outgassing, tests were also performed using humidified zero air. Based on the potential effect of outgassing on hydrocarbon measurements and CO_2 permeation on fuel economy measurements, material performance criteria are recommended. Tentative criteria are that hydrocarbon outgassing should be less than 15 $ppbC_1$ at 60 minutes using both dry and humidified zero air as source gases, and CO_2 permeation should be less than 0.2% at 60 minutes using a 2% CO_2 source gas. Of the material tested, only baked Tedlar met both criteria. Some of the other materials tested may meet criteria using alternative treatments of the material.

Since a precise and accurate hydrocarbon gas analyzer is needed for this testing, this paper also describes the calibration and determination of the limit of detection (LOD) of the Horiba 726LE Flame Ionization Detector (FID) hydrocarbon analyzer in the Horiba MEXA LE emission analyzer system. The LOD was calculated to be 3 $ppbC_1$. Such precision supports the reliability of the hydrocarbon outgassing data.

INTRODUCTION

In order to reduce air pollution from motor vehicles, the United States Environmental Protection Agency (USEPA) and the California Air Resources Board (ARB) regulations require vehicles to meet certain exhaust emission standards. These regulations require measurement of hydrocarbons, carbon monoxide, nitric oxides and carbon dioxide exhaust emissions using a Constant Volume Sampling (CVS) system. This system utilizes plastic sample bags to collect dilute vehicle exhaust.

The emission sample bags are commonly made of a plastic film called Tedlar, which is the trade name of polyvinylfluoride (PVF) film manufactured by DuPont. More specifically, the grades of the Tedlar film being used are TST20SG4 or TTR20SG4 [1]. Both film grades are made by the same extrusion process and biaxial-orientation or stretching, except that TST20SG4 passes a stricter visual defect inspection. A small amount of N,N- dimethylacetamide (DMAc) is added during the extrusion to aid processing [2]. DMAc, a liquid at room temperature with a boiling point of 166 degrees Celsius, outgasses from Teldar at room temperature and interferes with vehicle exhaust hydrocarbon measurements. This problem is especially critical when testing vehicles meeting California's Super Low Emission Vehicle (SULEV) emission standards.

In the late 1980's, ARB identified DMAc as a contaminant from Tedlar bags used to collect exhaust samples for hydrocarbon speciation by gas chromatography (GC bags) [3]. At that time, a bag flushing procedure was developed to minimize DMAc outgassing so that it did not interfere with speciation. In 1995, when a particular batch of Tedlar film had higher than normal DMAc outgassing levels, ARB began baking the Tedlar film to remove the DMAc before making the film into GC bags.

Earlier studies by others also identified this problem and Teflon FEP film was tested as CVS sampling bags [4,5]. Teflon FEP has little outgassing, but has relatively high carbon dioxide permeation compared to other materials. Permeation is significant enough to affect the fuel economy results. Consequently, some use both Tedlar and Teflon bags for repeated emission testing. Tedlar bag results are used to calculate the carbon dioxide level, while Teflon bag results are used to calculate hydrocarbon levels. Such practice is nonetheless cumbersome. Obviously, a single bag material that can be used for both measurements is desirable.

Because hydrocarbon outgassing and CO2 permeation can cause significant errors in hydrocarbon and fuel economy measurements, respectively, the following tentative bag performance criteria were used for this study:

1) With the bag filled with dry zero air, hydrocarbon outgassing shall be less than 15 ppbC$_1$ measured at 60 minutes.
2) With the bag filled with humidified zero air, hydrocarbon outgassing shall be less than 15 ppbC$_1$ measured at 60 minutes.
3) With the bag filled with 2% CO_2 gas, the reduction in CO2 concentration measured at 60 minutes shall be less than 0.2% of the concentration of the source gas.

In addition to dry zero air, humidified zero air was used as a test gas because DMAc and other organic compounds are water-soluble, and also to simulate the moisture content of dilute vehicle exhaust. Outgassing was measured at 60 minutes and 24 hours. In addition to total hydrocarbons, methane was measured at 24 hours because methane from the ambient air could permeate into the bag and could affect outgassing results. After the 24 hour tests, the effectiveness of flushing the bags with dry zero air was evaluated.

Total Hydrocarbon (THC) outgassing was measured using both a Horiba FIA-200 analyzer in a Horiba 9000 bench, and a Horiba FIA 726 LE analyzer in a Horiba MEXA 7200LE bench. The Limit of Detection was determined for each analyzer using the Horiba mass flow controller type gas divider (GDC 703). This gas divider generates and verifies the calibration curve for single range operation on the FIA 726 LE THC analyzer [6, 7]. The GDC 703 gas divider can provide cut points down to 0.2% for calibration and verification.

For this study hydrocarbon outgassing and CO2 permeation criteria were established as minimum goals to achieve. The one-hour outgassing criterion of 15 ppbC$_1$ represents a 3 to 5% error (0.3 to 0.5 mg/mi) in hydrocarbon measurement due to outgassing contamination. This error assumes a worst case scenario where the contamination of the ambient bags is zero and the contamination of the sample bags is 15 ppbC. The percent error is based on the 10mg/mi SULEV NMOG Standard. The one-hour soak time was selected to ensure that the bags meet the criterion with good margin considering that the maximum soak time (from the start of filling the bag to final reading) is 35 minutes for phase 2 of the standard FTP vehicle exhaust emission test procedure. The 15 ppbC$_1$ criterion does not include any hydrocarbon contamination contributed by the sampling and analytical systems, which can also be significant depending on their cleanliness.

The one-hour CO2 permeation criterion of 0.2% represents a fuel economy error of 0.1 miles per gallon. This assumes zero permeation for the ambient bag because the CO2 concentration inside and outside of the bag is the same, and assumes a worst case permeation of 0.2% for the sample bag. The 2% CO2 source gas was selected because it is the typical concentration of the exhaust emission sample bag. The one-hour soak period was selected for the same reason as for outgassing.

LITERATURE SEARCH

A literature search was conducted to identify the fluorocarbon polymers and their permeation properties. AtoFina, supplier for Kynar, provided a table with the permeation properties of all major fluorocarbon polymer types [8]. Table 1 lists the major fluorocarbon polymers and their respective manufacturer method and permeation properties. Based on this data from AtoFina and additional information from the auto manufactures and test equipment suppliers, candidate fluorocarbon polymer materials were selected for testing.

EXPERIMENTAL

EQUIPMENT

Zero Air Supply

ARB uses an AADCO Model 737-15 pure air generator to supply compressed air to the emission analyzers for zero, purge and burner air for the FID analyzers. This generator has two alternating catalytic converters, which burn the hydrocarbons in ambient air. A weekly check with gas chromatography (GC) is routinely conducted.

Humidifier (bubbler)

Mr. Jim Hill of General Motors has kindly supplied a custom made humidifier. It is a sealed stainless steel cylindrical container that can be filled with water and in which zero air can be bubbled through.

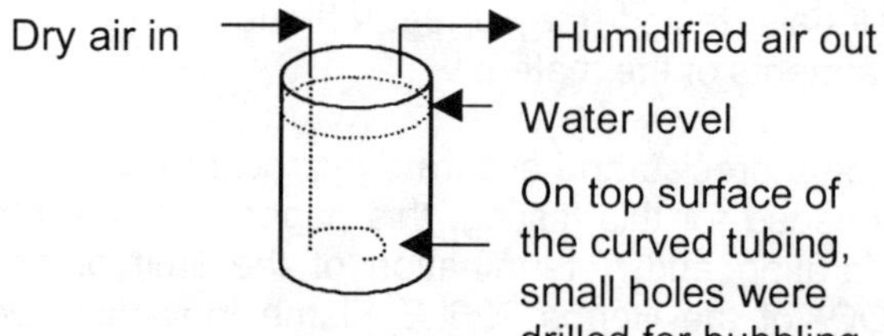

Figure 1: Sketch of Humidifier

See Figure 1 for its construction. The diameter of the cylinder is about 8.9 cm and the height is about 20.3 cm. At the top of the cylinder, there are three tapped holes for Swagelok adapters. The dry zero air was regulated at 5 psi and fed through one hole into a stainless tubing that extended into the vessel and is submerged in the water. The air bubbles up and then flows through an exit located at the second hole. The third hole is used for drain and fill. The stainless humidifier has been fired to remove any hydrocarbon contamination, according to Mr. Hill.

ASTM Type II water was used in the humidifier. The water was from an in-house system consisting of a double-mixed bed deionized water system and a Millipore Milli-QTM water system with four cartridges (carbon, 2-ion exchanges, and organic) and a 0.2 micron filter. After each fill, the humidifier was purged with dry zero air for a couple of hours to remove the hydrocarbons that entered the cylinder from ambient air during filling.

Hygrometer

A General Eastern Model E-2 hygrometer was used to measure the relative humidity of the humidified zero air.

CALIBRATION OF FID IN BAG BENCH

Analyzer Systems

A Horiba 9000 and a Horiba MEXA 7200 LE were used for this study. The Horiba 9000 was used for the early part of the study while the MEXA was being installed. The analyzer models used for hydrocarbon, methane, and carbon dioxide measurement are shown below.

	Horiba 9000	Horiba MEXA 7200LE
Hydrocarbon	FIA 222	FIA-726LE
Methane	NA	GFA-720LE
Carbon dioxide	NA	AIA-722

Gas Dividers

Two manual capillary gas dividers were used for the LOD study of the FIA-220. One has 10% cuts from 0% to 100% and the other one has 0%, 10%, 25%, 40%, 50%, 60%, 75%, 90%, and 100% cuts. They were put in series to generate 1%, 2.5%, 4%, and 5% cuts.

The Horiba GDC 703 Gas Divider was integrated in the MEXA bench. It has four mass flow controllers (MFC). MFC-1 controls zero gas flow, and MFC-2, 3 and 4 control span flow. The operating ranges of the MFCs are
0.2% to 2.4% for MFC-2, 2.6% to 12.4% for MFC-3, and 12.6% to 100% for MFC-4. The error specification is less than 1% reading scale (RS) for each point.

Analyzer Calibration and Verification

The FIA 220 FID hydrocarbon analyzer was calibrated for a 10-ppm range using a manual gas divider. Because a 1-ppm hydrocarbon standard is not readily available, the FIA 726LE FID hydrocarbon analyzer was calibrated for a 5-ppmC_1 range using the GDC-703 gas divider. Eleven cut points (at 10% intervals) were used to generate the first order non-weighted calibration curve. The methane response factor of this FID is 1.16. The methane analyzer was calibrated for a 5-ppmC_1 range and the CO2 analyzer was calibrated for a 4% range. The GDC-703 and calibration cylinder (1% primary standard) were used to verify calibration curve accuracy.

GDC 703 GAS DIVIDER CONFIRMATION

The gas divider was confirmed by verifying the NO_x calibration curve over the operational range of the gas divider, especially for the low cut points. The NO_x analyzer was calibrated for a 30-ppm range. A calibration line was generated with 11 points (10 percent cut). Since the NO_x analyzer is recognized as a linear analyzer, the curve was specified as first order without any weighted fit criteria. Then the calibration curve was verified using the gas divider for cut points at 100%, 80%, 60%, 40%, 20%, 16%, 12%, 8%, 4%, 1%, 08%, 0.6%, 0.4%, 0.2%, and 0%.

LOD DETERMINATION

According to Taylor, the LOD is defined as $3S_0$, where S_0 is the value of standard deviation as the concentration approaches zero [9].

ARB also has a method for LOD determination, which requires at least five repeats of a measurement close to LOD and not more than 6 times the LOD [10], which is confirmed after testing. In addition to these five measurements, a few other repeated measurements at higher concentration are also required to generate a least squares fit calibration curve. For a FID, this curve is first order, in other words, a straight line. The LOD is then calculated from the following equation:

$$LOD = (|b| + t * s) / m \quad \text{(Equation 1)}$$

where

s is standard deviation of the seven repeats at the lowest concentration.

t is from the t table for 99% confidence for a one-sided Student-t distribution (equivalent to 98% for two-sided) and t depends on degrees of freedom.

b is intercept of the least squares fit line. When the intercept is positive, this term may be neglected.

m is the slope of the least squares fit line.

BAG PREPARATION

Film and Bag Samples

Table II shows the materials selected for the study and associated bag information (film gauge, dimensions, film supplier etc.). Film suppliers provided the film samples for testing. Three variations of Tedlar film were selected for testing: new untreated Tedlar, new baked Tedlar as described in the film baking procedure, and aged baked Tedlar in which the bags were made and initially tested approximately four months before this study. ARB personnel made these film samples into bags. In addition, two grades of Kynar were selected: Kynar 2750 and 2800.

For this study, the standard large hub and Teflon tubing loops were not used. Instead, the bags were made using small hubs and without Teflon loops to minimize the voids and contamination from the internal bag components. However, these bags would experience increased stress around the hub when evacuating the bags. The detailed bag construction process is described in the following section.

The Kynar 2750, Teflon, and TFM TFE bags were purchased from Plastic Film Enterprises Inc. For these bags, the film material for one side of the bag underwent a crinkling process to reduce stress on the bag during evacuation by not creasing the material around the hub. One crinkled film sheet was sealed to a normal film sheet. For attaching lines to the bags, small Teflon fittings (1.25 inches OD) were installed in the center of the normal film sheet before sealing the bag. These fitting were used instead of the standard large internal Kynar hubs (5 inches OD) and Teflon tubing loops to minimize the dead space in the bag and minimize outgassing from the fitting materials.

Film Baking Procedures

Some bags were made from baked films. To maximize the exposed surface, the films were carefully hung in an air-circulating oven. Tedlar was baked at 120°C for 16 hours. Kynar Flex 2800 was baked at 100°C for 7 hours. As a safety precaution, the oven exhaust was vented outside the building.

Bag Construction Procedure

Except for the sample bags purchased from Plastic Film Enterprises, Inc. all other bags were constructed at ARB by sealing together two sheets of film using a thermal impulse heat sealing device made by Vertrod Corp. The heating can be adjusted through a knob scaled from 1 to 10. The typical setting for Tedlar is 5. Three edges of the bag are sealed first. Then a 9/16-inch hole is punched in the center of one side of the bag using a gasket punch and a Cajon connector fitting is installed. The Cajon fitting consists of two parts, a custom made 1 inch OD Teflon nut on the inside of the bag and a stainless male adapter (SS-4-TA-1-4ST) and rubber o-ring on the outside. Then the fourth edge of the bag is sealed. The bags constructed by Plastic Film Enterprise (PFE) utilized a 1.5 inch OD Kynar fitting (model 800). Swagelok SS-QC4-D-400 quick-connect fittings were connected to both the ARB and PFE constructed bags for filling and evacuating.

BAG TESTING PROCEDURES

Leak Check

Before testing, each bag was leak checked. The bag was connected to a vacuum pump, evacuated to a vacuum of greater than 29 inches of Hg, and then disconnected from the pump. If the bag held vacuum for 5 minutes, the bag was considered leak-free. Otherwise, the bag was not used for testing.

Dry Zero Air Tests

Each sample bag was filled and evacuated three times with dry zero air. A CAL (zero, span and zero-read) was performed on the FID. The bag was filled with dry zero air once more (the forth time), and then a THC reading was taken immediately after filling the bag and reported as the zero minutes read. For most bags, it generally took 5 minutes to fill the bag to 280 liters and less than 0.25 inches of water pressure using a line pressure of 25 psi. These bags were standard size CVS sample bags, which are normally filled to 100-140 liters for exhaust emission testing. CAL and read procedures were repeated at 15, 35, and 60 minutes. The number of measurements was reduced for the smaller sample bags.

Outgassing readings may be artificially low considering that the bags in this study were filled to twice the normal volume for exhaust emission testing, thereby diluting the outgassing concentrations. Outgassing level for typical fill volumes may be as much as double the values reported in this study.

Humidified Zero Air Tests

For the humidified zero air tests, baseline readings of the humidified zero air source were taken for relative humidity, temperature, methane, and THC. From the relative humidity and temperature, the absolute humidity was calculated. Usually, the temperature in the laboratory fluctuates less than 5 degrees F within 24 hours. Humidified zero air readings may differ from dry zero air, which is used to calibrate and zero the THC analyzer, due to effects of humidity on the system and the analyzer. Humidified zero air could potentially clean the FID and sampling system, and the added water vapor could affect sample flow by changing the viscosity of the source gas. The moisture may also lower the temperature of the heated FID.

Baseline readings were taken before testing each bag. Then each bag was filled and evacuated three times with dry zero air. Then the bag was filled with humidified zero air. Humidity and THC readings were taken immediately after filling the bags and reported as the zero minute reading. To achieve the desired humidity with the limitation of the humidifier apparatus, the line pressure for dry zero source gas was adjusted to only 5 psi, and thus required 30 minutes to fill the bags to 280 liters. CAL, THC, and bag reading (humidity and THC) procedures were repeated at 15 minutes, 35 minutes, 60 minutes, and 24 hours. For the 24-hour readings, methane was also read and THC was read continuously until the bag was completely evacuated.

After the 24 hour test, the effectiveness of flushing the bag with dry zero air was evaluated by filling and evacuating the bag three times with dry zero air. Then the bag was filled with humidified zero air and readings were taken at 0, 15, 35, and 60 minutes. At 60 minutes, THC was read continuously until the bag was evacuated.

Carbon Dioxide Permeation

A cylinder of 1.8% carbon dioxide (CO_2) in nitrogen was used to determine the carbon dioxide permeation rate of the sample bags. The sample bag was filled and evacuated three times with dry zero air and then filled with CO_2 gas. CAL and CO_2 read procedures were performed at 0 minutes, 15 minutes, 30 minutes, and 60 minutes.

RESULTS AND DISCUSSION

GDC 703 GAS DIVIDER CONFIRMATION

Calibration Results

Table 3 lists the NOx analyzer calibration curve results. All calibration points except for the 10% point have less than a +/-0.1% error. The 10% cut point has a –0.7% error.

Curve Confirmation Results

Table 4 lists the curve confirmation results. Cut points 16% and above have less than +/-0.2% error. Cut points below 16% have less than +/-1.0% error except for the 0.2% point which has a 4.9% error or a 3 $ppbC_1$ absolute error for a 0.054 $ppmC_1$ (54 $ppbC_1$) reading

LOD DETERMINATION RESULTS

LOD of FIA 220

Two manual gas dividers connected in series and a 0.97 $ppmC_1$ propane bottle were used to generate low concentration points. Five measurements were made for each concentration point. At zero concentration, the average of five readings is 0.3 $ppbC_1$ and the standard deviation is approximately 6 $ppbC_1$. At 0.0097 $ppmC_1$ (9.7 $ppbC_1$), the lowest concentration point generated, the average is 10.2 $ppbC_1$ and the standard deviation is 3.9 $ppbC_1$. The calculated LOD is 18 $ppbC_1$ for both the Taylor and ARB methods described in the experimental section.

LOD of FIA 726 LE

The LOD was determined using the internal gas divider and a 3.03 $ppmC_1$ propane bottle. Table 5 shows the instrument counts for each concentration measured. Instrument count is the internal instrument response to the gas concentration being measured. The LOD was determined from the linear regression of instrument counts versus concentration in $ppmC_1$. At 0.0061 $ppmC_1$ (6.1 $ppbC_1$), the lowest concentration generated, the average count is 10814 and the standard deviation is 4.3 counts. The calculated LOD is 3 $ppbC_1$.

DRY ZERO AIR SOAK TEST RESULTS

Table 6 lists the dry zero air test results. Usually readings below the LOD would be reported as "below LOD," but for information only all readings are shown. Most of the data were averages of two identical sample bags, unless a duplicate is either not available, as for Halar, or not necessary, as for untreated Tedlar. Table VI shows that untreated Tedlar had the highest outgassing at 0.38 $ppmC_1$ at 15 minutes. DMAc is believed to be the primary outgassing compound. Since untreated Tedlar is known for DMAc outgassing and may contaminate the new FID, untreated Tedlar was not tested to the extent of the other materials. Kynar Flex 2750 has the second highest outgassing at 0.14 $ppmC_1$ measured at 15 minutes. This relative high result indicates that a plastic film free of DMAc may still outgas significantly. For the materials tested at 35 minutes, TFM TFE has the highest result at 0.030 $ppmC_1$, which increased to 0.062 $ppmC_1$ at 60 minutes. However, as discussed in humidified zero air results, the 60–minute result is only 0.013 $ppmC_1$. Halar had the second highest 60-minute result at 0.053 $ppmC_1$. Halar was retested using the MEXA FID and the result was 0.032 $ppmC_1$.

TFM TFE should have little impurities since it is made by skiving a compression-molded billet. Possible sources of contamination may be from its manufacturing processes, including compression molding, skiving, and crinkling.

Figure III compares the triangle plots for all the materials with 60-minute results below 0.030 $ppmC_1$. The triangle for each material represents two measurements and the average. Only Teflon FEP, baked Tedlar, and aged baked Tedlar are less than 15 $ppbC_1$.

Expecting Teflon FEP bags to have very low results, they were tested a second time with dry zero air (after testing with humidified zero air). 60-minute results decreased from 0.018 to 0.008 $ppmC_1$. The FIA 220 was used for the first round of testing and FIA 726 was used for the second. It is possible that repeated testing with humidified zero air reduced the contamination level of the bag, which may or may not be attributed to outgassing from the material.

HUMIDIFIED ZERO AIR SOAK TEST RESULTS

60-Minute Results

Table 7 lists the humidified zero air results. The 60-minute results for Baked Tedlar, Aged Baked Tedlar, Teflon FEP, and TFM TFE are less than 15 ppbC$_1$. Apparently for the TFM TFE and Teflon, repeat testing with humidified zero air was effective in reducing the contamination level.

The Kynar Flex 2750 60-minute humidified zero air result is 0.023 ppbC$_1$. Although it is higher than the desired 0.015 ppbC$_1$ criterion, it is a significant improvement from the brand new bag reading of 0.14 ppmC$_1$, using dry zero air. Again, repeated purging during testing appears to clean the bag. The lower limit of cleaning was not determined nor in the scope of the study.

Figure 4 compares the triangle plots of the 60-minute results. For TFM TFE, the two data points are the same so no triangle plot shows. For Kynar Flex 2800, there is only one data point. Figure 5 shows the trends of interval readings for the 60-minute tests. Baked Tedlar and Teflon readings are very stable whereas TFM TFE and Tefzel readings show increasing trends and increase 7 to 8 ppbC$_1$ over 60 minutes.

24 Hour Results

Table 7 also shows the 24-hour humidified zero air test results. THC and methane concentrations were measured at 24 hours to determine the long term outgassing characteristics (whether levels stabilized or continue to increase after 60-minutes). Methane was measured to measure methane permeation into the bag from ambient air. Considering the goal of near zero emission measurements, methane permeation is not desirable for a bag material. Only Teflon FEP, TFM TFE, and Kynar Flex 2750 had measurable methane concentrations.

It is interesting to note that the humidity in Teflon FEP and TFM TFE bags almost remained constant, while the humidity in baked Tedlar, Kynar Flex 2750, baked Kynar Flex 2800, and Tefzel decreased more significantly after 24 hour soak. This findings suggests that moisture may have adsorbed on the surfaces of the latter materials, while Teflon FEP and TFM TFE, known for their non-stick properties, are less likely to adsorb moisture.

Figure 6 compares the 24-hour triangle plots. The results in the plots have been corrected for methane permeation. For the materials that had a positive methane reading, the total hydrocarbon readings were corrected for the FID's over response to methane and methane was subtracted from the total hydrocarbon reading. Except for the baked Tedlar bags, the contamination level of all the other bags increased substantially from 1 to 24 hours, especially for Tefzel and TFM TFE. Although a 24-hour outgassing criterion was not established, the baked Tedlar bags have the lowest 24-hour result at 10 ppbC$_1$. All other materials are above 30 ppbC$_1$.

Repeat 60 Minute Soak Results

After the 24-hour soak test, the 60-minute soak test was repeated to evaluate the effectiveness of purging the bag three times with dry zero air for reducing bag contamination levels. Table 8 and Figure 7 show the 60-minute results. The four materials that passed (below 15 ppbC$_1$) the first 60-minute test also passed the second 60-minute test. Kynar Flex 2750 and Tefzel did not pass the first or second 60-minutes tests. For all materials, purging with dry zero air was effective in reducing contamination levels to initial levels.

Continuous Read Results

Continuous read results after the 24-hour and repeat 60-minute tests, varied depending on the material. The maximum reading ranged from 8 to 15 minutes depending on bag size.

For baked Tedlar, THC levels decreased about 20 ppbC$_1$ over the first 2 minutes and stabilized somewhat after 3 minutes (still slightly decreasing). See Figure 8A. The Teflon results were similar, but readings were more stable after about 3 minutes. See Figure 8B. However for the 60-minute test, levels increased slightly (4 ppbC$_1$) over the last 2 minutes of reading.

For Tefzel, the 24-hour test stabilized at 2 minutes and then increased about 5 ppbC$_1$ over the last 2 minutes of reading. See Figure 8C. For the 60-minute test, the reading stabilized somewhat at about 3 minutes and then increased about 10 ppbC$_1$ over the last 5 minutes of reading.

For TFM TFE after the initial 2-minute stabilization period, the reading decreased very slightly (~1 ppbC$_1$) for the remainder of the reading. See Figure 8D.

Kynar was the only material that showed a continuous and steady increase during the reading. See Figure 8E.

The point to note is that for the continuous read tests, none of the materials showed a large increase in reading potential due to saturation of the system with the contaminant. In addition, none of the readings truly stabilized; however, the variations were very small for the 60-minute tests.

Baseline Humidified Zero Air Results

Table 7 also shows the baseline results for humidified zero air. Baseline readings were taken to determine any potential offset of humidified zero air due to sample system contamination or humidity effects on the FID analyzer. Baseline readings were taken immediately after performing a CAL on the FID and ranged from -0.004 to 0.004 $ppmC_1$ for a typical reading time of 200 seconds. Theoretically bag readings could be adjusted for the baseline levels, but they were not adjusted because the levels were so low and in most cases below the LOD.

Hydrocarbon Contamination in the FID

Another issue of concern in this study is system contamination. At the beginning of the study, the dry zero air generated over 11000 counts for the FIA 726 LE analyzer. At the beginning of the first round of humidified zero air testing, the counts for zero air decreased to 8000. By the end of testing, the counts decreased further to 6700. This reduction in counts suggests that humidified zero air cleaned the system. To minimize the effect of system cleaning on bag readings, the analyzer was zeroed before each measurement. This procedure was effective, considering that the repeat 60-minute results with a cleaner system were very similar to the initial 60-minute results. However, a stable baseline for dry zero air and humidified zero air is desirable and recommended before testing to eliminate any potential measurement error due to instability of the sampling and measurement system.

CARBON DIOIXDE PERMEATION

Table 9 and Figure 9 show the carbon dioxide permeation results. All materials passed the criterion (less than 0.2% reduction from the initial CO2 concentration of the bag) except for Teflon FEP, TFM TFE bags and Halar. Teflon is known to have relatively high CO2 permeation. Only one test was performed for Halar.

To determine the resolution of the CO2 analyzer, the 1.8% CO2 gas was measured six times. A CAL (zero and span adjust) was performed before each measurement. The average reading was 1.8111% with a standard deviation of 0.0002%, which is 0.01% of the average reading. Therefore, there is confidence in using the CO2 analyzer to test a criterion of 0.2% reduction (in bottle concentration).

LEAKING BAGS

During testing, a leak was found in one of the aged Tedlar bags (made and initially tested four months before this study), one of the Kynar 2800 bags and the Halar bag. The leak was determined by failure to hold vacuum. All three bags were made in-house with normal non-crinkled film, and pinhole leaks were detected around the connector area.

As mentioned in the bag construction section, a typical emission sample bag contains a large distribution hub and two loops of Teflon tubing. This assembly is designed to collect a distributed sample from the bag, assist in completely evacuating the bag during purging, and minimize stress on the bag material during evacuation, especially at the connector. Since none of the bags tested in this study contained the hub and tube assembly, they experienced higher than normal stress around the connector during evacuation.

CONCLUSIONS

1. Near zero hydrocarbons measurement has been demonstrated using the Horiba FIA-726LE analyzer and GCD 703 gas divider. Based on the LOD of 3 $ppbC_1$, there is confidence in the ppb level results.
2. Both the dry and humidified zero air results were repeatable for each of the two separate bags for each material tested. For humidified zero air, readings from 0 to 60 minutes were stable for the baked Tedlar and Teflon bags. The Kynar, Tefzel, and TFM TFE bags readings increased 4-8 $ppbC_1$.
3. Except for baked Tedlar, outgassing increased substantially from 1 hour to 24 hours, which indicates the outgassing is occurring for these materials or the bag fittings.
4. Based on the procedure used in this study, only the baked Tedlar bags passed the 15 $ppbC_1$ outgassing and the CO2 permeation criteria. This material also met the criteria after aging for four months. However, considering that one of the aged baked Tedlar bags developed a leak during testing, the longevity or durability of these bags needs to be evaluated further.
5. No bag passed the outgassing criterion without a purifying process, such as baking the material prior to making the bag or repeated purging of the bag. Purging the bags three times with dry zero air was effective in the reducing contamination level, but humidified zero air may be more effective.
6. The Teflon FEP and TFM TFE bags with one side crinkled failed the CO2 permeation criterion, which may be partially due to the crinkling process. Non-crinkled TFM TFE should be tested further.
7. The Tefzel, Kynar 2750, and baked Kynar 2800 bags did not meet the 15-$ppbC_1$ criterion and levels ranged from 18 to 21 $ppbC_1$. Further treatment of these materials may improve their outgassing performance to meet the criteria. All of these materials met the CO2 permeation criterion.

SUGGESTIONS FOR ADDTIONAL STUDY

1. Future studies should determine the LOD of the THC analyzer to ensure accuracy at ppb level measurements.
2. Before testing, stable baseline contamination levels of the sample system and analyzer should be measured for both dry and humidified zero air.
3. Kynar and TFM TFE bags using non-crinkle films should be tested for outgassing and CO2 permeation for the crinkling process may be contaminating and creating thin spots.
4. The outgassing of the various bag fittings should be investigated.

5. The longevity of the bags with small fittings should be tested, especially for materials that have been baked.
6. The humidity of the humidified zero air should be more tightly controlled.
7. A larger capacity humidifier is recommended for a faster fill rate.
8. Emission carry-over or hydrocarbon retention of the bags and sample system needs to be studied.
9. New bags should undergo a standard conditioning procedure (several flushes with humidified zero air) to stabilize outgassing before initial testing. For the early tests, most bags in this study had substantial contamination levels that generally decreased with subsequent testing.
10. Bag size and bag fill volume should be standardized to represent typical emission testing practices.

ACKNOWLEDGMENTS

The authors would like to thank Mr. Dave Peterson, Mr. Huy Khou, Mr. Saul Ortega, and Mr. George Gatt of Cell 1 at ARB for performing the testing for this study and their dedication to precision measurements. We also appreciate Mr. Michael O'Connor, our colleague at ARB, for his contribution to the study. Mr. Jim Hill of GM Powertrain kindly provided the humidifier used to generate humidified zero air as well as his experience in bag testing. The authors appreciate Mr. Kimikazu Yoda and Mr. Mike Lord of Toyota, Mr. Michael Sherman and Mr. Michael Vaillancourt of Ford Correlation Engineering, and Mr. Kazuhiro Murata and Mr. Robert Cassidy of Nissan for their valuable insights and discussions on bag material selection and emission testing. We also extend our appreciation to Mr. Mark Weber, Mr. Terry Sutherland, Mr. Karl Ostergaard, Mr. Hitoshi Hirai, Mr. Kazuya Tsurumi, Mr. Makoto Ito, Mr. Don Dennis, and Mr. Paul Thomas of Horiba Instruments Incorporated for their assistance in calibrating the MEXA 7200 LE Bag Bench. The authors also thank DuPont, Westlake, Honeywell (Allied Signal), Ausimont and Dyneon for providing product information and film samples.

REFERENCES

1. Tedlar Film Designation Guide, DuPont Tedlar.
2. Private Communication with DuPont.
3. Unpublished results, McMahon.
4. Sawano et. al., "A Study of Low Emissions Measurement Techniques," JSAE Paper No. 9631010, 1996
5. Tayama, et. al., "A Study of Gasoline-Fueled Near-Zero-Emission Vehicle Using an Improved Emission Measurements System," SAE Paper No. 982555, 1998
6. Sherman et. Al., "Evaluation of Horiba MEXA 7000 Bag Bench Analyzers for Single Range Operation," SAE Paper No. 1999-01-0147, 1999
7. Sherman et. al., "Evaluation of Mass Flow Controlle Gas Divider For Linearization Emission Analytica Equipment," SAE Paper No. 1999-01-0148, 1999
8. Kunststoffe, "Fluorocarbon Films – Present Situation and Future Outlook," 1980
9. Taylor, "Quality Assurance of Chemica Measurements," Lewis Publishers, Inc. 1987
10. "Procedure for LOD Determination for Speciated Hydrocarbon Analysis (MLD Method 1002/1003)," California Air Resources Board, 1999

CONTACT

For additional information, please contact:

Edward I Sun
Air Resources Engineer
California Air Resources Board
Mobil Source Control Division
Engineering Testing Section
9528 Telstar Avenue
El Monte, CA 91731
USA
Phone: (626)575-6650
Fax: (626)575-6633
E-mail address: esun@arb.ca.gov

DISCLAIMER

This paper has been reviewed by the staff of the California Air Resources Board and approved for publication. Approval does not signify that the contents necessarily reflect the views and policies of the Air Resources Board, nor does the mention of trade names or commercial products constitute endorsement or recommendation for use.

DEFINITIONS, ACRONYMS, ABBREVIATIONS

ARB:	California Air Resources Board
CAL:	Zero/Span and zero read
CO_2:	Carbon Dioxide
CVS:	Constant Volume Sampler
DMAc:	N,N- dimethylacetamide
FID:	Flame Ionization Detector
GC:	Gas Chromatography
HC:	Hydrocarbon
LEV:	Low Emission Vehicle
LOD:	Limit of Detection
MFC:	Mass Flow Controller
PVF:	poly-(vinyl-fluoride)
QC:	Quality Control
RH:	Relative Humidity
RS:	Reading Scale, also known as Percent Point or percent error based on the reading value
SULEV:	Super Ultra Low Emission Vehicle
THC:	Total Hydrocarbon
ULEV:	Ultra Low Emission Vehicle
USEPA:	United States Environmental Protection Agency

APPENDIX

TABLE 1 MAJOR FLUOROCARBON POLYMERS AND THEIR PERMEATION PROPERTIES

Resin Manufacturer	Resin Trade Name	Grade	Homo, Co, Ter-polymer	Polymer	Film Manufacturer	Film Process	Film Grade	Water Vapor * g/m2.d.bar	Air * cm3/m2.d.bar	Oxygen * cm3/m2.d.bar	Nitrogen * cm3/m2.d.bar	CO2 * cm3/m2.d.bar
Honeywell	Aclar**	11 A, 11 C	Homo	Chloro- trifluoro Ethylene	Honeywell	Melt Extrusion	11A, 11C	1	-	60	10	150
Honeywell	Aclar**	33 C	Co	Chloro- trifluoro Ethylene + Difluoro Ethylene	Honeywell	Melt Extrusion	33C					
AtoFina	Kynar	740	Homo	PVdF (Poly vinylidene Fluoride)	Westlake	Melt Extrusion		2	7	20	30	100
AtoFina	Kynar Flex	2800	Co	PVdF + Hexafluoro-propylene	Westlake	Melt Extrusion						
AtoFina	Kynar Flex	2750	Co	PVdF + Hexafluoro-propylene		Melt Extrusion						
Ausimont	Halar		Co	Ethylene + Chloro- trifluoro Ethylene	Westlake	Melt Extrusion		2	40	100	40	400
DuPont	Teflon PTFE ***		Homo	Polytetrafluoro Ethylene	DuPont	Compression Molding & Skiving		5	2000	1500	500	15000
DuPont	Teflon PFA ***		Co	Tetrafluoro Ethylene + 3-4% Perfluoroalkoxy	DuPont	Melt Extrusion		8	1150	-	-	7000
DuPont	Teflon FEP	Type L	Co	Tetrafluoro Ethylene + Hexafluoro propylene	DuPont	Melt Extrusion	Type-L (high temperature)	1	600	2900	1200	4700
DuPont	Tedlar		Homo	Polyvinyl Fluoride	DuPont	Melt Extrusion with DMAc	TTR 20SG4 or TST 20SG4	7	50	12	1	60
DuPont	Tefzel		Co	Ethylene + Tetrafluro Ethylene	DuPont	Melt Extrusion	500 LZ	2	175	350	120	1300
Dyneon	TFM PTFE	1700	Homo ****	Polytetrafluoro Ethylene + <1% Perfluoroalkoxy		Compression Molding & Skiving						

* Permeation data published in 1980 Kunststoffe paper entitled "Fluorocarbon Films - Present Situation and Future Outlook" based on generic polymers [8].
** Aclar was not tested for difficulties in sealing.
*** Teflon PTFE and PFA were not tested for their high permeation properties.
**** With less than 1% co-monomer, the TFM is considered a homopolymer by standard definition.

TABLE 2 INFORMATION OF THE SAMPLE BAGS

Bag Materials	**# of Bags Tested**	**Bag ID**	**Gauge**	**Bag Width (cm)**	**Bag Length (cm)**	**Film Suppliers**	**Bag Supplier**
Tedlar	1	18	2	102	152	DuPont	-
Baked Tedlar	2	41,42	2	102	152	DuPont	-
Aged Baked Tedlar*	2	36,37	2	102	152	DuPont	-
Kynar Flex 2750 **, ***	5	5,6,7,8,9,10	2	91	132	-	PFE
Baked Kynar Flex 2800	1	40	3.5	64	84	Westlake	-
Teflon FEP (L)***	2	13,14	5	91	132	-	PFE
Tefzel (500LZ)	2	38,39	5	102	152	DuPont	-
TFM TFE***	2	43,44	5	69	107	-	PFE
Halar	1	35	3	28	102	Westlake	-

* Samples were aged for 4 months after initial testing.
** Kynar Flex 2750 Bag samples were purged with humidified zero air extensively.
*** One side of the bags has been crinkled so that the bags can be evacuated smoothly without stress.

TABLE 3 A NON-WEIGHTED FIRST ORDER CALIBRATION CURVE FOR NO_X AT 30 PPM RANGE

Current Data Set				Current Curve	
Point #	Z/S Adjusted Counts	Gen. Conc. (ppm)	Cut (%)	Meas. Conc. (ppm)	Error (%)
1	124351	29.700	100.0	29.682	-0.06
2	112019	26.730	90.0	26.722	-0.03
3	99723	23.760	80.0	23.771	0.05
4	87355	20.790	70.0	20.802	0.06
5	74984	17.820	60.0	17.833	0.07
6	62588	14.850	50.0	14.858	0.05
7	50162	11.880	40.0	11.875	-0.04
8	37801	8.910	30.0	8.909	-0.01
9	25427	5.940	20.0	5.938	-0.03
10	12972	2.970	10.0	2.949	-0.70
11	724	0.000	0.0	0.009	-

TABLE 4 GAS DIVIDER GDC 703 CONFIRMATION ON A FIRST ORDER CURVE OF NO_X AT 30 PPM RANGE

Candidate Data Set				Current Curve		MFC Used
Point #	Counts	Gen. Conc. (ppm)	Cut (%)	Meas. Conc. (ppm)	Error (%)	
1	111983	27.000	100.0	26.975	-0.09	4
2	89682	21.600	80.0	21.605	0.03	1,4
3	67306	16.200	60.0	16.218	0.11	1,4
4	44791	10.800	40.0	10.798	-0.02	1,4
5	22389	5.400	20.0	5.405	0.09	1,4
6	17917	4.320	16.0	4.328	0.19	1,4
7	13298	3.240	12.0	3.216	-0.74	1,3
8	8854	2.160	8.0	2.146	-0.64	1,3
9	4401	1.080	4.0	1.074	-0.55	1,3
10	1072	0.270	1.0	0.273	0.95	1,2
11	843	0.216	0.8	0.217	0.68	1,2
12	618	0.162	0.6	0.163	0.75	1,2
13	390	0.108	0.4	0.108	0.28	1,2
14	175	0.054	0.2	0.057	4.94	1,2
15	-50	0.000	0.0	0.002	-	1

TABLE 5 COUNTS OF LOD STUDY OF FIA 726 LE

Conc. ($ppmC_1$)	1	2	3	4	5	6	7
0.1517	11441	11447	11437	11433	11437	11437	11435
0.0303	10918	10927	10922	10916	10919	10922	10916
0.0243	10894	10898	10890	10888	10894	10895	10896
0.0121	10837	10851	10838	10833	10846	10845	10843
0.0061	10815	10816	10816	10807	10818	10813	10816

The t (=3.1) is found from table with degree of freedom equal to 6.
The s (=4.3124) is the standard deviation of seven repeat tests at 0.0061 $ppmC_1$.
The m (=4276.529) and b (10789.48) are the slope and intercept of a least square fit line based on all the points.
The original LOD equals (|b| + t*s)/m if b is negative. Since b is positive, LOD equals (t*s)/m = 0.0031 $ppmC_1$.

TABLE 6 DRY ZERO AIR SOAK RESULTS

Material	**Bag ID**	**New Bag**	**FID**	**0 min**	**15 min**	**35 min**	**60 min**
					($ppmC_1$)		
Tedlar	18	Yes	FIA-200	-	**0.38**	-	-
Baked Tedlar	41	Yes	FIA 726 LE	0.006	0.005	-	0.006
	42	Yes	FIA 726 LE	0.003	0.004	-	0.005
		Average		**0.005**	**0.005**	-	**0.006**
Aged Baked Tedlar	36	No	FIA 726 LE	0.010	0.012	-	0.009
	37	No	FIA 726 LE	0.005	0.007	-	0.010
		Average		**0.008**	**0.010**	-	**0.010**
Kynar Flex 2750	9	Yes	FIA-200	-	0.17	-	-
	10	Yes	FIA-200	-	0.10	-	-
		Average		-	**0.14**	-	-
Baked Kynar Flex 2800	40	Yes	FIA 726 LE	**0.014**	-	-	**0.026**
	40	No	FIA 726 LE	0.011	-	-	0.018
Teflon FEP (L)	13	Yes	FIA-200	-	0.016	-	0.013
	14	Yes	FIA-200	-	0.023	-	0.023
		Average		-	**0.020**	-	**0.018**
	13	No	FIA 726 LE	0.005	0.005	0.006	0.006
	14	No	FIA 726 LE	0.005	0.005	0.005	0.009
		Average		**0.005**	**0.005**	**0.006**	**0.008**
Tefzel	38	Yes	FIA 726 LE	0.013	0.018	0.018	0.024
	39	Yes	FIA 726 LE	0.013	0.013	0.018	0.023
		Average		**0.013**	**0.016**	**0.018**	**0.024**
TFM TFE	43	Yes	FIA 726 LE	0.024	-	0.036	0.070
	44	Yes	FIA 726 LE	0.016	-	0.024	0.053
		Average		**0.020**	-	**0.030**	**0.062**
Halar	35	Yes	FIA-200	-	-	-	**0.053**
	35	No	FIA 726 LE	**0.010**	-	-	-
	35	No	FIA 726 LE	-	-	-	**0.032**

Note: FIA 200 stands for FIA 200 series FID, which may be FIA-220 or FIA-222.

TABLE 7 WET ZERO AIR SOAK RESULTS

Material	Bag ID	Analysis	Before Test	ppmC_1					Note
				0 min	15 min	35 min	60 min	24 hrs	
Baked Tedlar	41	THC	0.001	0.005	0.003	0.001	0.004	0.011	
		CH4	-0.004	-	-	-	-	-0.002	
		Grains/lb	48.2	47.7	47.3	48.3	47.9	34.8	
	42	THC	0.001	0.002	0.001	0.000	0.001	0.009	
		CH4	-0.002	-	-	-	-	-0.002	
		Grains/lb	54.0	50.6	50.6	51.9	50.5	34.7	
	THC Average		**0.001**	**0.004**	**0.002**	**0.001**	**0.003**	**0.010**	
Aged Baked Tedlar	36	THC	-0.004	0.001	-0.002	0.001	0.000	-	0.021 after 28 hrs
		CH4	-0.001	-0.001	-	-	-	-	-0.002 after 28 hrs
		Grains/lb	60.5	54.3	54.3	53.7	53.2	-	25.8 after 28 hrs
	37	THC	0.004	0.005	0.005	0.004	0.004	-	0.013 after 96 hrs
		CH4	-0.001	-	-	-	-	-	-0.002 after 96 hrs
		Grains/lb	56.4	51.8	51.5	51.4	51.0	-	47.6 after 96 hrs
	THC Average		**0.000**	**0.003**	**0.002**	**0.003**	**0.002**	**-**	
Kynar Flex 2750	5	THC	-0.001	0.011	0.011	0.014	0.015	0.059	going up/high zero
		CH4	-0.001	-	-	-	-	0.012	
		Grains/lb	69.2	65.7	65.2	65.6	64.7	53.6	
	10	THC	0.001	0.016	0.015	0.018	0.021	0.062	going up/high zero
		CH4	0.000	-	-	-	-	0.011	
		Grains/lb	59.4	59.0	58.1	57.4	56.6	53.8	
	THC Average		**0.000**	**0.014**	**0.013**	**0.016**	**0.018**	**0.061**	
Baked Kynar Flex 2800	40	THC	**0.003**	**0.011**	-	-	**0.021**	-	0.076 after 96 hrs
		CH4	-0.003	-0.001	-	-	-	-	0.000 after 96 hrs*
		Grains/lb	58.3	52.1	-	-	50.3	-	46.9 after 96 hrs
Teflon FEP (L)	13	THC	0.000	0.005	0.003	0.005	0.006	0.055	
		CH4	-0.001	-	-	-	-	0.022	
		Grains/lb	63.9	62.5	59.9	60.0	60.4	59.0	
	14	THC	-0.002	0.002	0.003	0.003	0.004	0.054	
		CH4	0.000	-	-	-	-	0.019	
		Grains/lb	62.6	61.5	61.3	61.3	61.9	58.5	
	THC Average		**-0.001**	**0.004**	**0.003**	**0.004**	**0.005**	**0.055**	
Tefzel (500LZ)	38	THC	0.003	0.013	0.012	0.014	0.018	0.107	
		CH4	-0.002	-	-	-	-	0.000	
		Grains/lb	60.6	57.8	57.6	58.0	58.6	49.6	
	39	THC	-0.001	0.007	0.009	0.013	0.016	0.151	
		CH4	-0.001	-	-	-	-	0.001	
		Grains/lb	82.2	82.2	74.2	73.8	75.8	59.4	
	THC Average		**0.001**	**0.010**	**0.011**	**0.014**	**0.017**	**0.129**	
TFM TFE	43	THC	0.001	0.004	-	-	0.013	0.137	
		CH4	-0.001	-0.001	-	-	-	0.020	
		Grains/lb	74.3	72.5	-	-	71.1	72.1	
	44	THC	0.001	0.005	-	-	0.013	0.209	
		CH4	-0.003	-	-	-	-	0.025	
		Grains/lb	71.6	65.3	-	-	68.5	68.1	
	THC Average		**0.001**	**0.005**	-	-	**0.013**	**0.173**	
Halar	35	THC	0.002	0.022	-	-	-	-	Leak
		CH4	-0.003	-	-	-	-	-	
		Grains/lb	75.2	71.7	-	-	-	-	

* There was an instantaneous 0.014 ppmC_1 reading for methane. The reading drops to zero afterwards.

TABLE 8 REPEATED WET ZERO AIR SOAK RESULTS

Material	Bag ID	Analysis	Before Test	ppmC_1				Note
				0 min	15 min	35 min	60 min	
Baked Tedlar	41	THC	0.002	0.003	0.001	0.002	0.002	
		CH4	-0.001	-	-	-	-	
		Grains/lb	51.9	44.6	44.7	44.6	43.6	
	42	THC	-0.001	0.001	-0.001	-0.001	-0.001	
		CH4	-0.002	-	-	-	-	
		Grains/lb	56.5	51.0	51.0	50.9	50.2	
	THC Average		**0.001**	**0.002**	**0.000**	**0.001**	**0.001**	
Aged Baked Tedlar	36	THC	**0.000**	**0.003**	**0.002**	**0.003**	**0.003**	
		CH4	-0.001	-0.001	-	-	-	
		Grains/lb	70.0	64.1	63.6	63.6	63.7	
	37	THC	-	-	-	-	-	Leak
		CH4	-	-	-	-	-	
		Grains/lb	-	-	-	-	-	
Kynar Flex 2750	5	THC	-0.004	0.007	0.011	0.013	0.019	
		CH4	-0.001	-	-	-	-	
		Grains/lb	71.6	68.9	69.1	68.7	68.6	
	10	THC	0.001	0.017	0.019	0.022	0.026	
		CH4	-0.001	-	-	-	0.002	
		Grains/lb	68.3	66.8	66.3	65.4	66.4	
	THC Average		**-0.002**	**0.012**	**0.015**	**0.018**	**0.023**	
Baked Kynar Flex 2800	40	THC	-	-	-	-	-	Leak
		CH4	-	-	-	-	-	
		Grains/lb	-	-	-	-	-	
Teflon FEP (L)	13	THC	0.000	0.004	0.004	0.005	0.007	
		CH4	-0.004	-	-	-	-0.001	
		Grains/lb	68.7	66.7	67.8	68.9	67.8	
	14	THC	-0.003	0.001	0.002	0.003	0.005	
		CH4	-0.002	-	-	-	-0.001	
		Grains/lb	72.8	71.0	70.7	71.9	71.3	
	THC Average		**-0.002**	**0.003**	**0.003**	**0.004**	**0.006**	
Tefzel (500LZ)	38	THC	0.001	0.008	0.009	0.012	0.016	
		CH4	-0.001	-	-	-	-0.001	
		Grains/lb	80.0	75.6	76.3	75.7	76.0	
	39	THC	-0.004	0.010	0.014	0.016	0.018	
		CH4	-0.002	-	-	-	-0.004	
		Grains/lb	82.9	75.9	75.9	76.2	76.3	
	THC Average		**-0.002**	**0.009**	**0.012**	**0.014**	**0.017**	
TFM TFE	43	THC	-0.004	0.003	-	0.006	0.009	
		CH4	0.000	-	-	-	-	
		Grains/lb	68.3	67.1	-	65.4	64.2	
	44	THC	-0.001	0.004	-	0.010	0.016	
		CH4	-0.001	-	-	-	-	
		Grains/lb	85.5	79.1	-	78.9	74.1	
	THC Average		**-0.003**	**0.004**	-	**0.008**	**0.013**	
Halar	35	THC	-	-	-	-	-	Leak
		CH4	-	-	-	-	-	
		Grains/lb	-	-	-	-	-	

TABLE 9 CARBON DIOXIDE REDUCTION THROUGH TIME

Material	Bag ID	Time (minutes)	0	15	30	60
Baked Tedlar	41	Counts	18125.9	18114.3	18103.6	18106.8
		% Reduction	-	-0.064	-0.123	-0.105
	42	Counts	18126.0	18120.8	18116.3	18101.9
		% Reduction	-	-0.029	-0.054	-0.133
	Average % Reduction			**-0.046**	**-0.088**	**-0.119**
Aged Baked Tedlar	36	Counts	18059.4	18056.1	18053.4	18045.8
		% Reduction	-	-0.018	-0.033	-0.075
	37	Counts	18053.5	18055.1	18052.7	18047.1
		% Reduction	-	0.009	-0.004	-0.035
	Average % Reduction			**-0.005**	**-0.019**	**-0.055**
Kynar Flex 2750	5	Counts	18106.6	18104.9	18097.2	18090.1
		% Reduction	-	-0.010	-0.052	-0.091
	7	Counts	18111.4	18113.3	18104.5	18085.2
		% Reduction	-	0.010	-0.038	-0.145
	Average % Reduction			**0.000**	**-0.045**	**-0.118**
Baked Kynar Flex 2800	40	Counts	17983	-	-	17970
		% Reduction	-	-	-	-0.072
	40	Counts	18016	-	-	18001
		% Reduction	-	-	-	-0.083
	Average % Reduction			-	-	**-0.078**
Teflon FEP	13	Counts	18114.2	18098.2	18069.0	18038.5
		% Reduction	-	-0.088	-0.250	-0.418
	14	Counts	18124.3	18090.6	18077.8	18034.3
		% Reduction	-	-0.186	-0.256	-0.497
	Average % Reduction			**-0.137**	**-0.253**	**-0.457**
Tefzel	38	Counts	18005.6	18008.7	18005.3	17981.6
		% Reduction	-	0.017	-0.002	-0.133
	39	Counts	18027.4	18015.1	18008.5	18001.5
		% Reduction	-	-0.068	-0.105	-0.144
	Average % Reduction			**-0.026**	**-0.053**	**-0.139**
TFM TFE	43	Counts	18100.1	-	18066.1	18031.0
		% Reduction	-	-	-0.188	-0.382
	44	Counts	18087.1	-	18062.4	18013.1
		% Reduction	-	-	-0.137	-0.409
	Average % Reduction			-	**-0.162**	**-0.396**
Halar	35	Counts	18038	-	-	17999
		% Reduction	-	-	-	**-0.266**

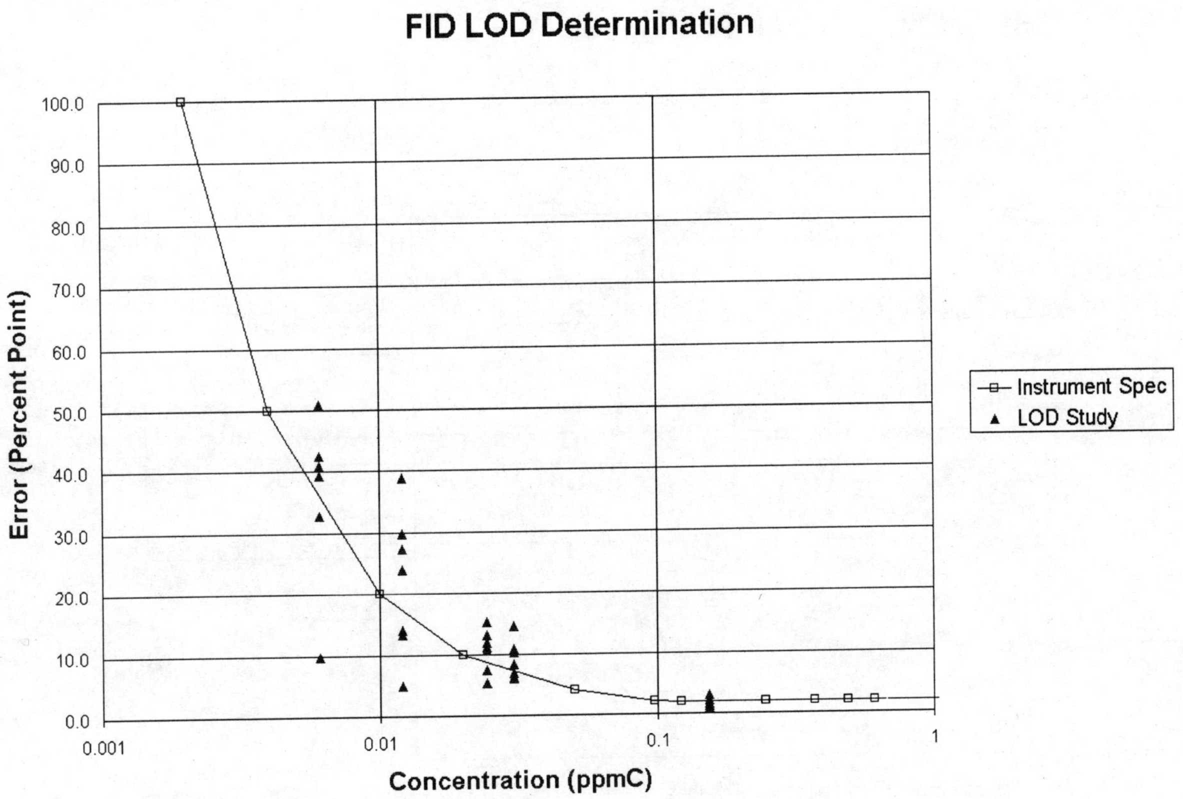

FIGURE 2 LOD DETERMINATION

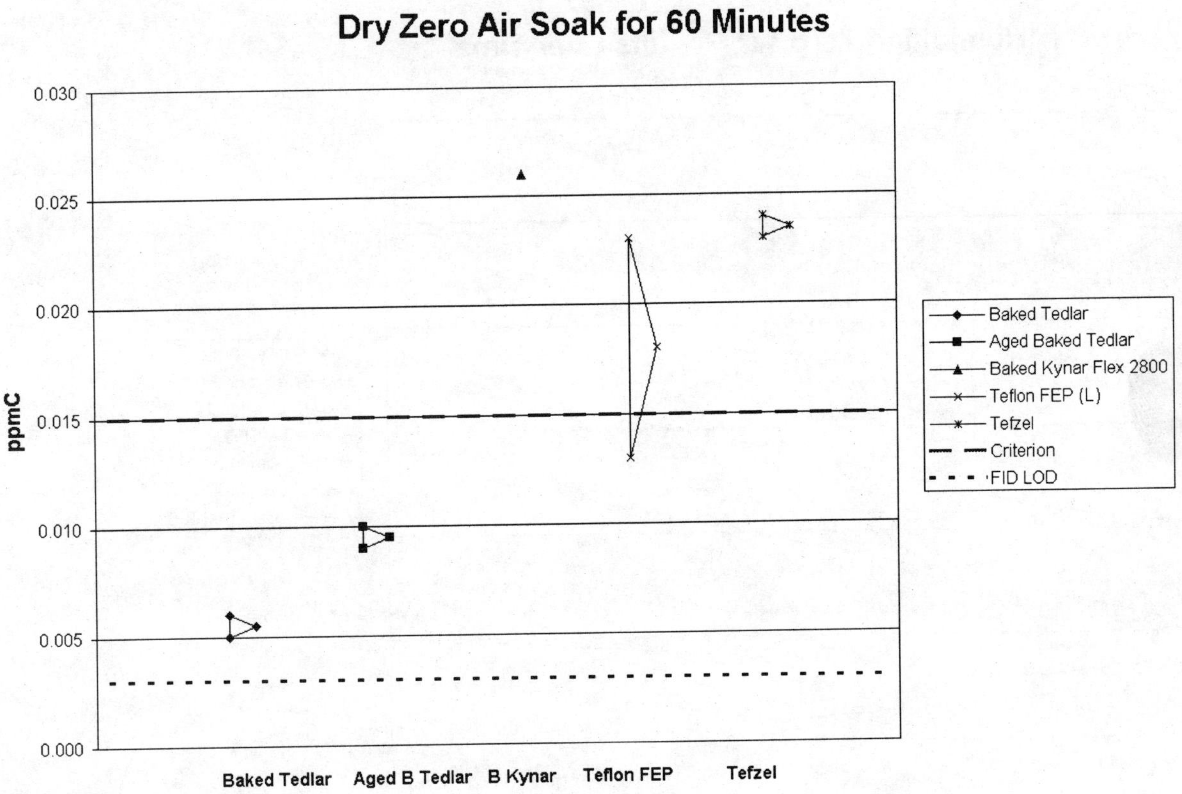

FIGURE 3 DRY ZERO AIR SOAK FOR 60 MINUTES

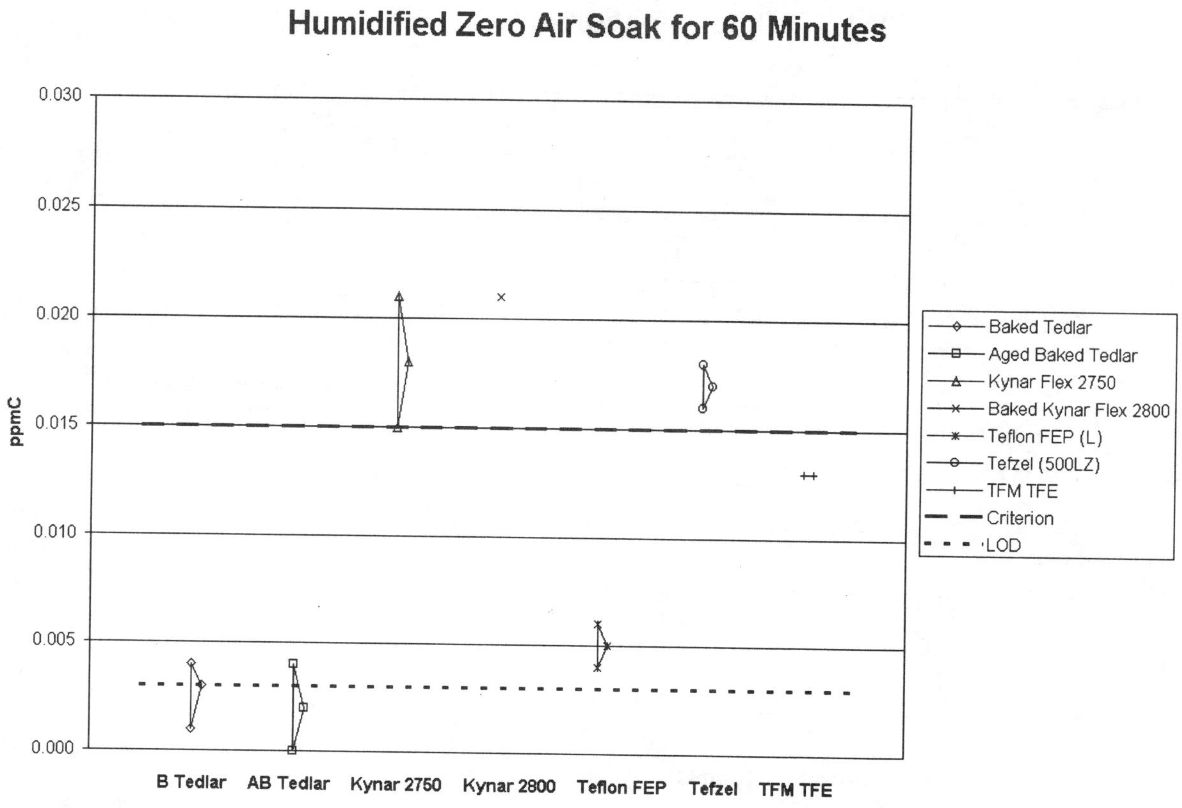

FIGURE 4 HUMIDIFIED ZERO AIR SOAK FOR 60 MINUTES

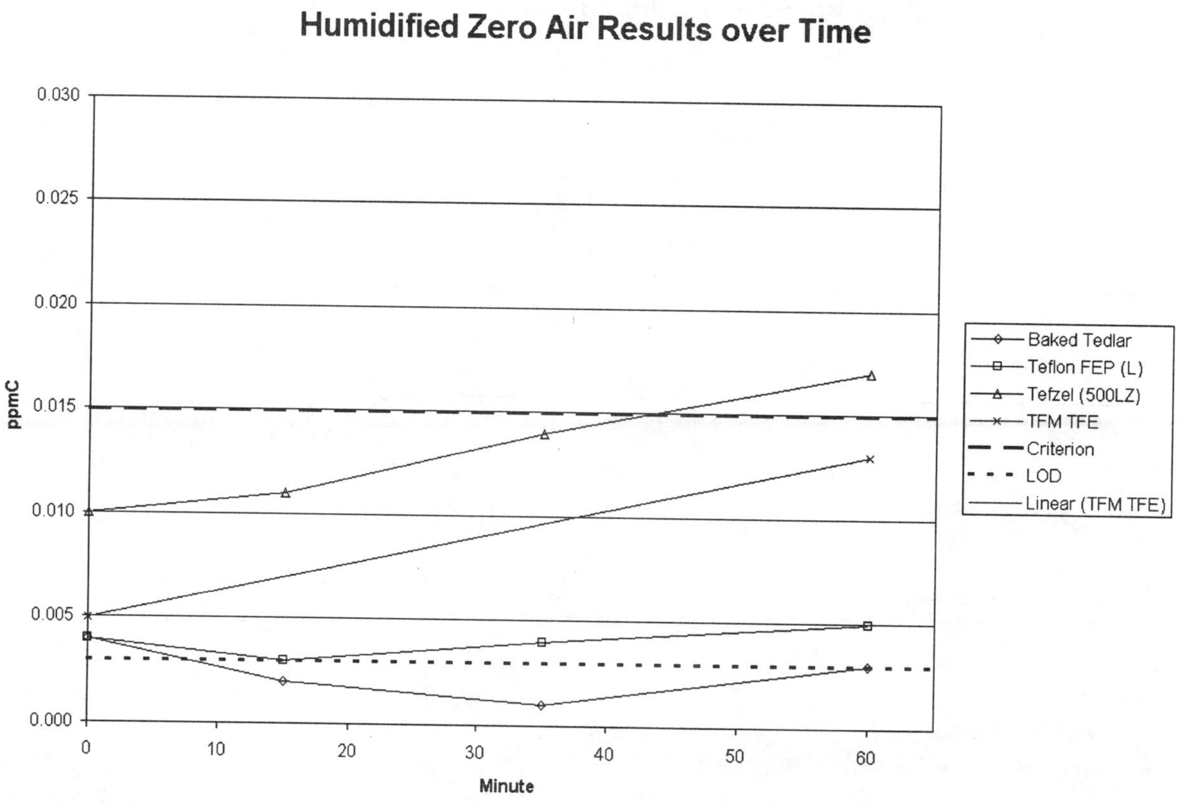

FIGURE 5 HUMIDIFIED ZERO AIR RESULTS (0-60 MINUTES)

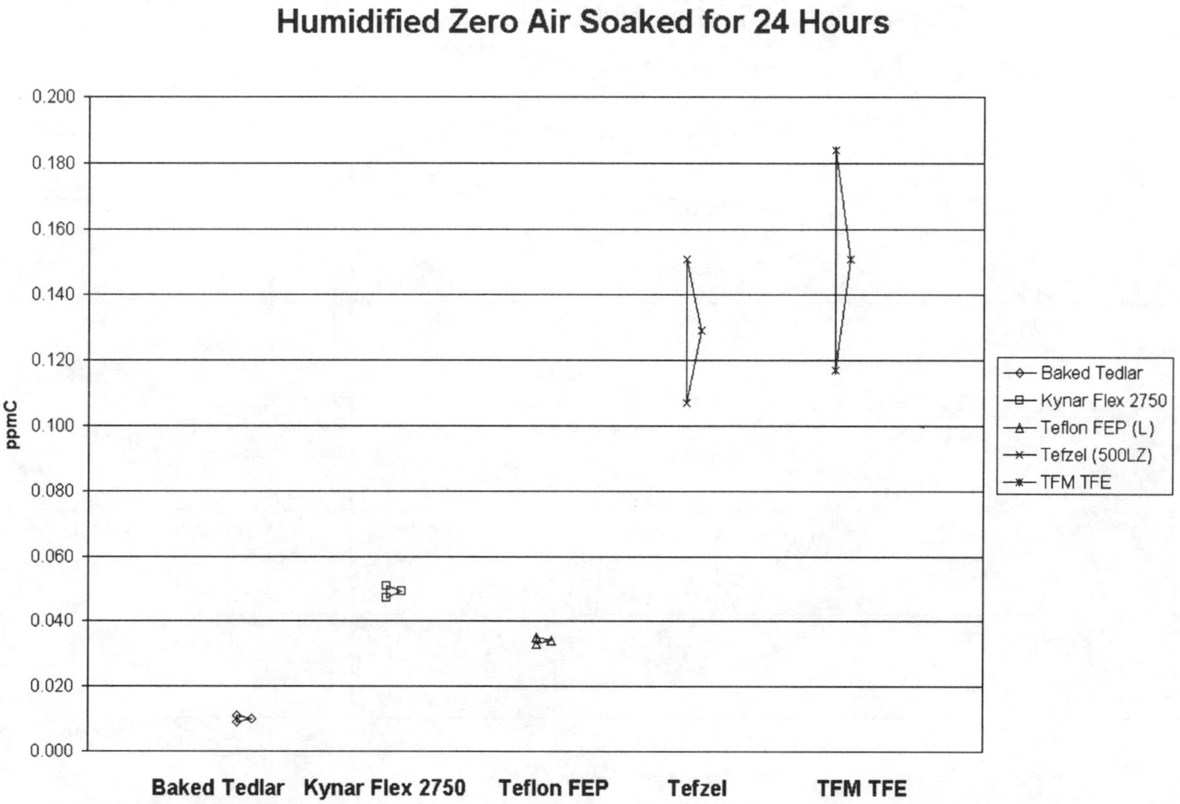

FIGURE 6 HUMIDIFIED ZERO AIR SOAK FOR 24 HOURS (THC DEDUCT METHANE RESPONSE)

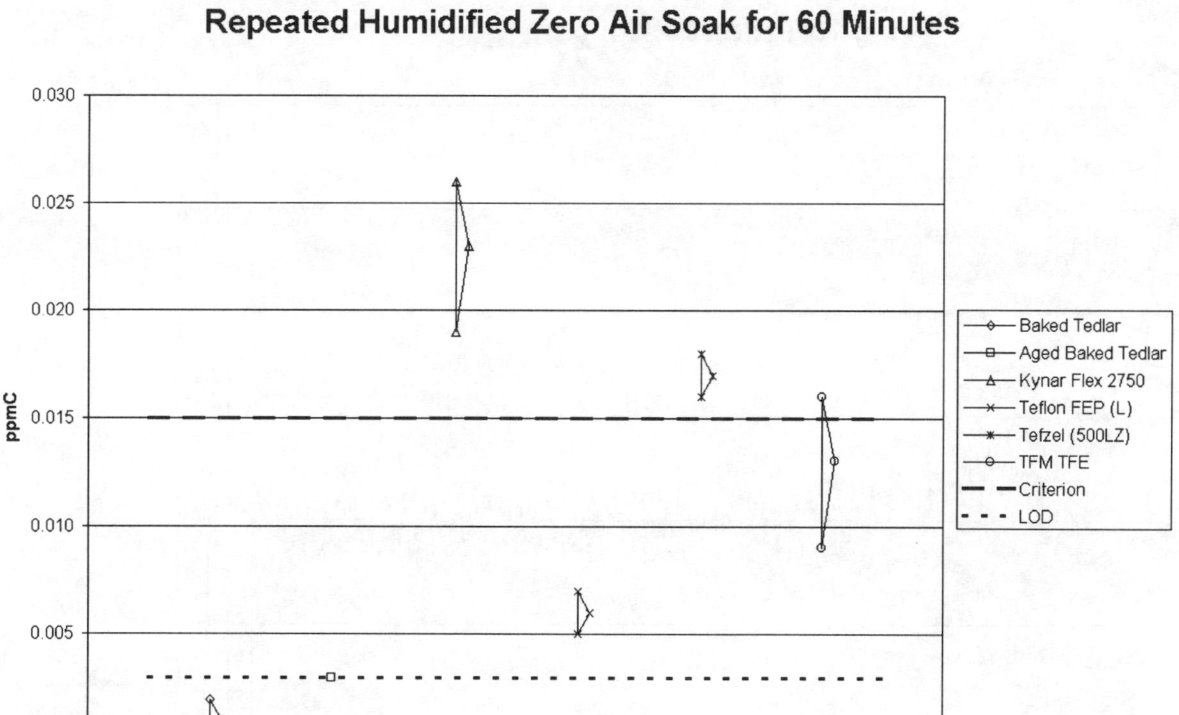

FIGURE 7 REPEATED HUMIDIFIED ZERO AIR SOAK FOR 60 MINUTES

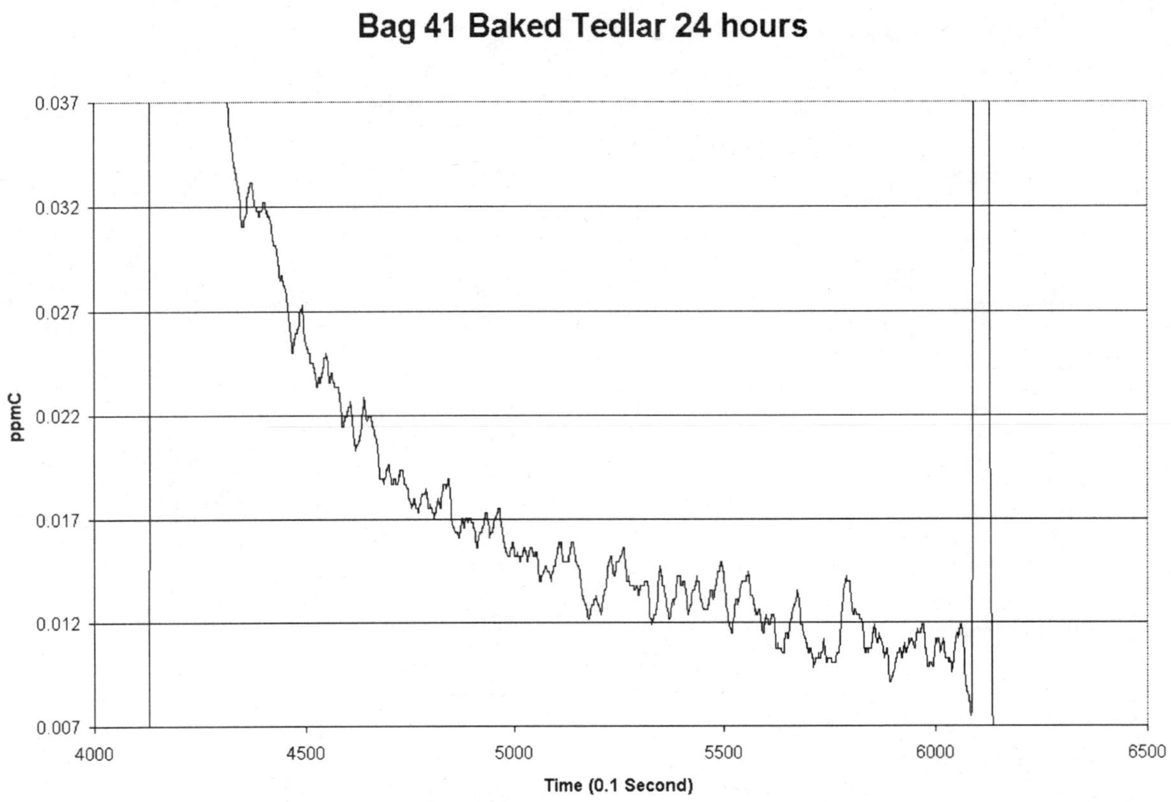

FIGURE 8A BAKED TEDLAR WITH HUMIDIFIED ZERO AIR SOAKED FOR 24 HOURS

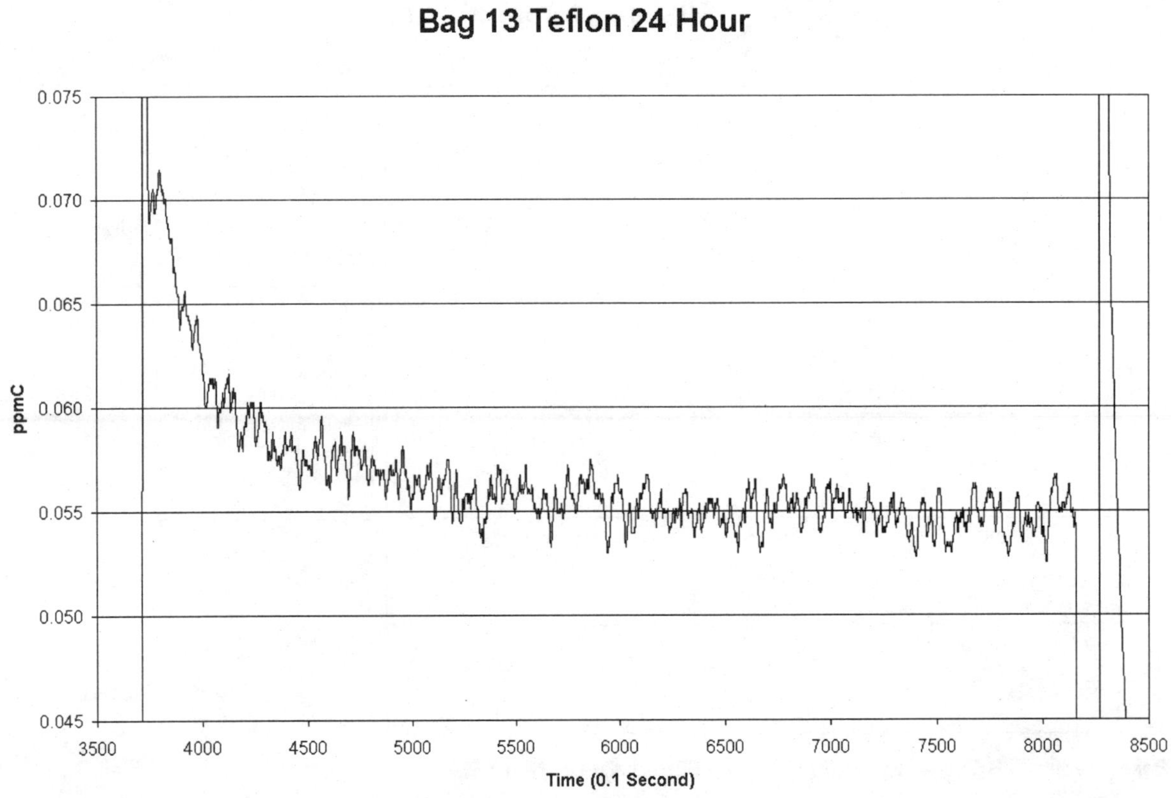

FIGURE 8B TEFLON FEP WITH HUMIDIFIED ZERO AIR SOAKED FOR 24 HOURS

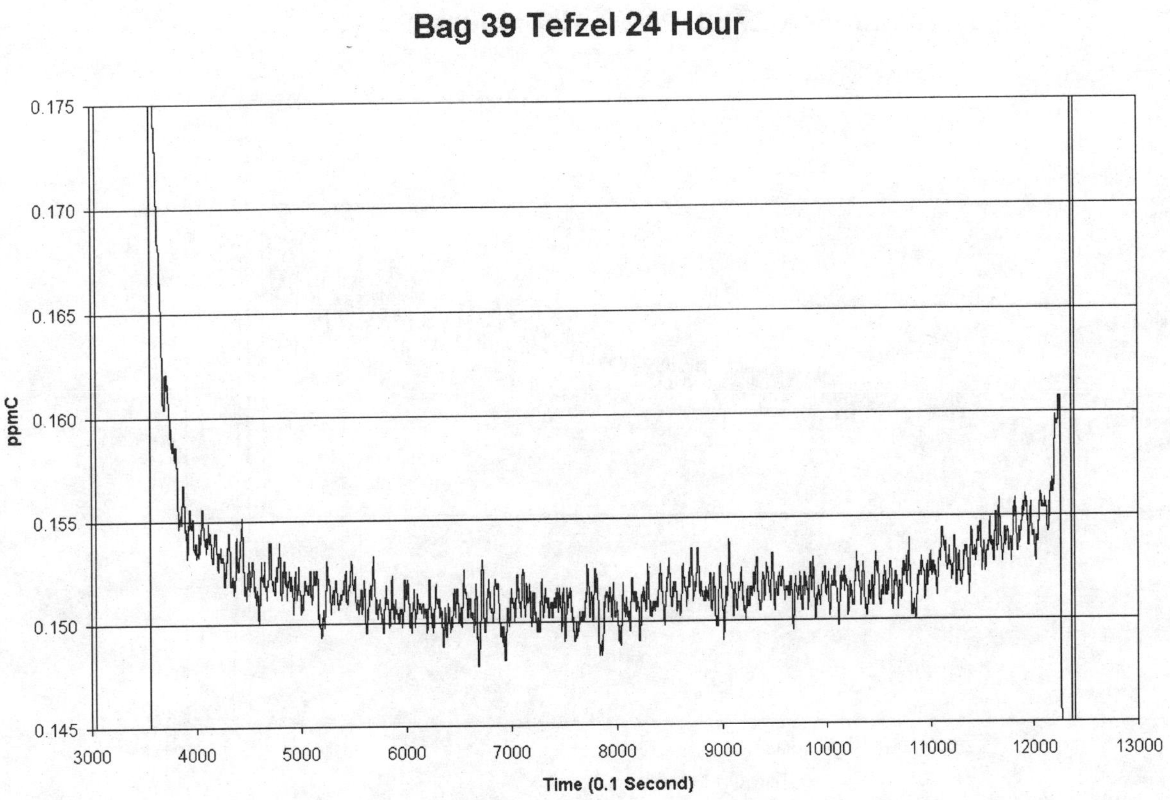

FIGURE 8C TEFZEL WITH HUMIDIFIED ZERO AIR SOAKED FOR 24 HOURS

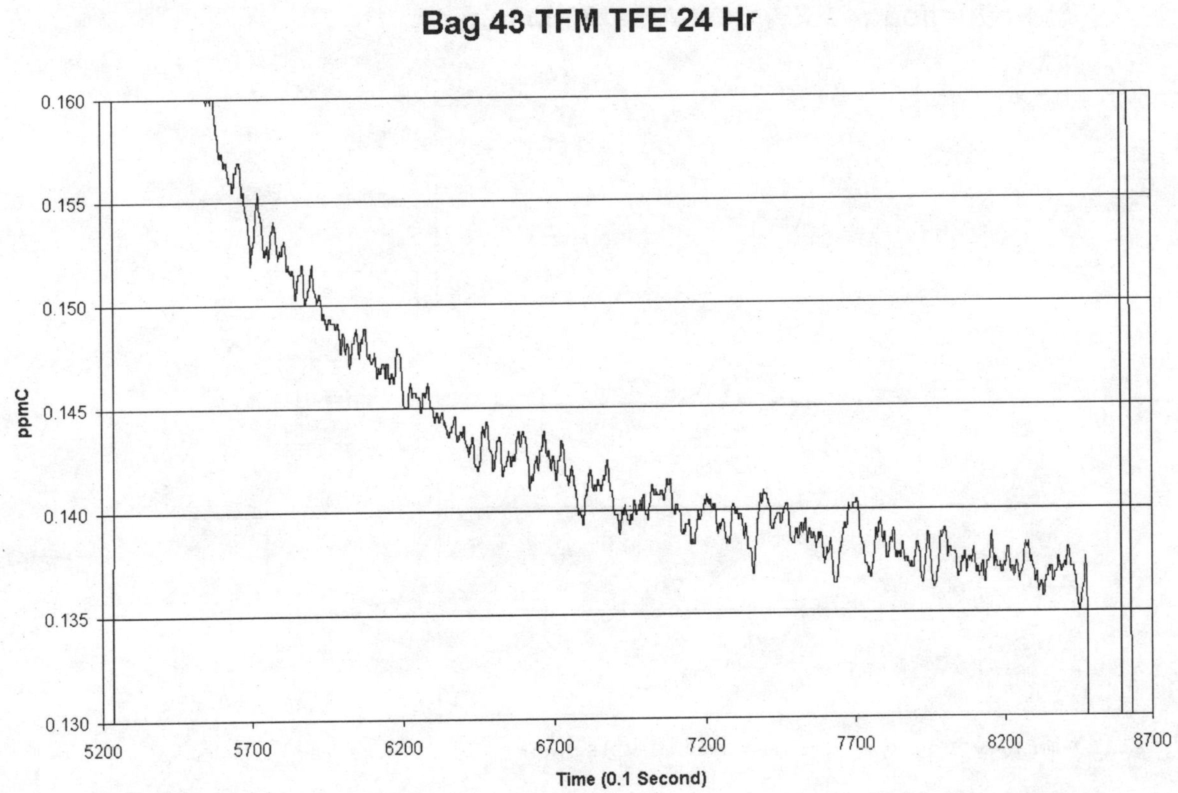

FIGURE 8D TFM TFE WITH HUMIDIFIED ZERO AIR SOAKED FOR 24 HOURS

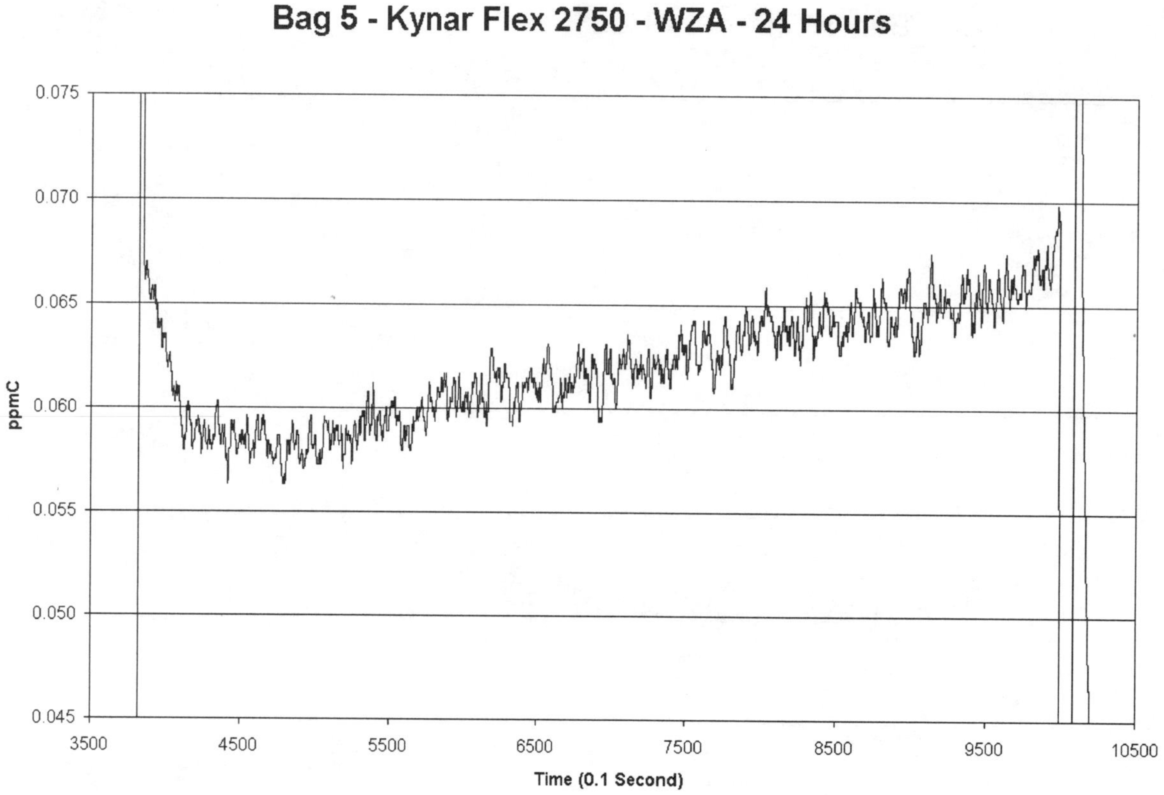

FIGURE 8E KYNAR FLEX 2750 WITH HUMIDIFIED ZERO AIR SOAKED FOR 24 HOURS

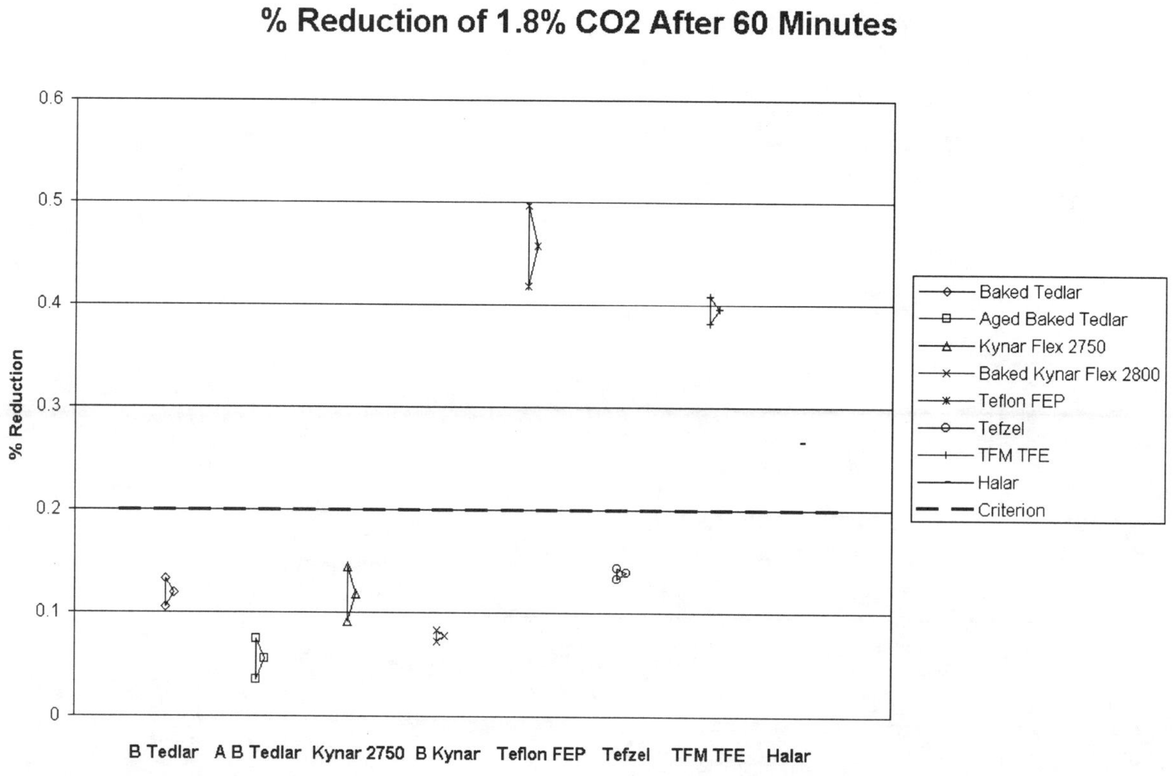

FIGURE 9 CO_2 REDUCTION AFTER 60 MINUTES

2002-01-0046

Advanced Emissions Test Site for Confident PZEV Measurements

Mark Guenther, Michael T. Sherman and Mike Vaillancourt
Correlation Engineering, Ford Motor Company

Dan Carpenter
Product Analysis and Verification, Ford Motor Company

Rick Rooney and Scott Porter
Horiba Instruments, Inc.

ABSTRACT

As automakers begin to develop and certify vehicles that meet the California Air Resources Board LEV II and Environmental Protection Agency Tier II Regulations, emissions test cells must be designed and implemented that are capable of accurate low-level measurements. A new test cell has been installed at Ford Motor Company for use in testing vehicles that meet the stringent Partial Zero Emission Vehicle tailpipe requirements (NMOG = 10 mg/mile, NOx = 20 mg/mile). This test cell includes a redesigned Bag Mini-Diluter (BMD), improved analytical benches, an ultrasonic exhaust flow meter with an integrated tailpipe pressure control system, a conventional constant volume sampler (CVS), and a moveable electric dynamometer. The Bag Mini-Diluter will be used as the primary sampling system for the tailpipe measurements. The moveable electric dynamometer enables the test cell to be configured so that the vehicle is moved to the test equipment rather than moving the test equipment to the vehicle.

INTRODUCTION

A new test site has been installed at Ford Motor Company to test vehicles designed to meet stringent California Partial Zero Emission Vehicle (PZEV) tailpipe emissions requirements [1]. The success of this project and return on investment will be measured by the increased confidence in exhaust emissions measurements below PZEV levels, i.e., below 10 mg/mile NMOG (Non Methane Organic Gases) and 20 mg/mile NOx (Oxides of Nitrogen). Several upgrades were necessary to each of our traditional systems that make up the test installation in order to specifically address the difficulties in making measurements at these low levels.

To help visualize the task, exhaust emissions measurements are normally made with diluted samples in the parts per million regime. These concentrations are then resolved in accordance with the regulations for final gram per mile determinations. For the new PZEV requirements, however, it is now more convenient to reference part per billion (ppb) concentrations and milligram per mile determinations to better communicate the low-level performance and accuracy necessary. For example, a systemic error of 50 ppbC applied to all three bags of the FTP (Federal Test Procedure) could amount to an error of 1 mg/mile to the final determination. To ensure accurate measurements, the requirement for a "clean" sampling system was determined to be a system contribution of no more than 30 ppbC to any measurement.

When using the conventional CVS (Constant Volume Sampling) system and testing PZEV units, it is not uncommon for the tailpipe exhaust to be cleaner than the ambient air used in the dilution process. This creates a problem for the laboratory and for those of us concerned with measurement accuracy and variability. Significant upgrades were needed to the traditional systems and methodologies normally used for emissions testing to ensure confident measurements for NMOG emissions below 10 mg/mile and NOx emissions below 20 mg/mile. For example, a Bag Mini-Diluter (BMD) was selected as the optimal sampling system for the PZEV test cell.

Several other considerations in the exhaust measurement process were also examined and refined. Apart from the test vehicle, the variability or uncertainty of an exhaust emissions test can be grouped for consideration in five different categories. These were evaluated for cost effective improvement with our suppliers and PZEV performance specifications were jointly established. Our considerations for the dynamometer, bag bench (gas analyzers used to measure diluted exhaust gas collected in sample bags), sampling system, facilities, and testing processes are presented here.

PZEV MEASUREMENT CONSIDERATIONS

DYNAMOMETER - Standard electric single roll dynamometer installations are required to meet the industry acceptance procedures established by expert representatives of AAM (Alliance of Automobile Manufacturers), AIAM (Association of International Automobile Manufacturers), EPA and ARB [2]. Additionally, for the PZEV Test Site, a movable dynamometer was specified to allow consistent vehicle positioning relative to the cooling fan, climate control system, and emissions measurement equipment. For additional accuracy, we specified a ½pound (2.2 N) tolerance for simulated load throughout the speed range. This was achieved by incorporating a high precision load sensor with a new mounting system, a redesigned motor bearing lubrication system, and improved dynamometer control algorithms.

GAS ANALYZERS (BAG BENCH) - The basic requirements for low-level emissions measurements are met with the low emission (LE) type instrumentation [3]. For this PZEV test cell application, additional improvements were made to the FID (Flame Ionization Detector) for total hydrocarbon measurement, GC (Gas Chromatography) for methane measurement, Chemi-luminescent detector for NOx measurement and NDIR (Non-Dispersive Infrared) for CO measurement. Signal to noise ratio was improved in each case by a factor of two, except for the CO instrument, where an improvement factor of nearly 100 was realized.

Previously, the noise associated with the analyzer output signal near the zero point made ultra-low concentration measurements difficult as the signal to noise ratio influenced the repeatability of these measurements. Various methods were utilized to stabilize this signal. Stabilization of the high voltage power in the analyzer detector and heat isolation of electrical components resulted in the greatly improved zero signal noise levels.

The new series of instruments developed for this project is now available as super low emission (SLE) type instrumentation. A schematic configuration of this system is shown in Figure 1.

In addition to the technical refinements of the redesigned instruments, the bench was critically examined and engineered to minimize sample surface area and the potential for hydrocarbon adsorption. Where appropriate, Teflon was replaced with electropolished, chemically passivated stainless steel, and new materials were selected to reduce hydrocarbor desorption.

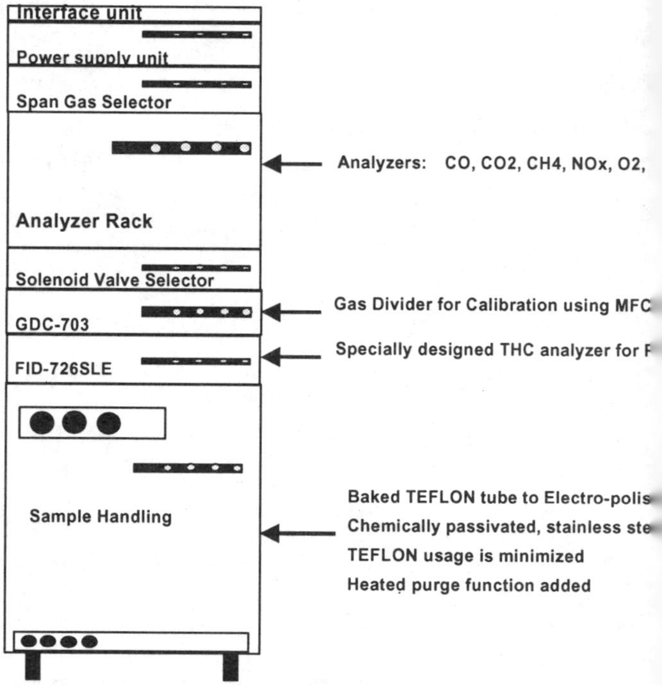

Figure 1. Bag bench for low-level emissions measurement

A heated purge function was added to address all components and lines in contact with the hydrocarbon sample stream. The sequence for the heated purge of the bag read line includes the gradual ramping of temperature to 100°C over a time interval of 11 minutes, while flowing high-purity nitrogen or zero air. This encourages the release of hydrocarbons from the internal surfaces of the tubing, valves and other components. As illustrated in Figure 2, the temperature is then gradually reduced to ambient temperature over a time period of 9 minutes. Rather than maintain a constant temperature of 100°C, the purge begins and ends at ambient temperature to allow the gas analyzer to monitor purge gas concentrations under conditions similar to actual testing.

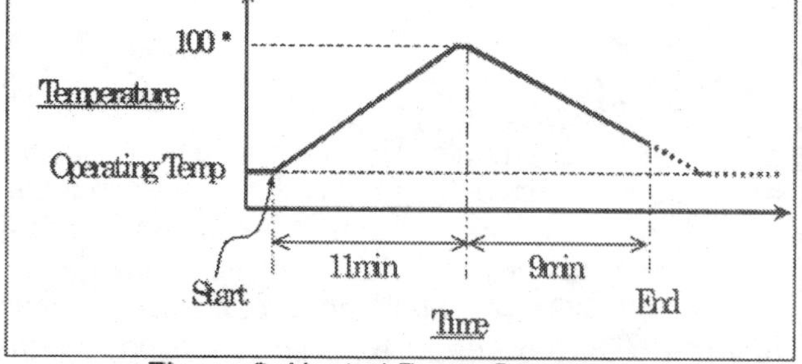

Figure 2. Heated Purge Sequence

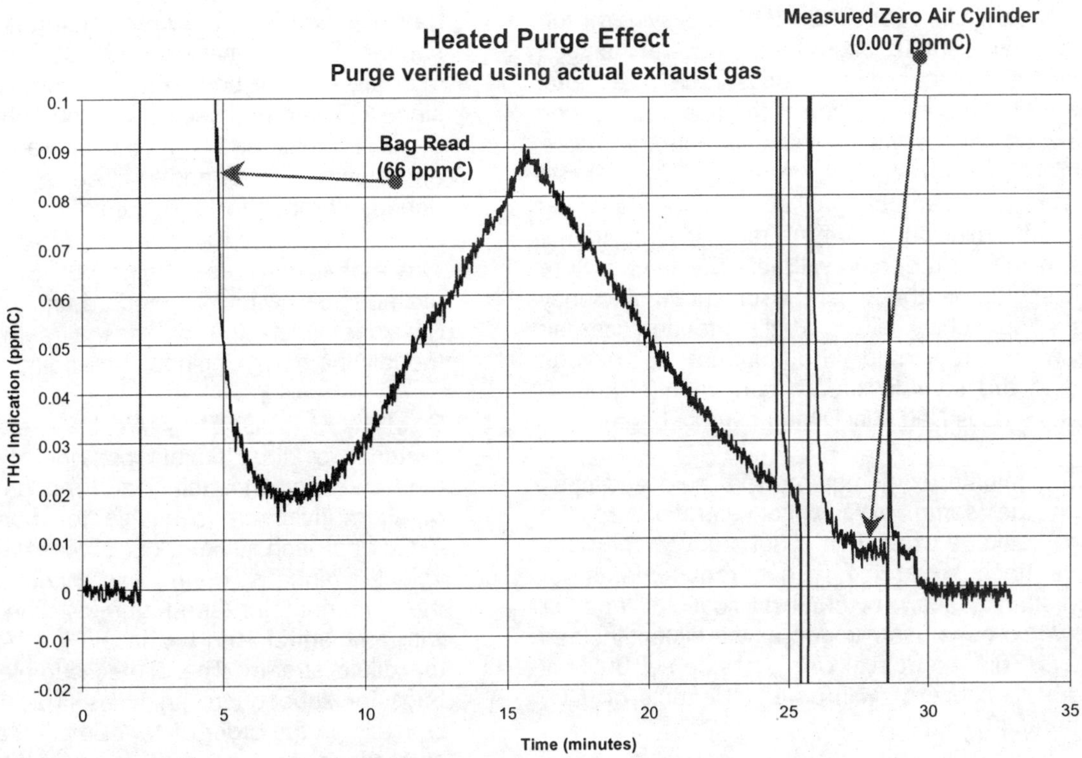

Figure 3. Heated Purge Effectiveness

Experimental tests have shown that above 100°C, there is minimal improvement in hydrocarbon reduction in the sample line. Purge gas temperatures up to 150°C showed negligible concentration differences when compared to concentration of purge gas temperatures at 100°C. The time to achieve sample line concentrations below 0.01 ppmC was approximately 240 seconds at purge gas temperatures of both 100°C and 150°C.

An experiment was performed to optimize the purge time and determine the effectiveness of the heated purge function. Two sample bags were filled; one with exhaust gas and the other with high-purity zero air. The THC concentration in the bag read line was monitored during the following sequence:

1. Zero the gas analyzer
2. Read the bag filled with diluted exhaust gas
3. Perform heated purge function
4. Initialize a system reset
5. Read the bag filled with zero-air
6. Zero the gas analyzer

Figure 3 summarizes the results of the heated purge experiment. A direct correlation exists between the temperature of the purge gas and the amount of hydrocarbons removed from the system tubing. As the temperature increases, the concentration of the gas sample increases proportionally until the hydrocarbons are purged from the system.

For this experiment, the exhaust gas in the sample bag consisted of the concentrations listed in Table 1.

Exhaust Gas Constituent	Sample Bag Concentration
THC	66.00 ppmC
CH_4	12.53 ppm
NOx	0.55 ppm
CO	48.90 ppm
CO_2	2.14%
O_2	17.96%

Table 1. Sample Bag Composition

All instruments in the SLE bag bench are operated on their high gain setting, which limits the full dynamic range but allows better resolution for low-level measurements. To better understand the sampling system performance and to support test site diagnostics, oxygen, SF_6 and humidity instruments compliment the standard bag bench analyzers. We are currently evaluating each of these instruments to determine whether they should be added to future bag benches on PZEV-capable test sites.

The oxygen instrument provides a Quality Assurance measure to monitor the dilution process and the integrity of the emissions test. The test site software allows the operator to select the oxygen instrument for bag measurements or to operate the system without the oxygen instrument.

Due to a lower dilution ratio than the CVS system, the Bag Mini-Diluter sample bags have a lower oxygen concentration in the sample bags. As the total hydrocarbon (THC) concentration in the bag approaches zero, the oxygen concentration in the sample bag has a greater effect on the analyzer reading. The oxygen concentration in the sample gas adds to the flame air supply to determine the air/fuel ratio in the flame. Variations in this air/fuel ratio will vary the temperature and size of the flame and affect the ionization efficiency producing a quenching effect. While the analyzer air supply can be regulated, the amount of oxygen introduced in the sample stream can vary (e.g. CVS sample bags versus Bag Mini-Diluter sample bags).

In order to eliminate oxygen quenching, zero and span gases with the same oxygen concentration as the sample gas could be used. This is not practical for many applications because of the unique requirements for these zero/span gases. For BMD applications, another approach would be to use nitrogen as the diluent instead of zero air. This approach was considered but not adopted due to concerns with potential effects on CO_2 measurements [6].

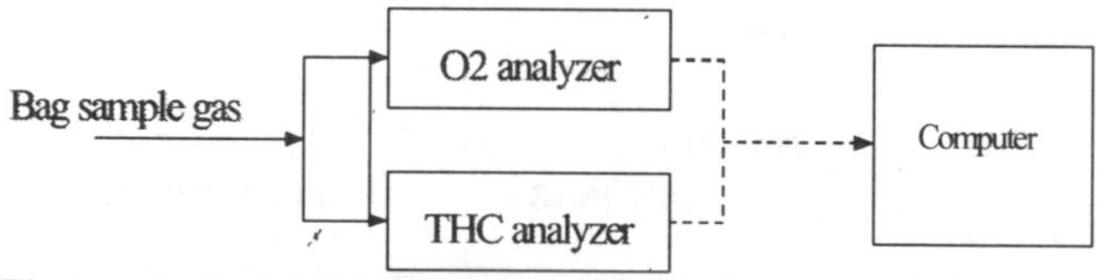

Figure 4. Total Hydrocarbon (THC) Compensation for O_2 Concentration

Figure 4 illustrates the method that was adopted to compensate the THC analyzer reading based on the oxygen content in the sample. Corrections for the oxygen quench can be classified into two categories, span and zero corrections. Recently, it was demonstrated that the O_2 quenching of the THC signal is not only significant for span gas quenching, but also for the zero reading.

The following formula is used to correct for the oxygen quenching effect:

THC indication = THC detected concentration - coefficient * O_2 detected concentration

Where the coefficient is represented as a "ZERO shift " and calculated from O_2 interference at zero.

Though it is recognized that changes in oxygen concentration may have a minor influence on low-level measurements [4], the current strategy is to monitor and control the range of oxygen manufactured by the zero air systems. Investigations are underway for an improved, dedicated, zero air generation system to compliment the unique demands of the BMD test site.

A relative humidity sensor has been adopted on the bench to determine the water content in each sample bag. The water and oxygen content is recorded for each sample bag to increase understanding of the NDIR interferences by both of these components [5]. Finally, since SF_6 can be injected into the vehicle tailpipe during an FTP test and not be altered by the gaseous emissions, it is planned to trace SF_6 through the system and into the bag for a "dynamic" mass recovery test [6].

The modal benches (engine out, mid-bed and tailpipe) are equipped with the standard analyzers except for two revisions: each line includes a methane analyzer and the tailpipe bench includes an additional SF_6 analyzer.

SAMPLING SYSTEM - We believe that the sampling system is the major contributor to hydrocarbon measurement uncertainty and recognize that a number of alternatives are available to improve the standard CVS approach. The CVS process gathers all of the vehicle's exhaust during a prescribed driving cycle on the dynamometer (see Figure 5). The exhaust is diluted with ambient air and a Critical Flow Venturi then meters the dilute stream. The diluted samples are collected in bags for subsequent analyses. This dilution process is typically on the order of 10:1 ambient air to exhaust, and corrections are made to account for the ambient air concentrations when determining the tailpipe exhaust emissions.

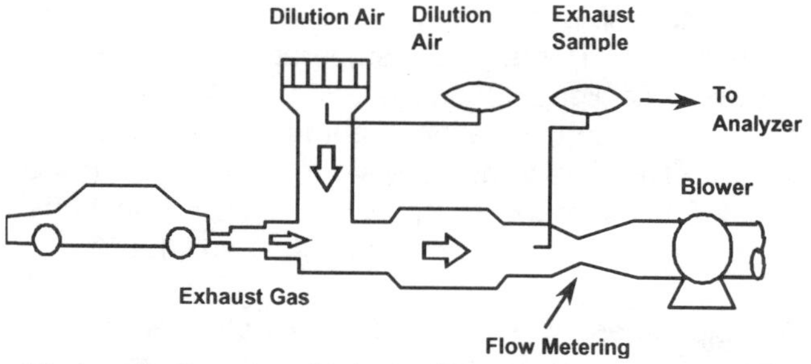

Figure 5. Constant Volume Sampler (CVS) Technique

The CVS dilution process is variable. Too much dilution takes place during idle conditions and too little occurs during accelerations. Furthermore, ambient air is no longer suitable as a diluent.

For this project, Dilution Air Refinement and heated CVS systems were considered. Both offer significant improvement and are attractive because they fit within the confines of the existing Code of Regulation. However, we elected the Bag Mini-Diluter approach as being the most cost effective, accurate, and capable technology with great potential for technical improvement in the future [7]. The Bag Mini-Diluter features a constant dilution process whereby dilution is the same for all portions of the test and is fixed by design (see Figure 6). Dry zero grade air is used for the diluent and ambient measurements are not required in the calculation of mass emissions.

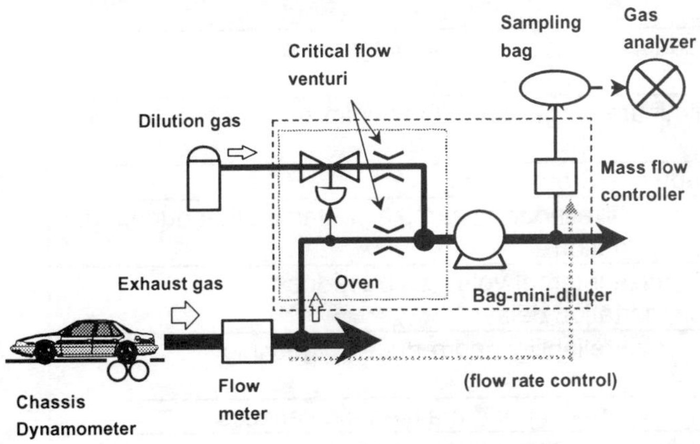

Figure 6. Schematic of Bag Mini-Diluter Technique

Bag Mini-Diluter Modifications - The new "production version" of the Bag Mini-Diluter (see Figure 7) includes many new features to enhance the quality of the PZEV measurements. The BMD improvements are summarized in Table 2.

For manufacturing of the new BMD, experience was gleaned from the semi-conductor industry to help define new manufacturing methods and to select better materials. Fittings were avoided wherever possible. Electro-polished, chemically passivated, stainless steel tubing with large radius bends was used to minimize microscopic fissures (and nucleation sites) internal to the tubing. Minimum use of Teflon was permitted only at pump connections, and Latex gloves were worn during the assembly process.

As a result this fastidious "attention to detail", the total hydrocarbon "outgas" specification of 30 parts per billion carbon (ppbC) was satisfied with no individual hydrocarbon compound in excess of 10 ppbC. This performance specification was verified by Gas Chromatography with a dry zero air bag fill. More stringent diagnostics are being evaluated.

Traditional Tedlar® sample bags, especially when new, are known to outgas N,N-dimethylacetamide and may not be suitable for low-level hydrocarbon applications. While investigations continue to identify better materials, for the interim, a TFM (Dyneon) material has been selected for the BMD sample bags. This material is a blend of two different Teflon materials that minimizes hydrocarbon contamination/outgassing and reduces permeation of CO_2. In our permeation experiments with conditioned bags at room temperature, bags were filled with zero air and with 2% CO_2. We then made measurements before and after a 1-hour soak. The Dyneon bag outgassing and permeation data is shown in Table 3.

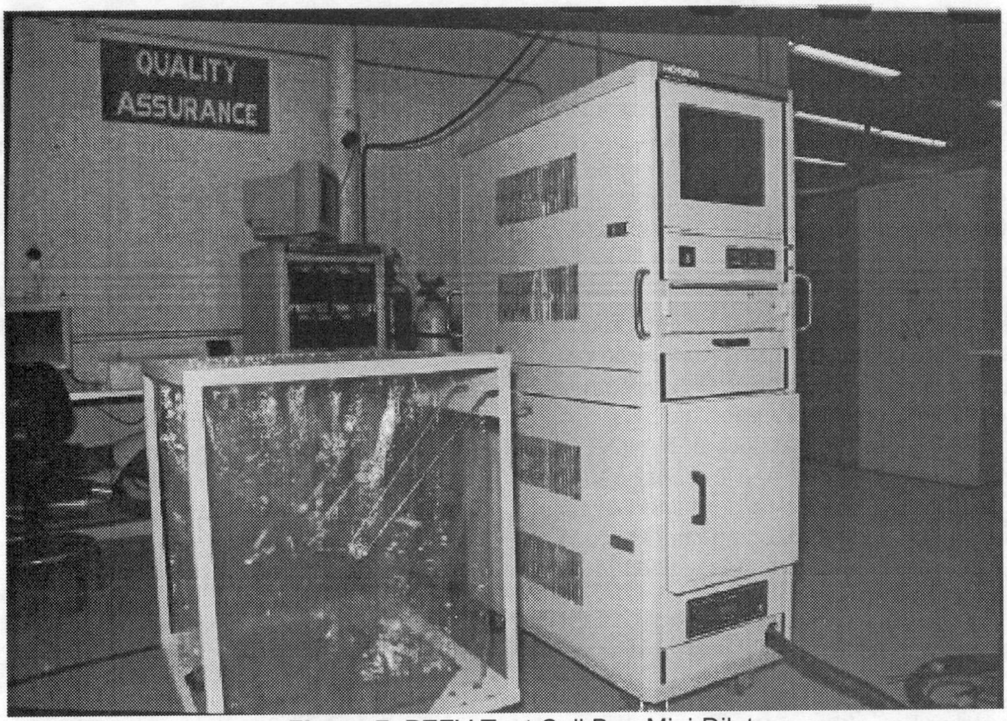

Figure 7. PZEV Test Cell Bag Mini-Diluter

BMD Improvement	Expected Benefit
1. Use semiconductor grade, electro-polished, 316 stainless steel (Ra<10)	Reduce HC hang-up and outgassing
2. Upgrade sample bags: 70 liter TFM crinkle w/o internal plumbing	Eliminate outgassing, and minimize CO2 permeation
3. Eliminate Viton	Reduce HC outgassing
4. Minimize internal volumes for sample stream	Reduce likelihood of sample contamination during gas transportation
5. Adopt miniature valve and manifolds	Minimize internal volume and reduce gas transportation delays
6. Use custom KNF Neuberger Pump w/ Teflon coated heads	Improve reliability and reduce outgassing
7. Use VCR fittings in critical areas	Improve sealing and maintenance access
8. Improve oven design	Reduce size and Improve access
9. Improve H_20 detector	Better response time (<10 seconds)
10. Increase heating coil	Improve temperature condition at CFV
11. Adopt PC control system	Improve compatibility and data storage
12. Adopt NI LabView Software	Incorporate user-friendly software
13. Adopt NI data boards	Increase to 16-bit resolution
14. Incorporate Windows 2000 System	Improve compatibility and ease-of-use
15. Eliminate process controllers	Improve compatibility and ease-of-use
16. Add diluent (zero air) bag	Ensure diluent purity and monitor contamination
17. Adopt 100% heated purge	Sweep residual HC's from BMD
18. Improve system and bag leak check	Improve system integrity
19. Add on-board read pump	Localize critical BMD components
20. Add venturi temperature thermocouples	Monitor CFV performance and improve ability to control dilution ratio at fixed rate

Table 2. PZEV Bag Mini-Diluter Upgrades

For the BMD application, the bags are not heated, and a consistent 5.5:1 dilution ratio is maintained. Further reduction in dilution is not practical due to gas composition effects and the desire to observe CO_2 correlation to the CVS technique. The PZEV Test Cell also includes a traditional CVS system for paired emissions measurements, ensuring that correlation is maintained between the BMD and CVS CO_2 measurements.

Dilution Ratio Accuracy - To maintain a consistent dilution ratio in the Mini-Diluter, the gas temperatures and pressures at the inlet to the critical flow venturis must be held constant and near the same value [7]. To improve the control of the gas temperature at the inlet to the critical flow venturis, the diluent and exhaust gases pass through coils of tubing which are contained in an oven in the Mini-Diluter prior to reaching the venturis (see Figure 8). The oven is maintained at 70°C to prevent condensation of the exhaust gas sample. One tube coil is used for the diluent gas and one is used for the exhaust gas.

As the gases pass through the coiled tubing, heat is transferred between the gases and the air in the oven, so that the temperature of the gas leaving the coil is near the oven temperature. Figure 9 shows the performance of the venturi inlet temperature control for the diluent and exhaust gases using as PZEV vehicle during phase 1 of an FTP test cycle. The exhaust (sample) gas temperature remains relatively constant throughout the test cycle, but the diluent gas temperature declines slightly throughout the phase. The slight decline in temperature is attributed to rapid expansion of the diluent gas through the inlet pressure regulator at high flow, thus slightly decreasing the diluent inlet temperature.

	Tedlar	Dyneon	Teflon PTFE
Outgassing ppmC	2.90	0.04	0.03
CO_2 Permeation (% change per hour)	-0.12	- 0.20	- 0.47

Table 3. Bag Material Comparison

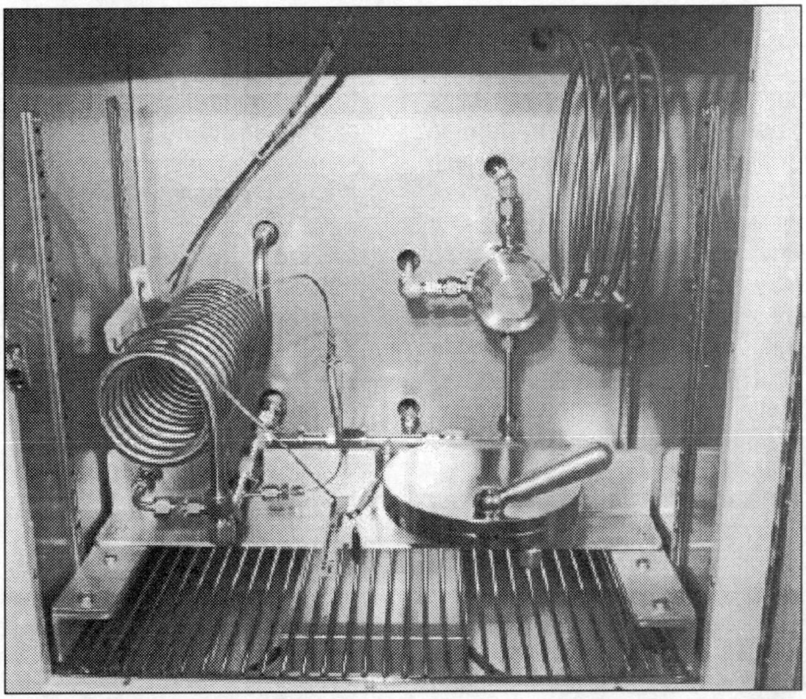

Figure 8. Heat transfer coils in BMD oven for stable venturi inlet temperatures

A faster relative humidity sensor (improved from the prototype BMD) with a T90 response time of less than 10 seconds was also implemented on the BMD. This relative humidity sensor is used to calculate the water content of the BMD sample. The BMD water content is then used to account for the slight influence of water on the sample venturi during the Cold Start portion of the FTP test cycle. As a result, a water adjustment factor (typically less than 2%) is commanded to the bag fill mass flow controller to compensate for the small change in the BMD dilution ratio during the Cold Start. A fast humidity sensor response time is important in order to implement compensation as close to "real-time" as possible.

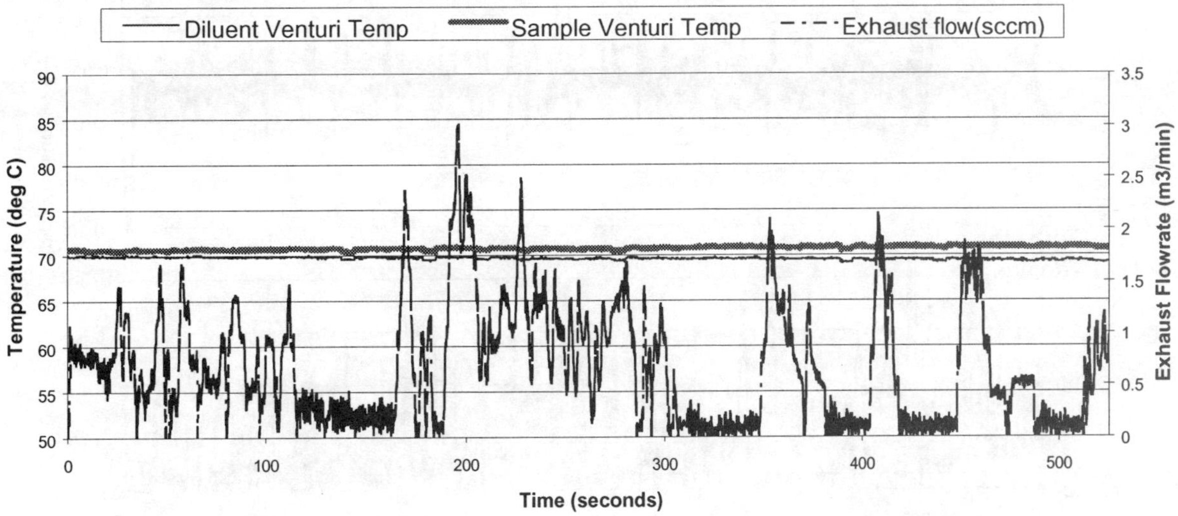

Figure 9. Venturi Inlet Temperatures during FTP, Phase 1 on Bag Mini-Diluter (ULEV Test Vehicle)

Diluent Bag - For the latest version of the Bag Mini-Diluter, a separate bag was added to the system to monitor the purity of the diluent gas that is mixed with the vehicle exhaust gas. Typically, subtraction of the low diluent gas concentrations are not required when using Bag Mini-Diluter technology, as is the case when using a CVS. The Diluent Bag may be used as a quality control tool to ensure on a test-by-test basis that the diluent gas is truly "zero" grade. This diluent bag is filled with the diluent gas at the same time as a sample bag during an emission test, and then analyzed separately to determine the purity level of the diluent gas. Improperly labeled gas bottles, excessive load on the zero air supply system, or contaminated sample read lines might be detected using this quality control method.

Bag Fill Proportionality - The BMD time delay [6] must be determined to account for the gas transport time and the response time of the bag-fill mass flow controller (MFC). A Fast Flame Ionization Detector (FID) was used to determine the time between hydrocarbon concentration changes during propane injections while a dynamic blower "stepped" between simulated exhaust flow rates. Using this technique, a time delay of 1.4 seconds was determined for the BMD system. This time delay, along with the MFC response time, is used in the BMD control software to ensure proportionality between the bag fill MFC and the exhaust flow signal. On the new BMD design, the mass flow controller demonstrated a T90 response time of less than 0.4 seconds.

Several revisions in the BMD control software have been adopted to improve MFC proportionality. The software has been modified to provide the actual flow curve from the MFC, as determined by a flow standard, and an auto-zero function has been incorporated to provide better accuracy. In addition, the internal MFC "auto-linearize" function has been turned off so that the actual flow throughout the full-scale range will be output by the MFC (rather that a flow output that was forced through an internal MFC linearization scheme).

100% Heated Purge - A six-stage heated purge sequence was introduced to the new BMD design to remove residual hydrocarbons from the bag fill and bag analysis lines. The purge sequence is performed during both pre-test and post-test operations. The six stages encompass 100% of the system sample and read bag flow circuits. These stages consist of the following sequences, whereby a purge gas flows through the system while being heated by the sample lines:

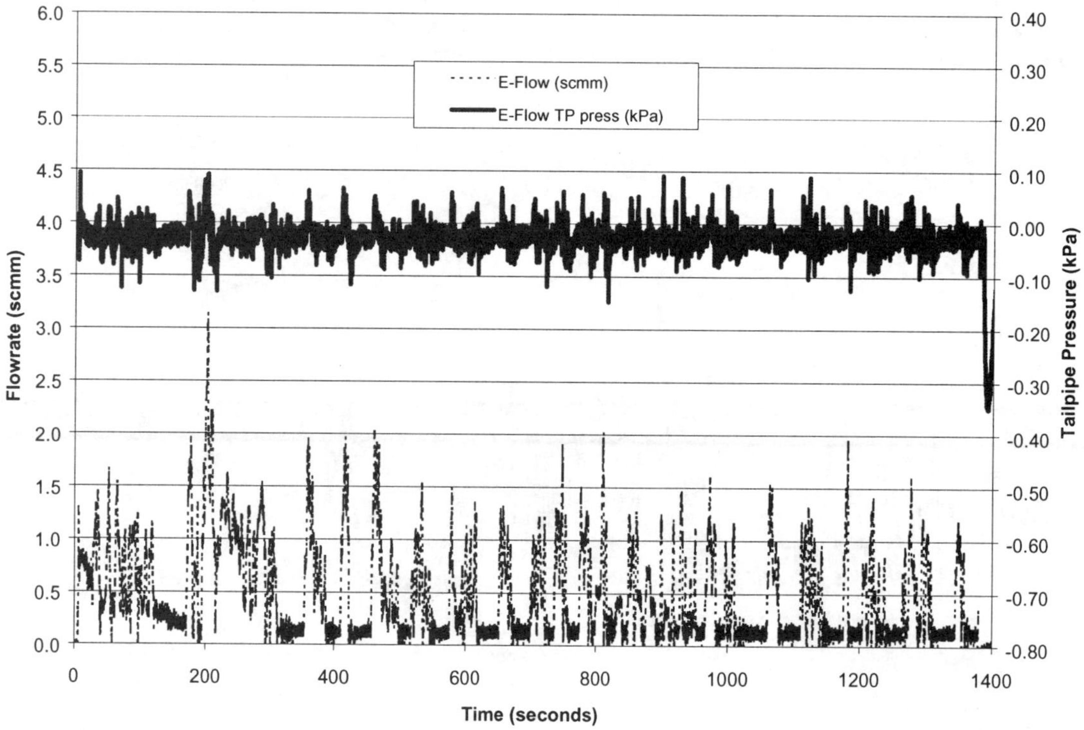

Figure 10. Tailpipe Pressure Control, 350 scfm (9.91 m^3/min), Phase 1 and 2 of FTP, ULEV Test Vehicle

1. Purge sample line (backflush).
2. Purge sample-fill line to the bag fill manifold.
3. Purge sample-fill valves to the samples bags.
4. Purge the read-bag manifold
5. Purge the read-bag valves
6. Purge bags

The system incorporates two user-configurable purge times for each stage. A shorter purge-time is currently used pre-test to perform a quick flush of the system to guarantee the system is clean prior to sampling. Another longer purge time is used post-test to remove residual hydrocarbons from the system. Effectiveness of the purge strategy is demonstrated by routine contamination checks (see data in "Process" section).

Ultrasonic Exhaust Flow Measurement - To complete the system, a modified E-Flow Direct Exhaust Volume Measurement System is provided with the ultrasonic exhaust measurement section aligned vertically to improve stratification under low flow conditions. The E-Flow also acts as a Mixing Tee for ambient air dilution to a standard multiple venturi CVS system connected in series. The CVS provides a selection of 14 flow rates for the diluted exhaust flow.

This flow meter contains an integral tailpipe pressure control system to maintain a range of ±1.0 inches W.C. (±0.25 kPa) during a Federal Test Procedure (FTP) test cycle. The pressure control system uses closed-loop control whereby the tailpipe pressure is measured by a transducer at the vehicle tailpipe. A loop controller processes the pressure transducer signal and provides a set point for a variable frequency drive, which in turn controls the motors for the dilution air blowers. Figure 10 demonstrates the capability of the pressure control system to maintain the pressure within the defined limits throughout flow rate excursions experienced during the first two phases of a FTP test using a PZEV test vehicle. For this test, the pressure set point was biased at -0.25 inches W.C (-0.06 kPa).

The response time of the E-Flow to a step change in flow rate was measured using a dynamic blower controller containing a hot-wire anemometer. The blower was used to inject air into the E-Flow while simultaneously varying the flow rate in step changes. Figure 11 demonstrates the E-Flow response compared to the hotwire anemometer under ascending flow conditions. The data was collected at 200 Hz, which provides ample resolution to determine the response of the E-Flow.

The initial response time of 0.390 seconds was recorded for ascending flow. Ascending flow was measured when the flow rate injected into the E-Flow was ramped from 3 scfm to 113 scfm (0.09 scmm to 3.20 scmm). Additionally, when the flow rate was ramped from 113 scfm to 4 scfm (3.20 scmm to 0.11 scmm), the initial descending flow response is slightly slower at 0.460 seconds (Figure 12).

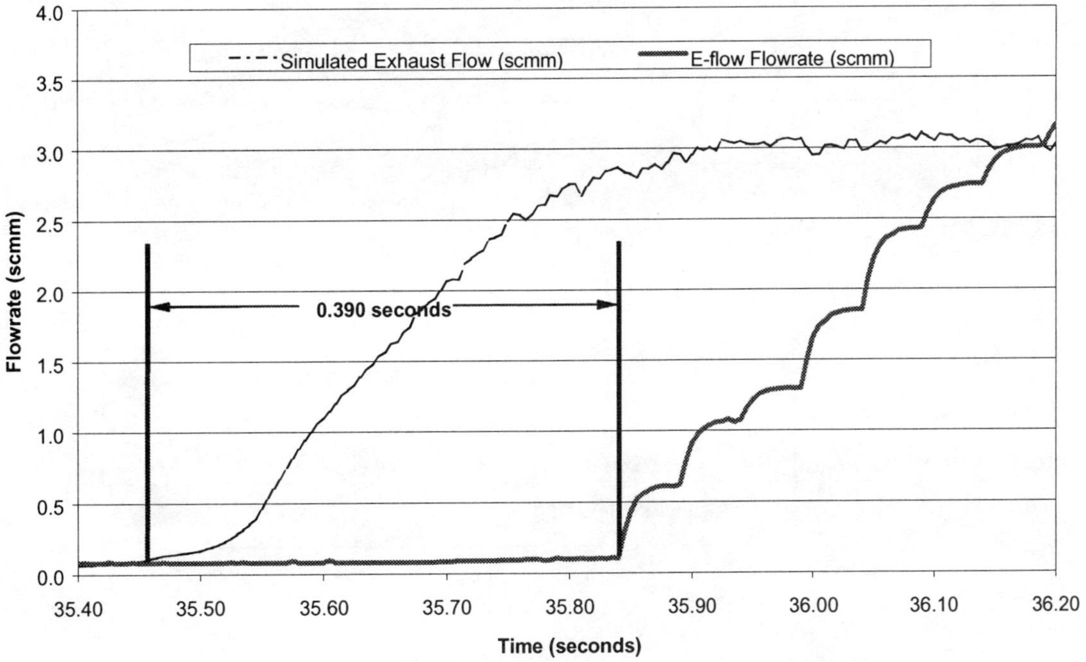

Figure 11. E-Flow Response Time – Ascending

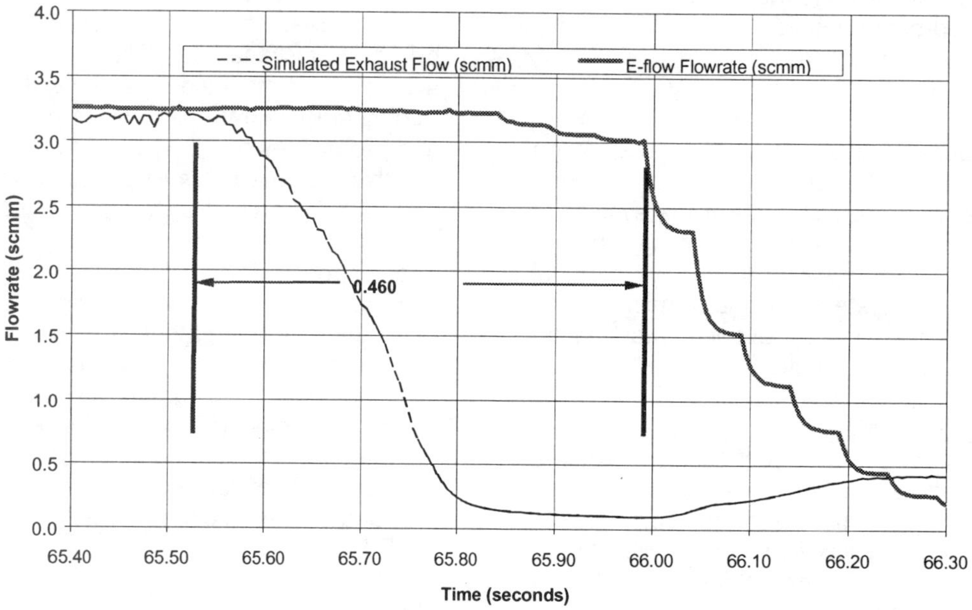

Figure 12. E-Flow Response Time – Descending

The T90 time was measured using the same technique. A step change in flow resulted in a 90% response time of 0.400 seconds (see Figure 13). This is less than the manufacturer's stated response of 0.500 seconds.

The critical factors for PZEV emissions measurement have been limiting contamination and improving instrument accuracy. However, for fuel economy, the key element is exhaust flow measurement. Prior to shipment, the E-Flow was calibrated by the manufacturer using a bell prover. Verification of the calibration was performed using Laminar Flow Elements (LFEs) and a blower to push ambient air into the inlet of the E-Flow.

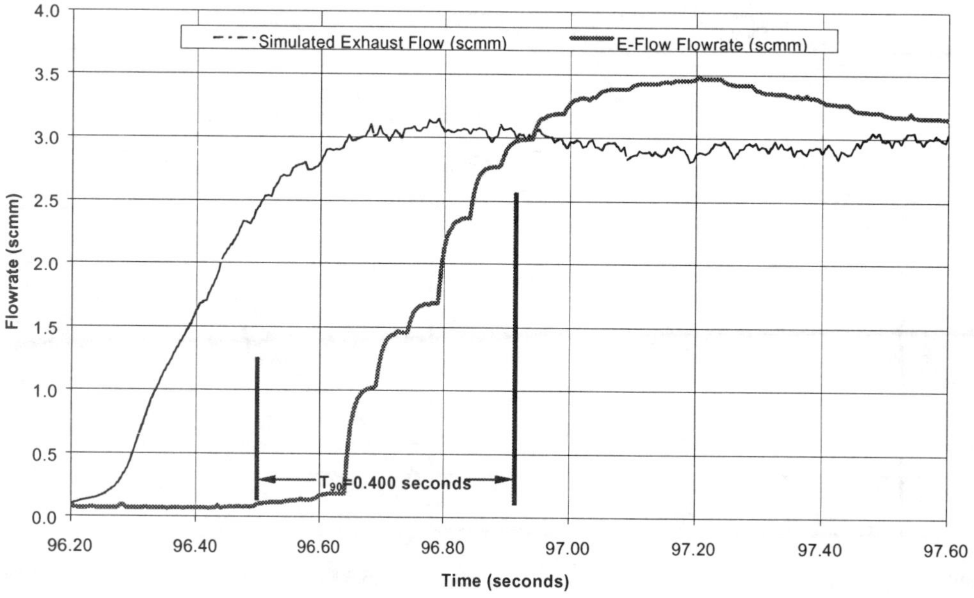

Figure 13. E-Flow T90 Response Time

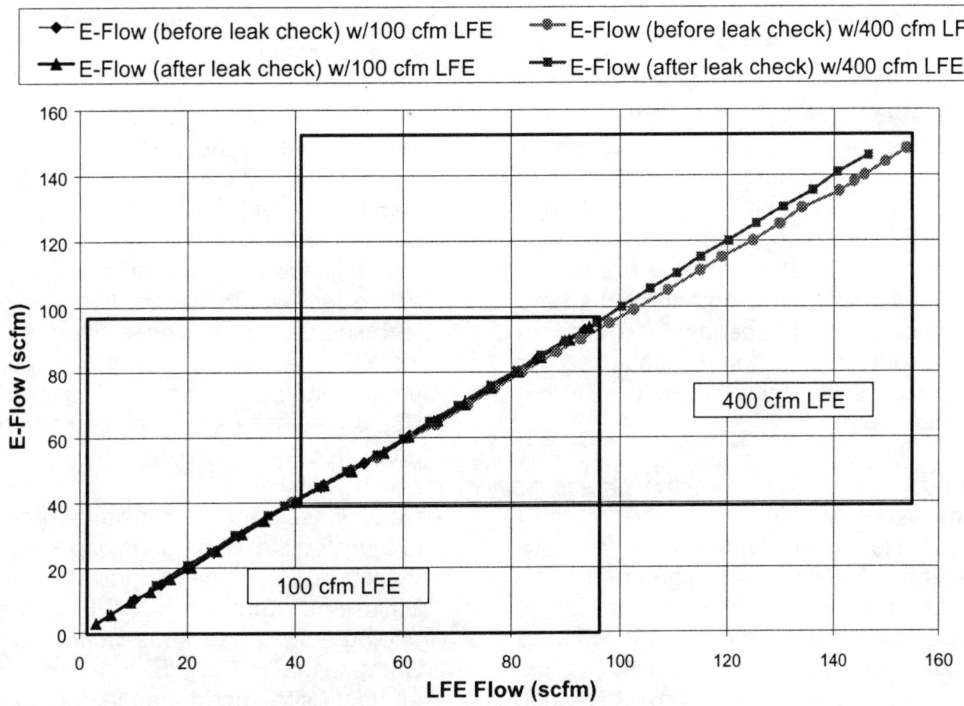

Figure 14. E-Flow Calibration Using "Overlapping" LFEs

Two different LFEs were used during this calibration in order to have the capability to span the full-scale range of the E-Flow (325 cfm or 9.20 cmm) while maintaining accuracy at the lower 20% of the range. A laminar flow element with a full-scale range of 100 cfm (2.83 cmm) was used for calibration of the E-Flow from 3 to 94 scfm (0.08 scmm to 2.66 scmm). A second LFE with a full-scale range of 400 cfm (11.33 cmm) was used to calibrate the E-Flow from 40 to 145 scfm (1.13 scmm to 4.11 scmm). This dual LFE method allowed for overlap of several data points to gain further confidence in the flow data (see Figure 14). Due to limitations on the blower supplying air to the E-Flow and LFE, 145 scfm was the maximum flow rate achieved for this calibration verification. The maximum flow rate of 145 scfm satisfies the flow range of PZEV vehicles currently under test.

Initial calibration data was collected with the E-Flow 'as-is'. Discrepancies between the factory calibration and the LFE calibration were discovered as the flow rate increased, indicating cause for investigation. Investigation of the interconnect piping from the calibration equipment to the E-Flow and internal components in the flow meter uncovered leaks at junction points. As the flow rate of the calibration blower increased, the pressure in the system piping increased producing more apparent leaks. The calibration procedure was repeated after these leaks were sealed, with improved results.

After repairing the leaks, data collected using the 100 cfm LFE demonstrated that the E-Flow calibration is accurate to within ±1.5% of point for flows between 3 and 30 scfm (0.08 and 0.85 scmm). Above 30 scfm, both the 100 cfm LFE and the 400 cfm LFE agreed with the E-Flow to within ±1.0% of point. Verification of the E-Flow calibration offers confidence in the measurement of the exhaust flow rate used to calculate the mass emissions of PZEV vehicles.

ENVIRONMENTAL OR FACILITY ISSUES - The PZEV Test Cell is temperature and humidity controlled to tight tolerances along with the aisle leading into the testing area. The cell is closed to the aisle during testing to help stabilize the ambient and to ensure front-to-rear airflow. A Road Speed Modulated (RSM) fan (45,000 cfm or 1,275 cmm) is provided along with the industry standard Hartzell. For US06 testing, the RSM fan is maintained at 15,000 cfm (425 cmm), which corresponds to an air speed of 20 mph (32 kph). Ample space in the test cell is needed to reduce variability due to air handling and airflow differences from car line to car line. The cell is approximately 1900 square feet by 19 feet in height (176.5 square meters by 5.8 meters in height) with 60 air exchanges per hour.

The layout of the equipment within the test site is for performance and not convenience. Plumbing lengths are kept to a minimum. The gas cylinders used for the zero, span and linearization functions are closely coupled to the bag bench by housing the cylinders in special cabinets within the test cell. The BMD sampling system and bag rack are within 8 feet of the bag bench, with a heated line between the BMD and bag bench. The moveable dynamometer allows the sampling equipment to remain stationary, while ensuring that the

BMD is no more than 12 feet from the vehicle's tailpipe at any time.

Also included in the "facility" category is the manufactured zero air generation system which is routinely tested for oxygen stability and hydrocarbon background level. The current system, though monitored, supplies 45 other test sites in the laboratory in addition to the PZEV facility. We believe there may be an opportunity for further improvement in this area and are experimenting with an independent system with part per billion (ppb) purification capabilities to determine the quantitative improvement to the testing data.

SAES Pure Gas manufactured the alternative on-site air purification system. Their MegaTorr Clean Air Supply (CAS) purifier is capable of providing zero air for BMD dilution with impurities at the ppb level. The purification is accomplished through three steps. First, gross moisture is removed using a water trap. This is followed by heated catalytic conversion of hydrocarbons, CO and H_2. Finally, NO_X, CO_2 and H_2O are removed through ambient adsorption. The purification process will also reduce the variability in oxygen content experienced with the current zero air supply system. Further testing, including continuous measurements of hydrocarbons, CO_2, NO_X, and oxygen, will be conducted over the next several months.

PROCESS - By "process" we refer to the operation involved within the test cell, the quality control of the equipment, and the change in culture needed to operate consistently at PZEV levels. While every precaution has been made to minimize the risk of hydrocarbon contamination, our intent is to further restrict the usage of the test cell to "clean" vehicles only. The suppliers, technicians, and skilled trades involved with the installations and operation of the test cell have an understanding of the "part per billion" influences on PZEV emissions. Of course, smoking is strictly prohibited.

Quality Assurance tools have been developed that are unique for the BMD sampling system, but perhaps the most significant development is the Vehicle Exhaust Emissions Simulator (VEES, See Figure 15 for photograph and subsequent data in "Test Site Comparison"). The VEES allows consistent monitoring of test site performance at the PZEV levels by simulating the vehicle's second by second mass emissions with gases injected directly into the BMD or CVS system [8].

Figure 15. Vehicle Exhaust Emissions Simulator (VEES)

Other quality assurance techniques are being evaluated for routine usage in the PZEV Test Cell, including dynamic propane injections, SF_6 injections, contamination (or "zero") checks, and dilution ratio checks.

Dynamic Propane Injections - During the Dynamic Propane Injection, propane is injected into the sampling system using a critical flow orifice (CFO) for metering. The propane is mixed with ambient air that follows a dynamic profile generated by a blower and a computer-controlled flow valve. A tolerance of ±2% is maintained between the mass of propane recovered by the BMD and the calculated mass from the CFO. This technique has proven to be a useful quality assurance tool for the BMD measurement system because it ensures accuracy of the exhaust flow meter and the system delay times.

Sulfur Hexaflouride Injections - To ensure the accuracy and measurement integrity of the test cell analysis equipment, a known mass of Sulfur Hexaflouride (SF_6) gas may be injected into the sample stream and collected in the BMD and CVS sample bags. A special gas analyzer calibrated to read SF_6 is used to read the sample bags and calculate the injected mass of the gas. SF_6 has several advantages over the traditional propane mass recovery method.

The first advantage is that SF_6 is not typically present in the test cell background air. Correction of background propane concentrations when calculating injected mass is a large source of variability and error in the propane recovery method. Because SF_6 is not present in the background air, subtraction of background concentrations is not necessary, thus eliminating this source of variability.

Another major advantage of the SF_6 mass recovery method is that the SF_6 gas may be injected during an actual vehicle test, without interfering with the emissions test results. This technique may then be used to verify the analysis system under actual operating pressures, temperatures, and flow rates. The accuracy of the Bag Mini-Diluter, exhaust flow measurement system, CVS, and gas analysis system may all be verified using this recovery technique. Due to time constraints, we have not collected sufficient data to report on the usefulness of SF_6 injections for quality assurance.

Contamination Checks - As reported previously [6 and 7], contamination checks are a valuable tool to determine the cleanliness of the measurement system. During the contamination check, no exhaust gas is introduced into either the BMD or CVS system. The BMD sample probe is flushed with high-purity zero air and the exhaust inlet to the E-Flow is capped while the timing of sample collection/analysis for a typical emissions test is duplicated. Three separate phases are performed for each contamination check to mimic the Federal Test Procedure.

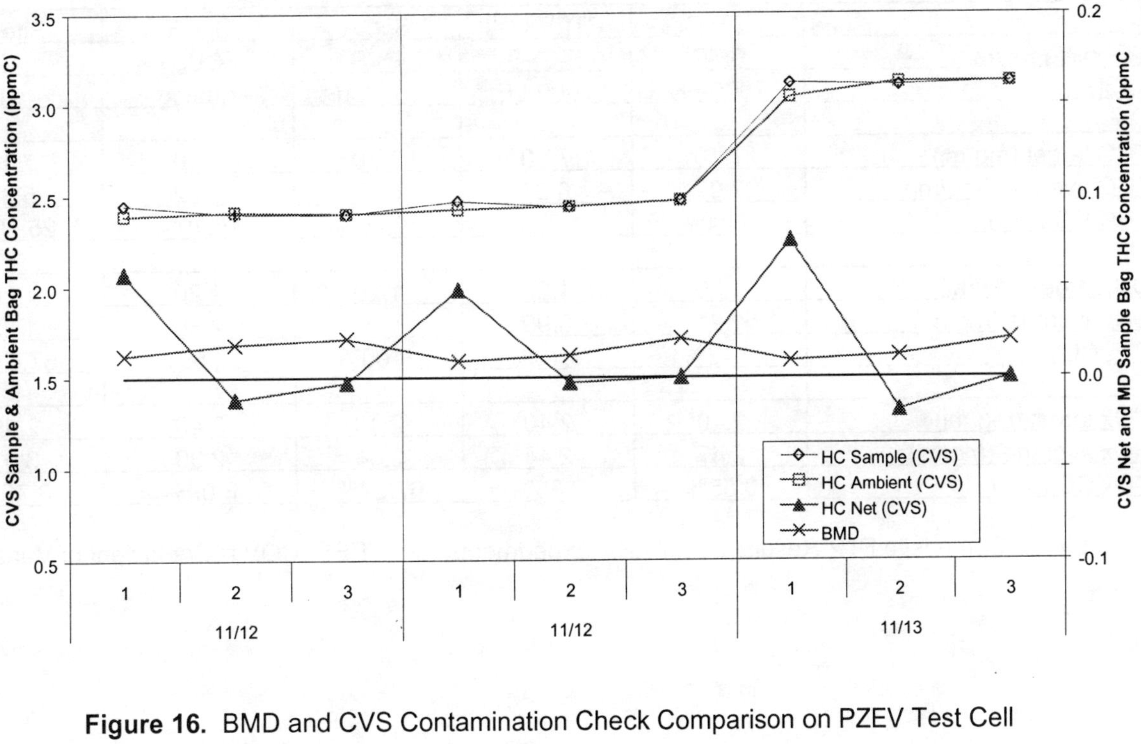

Figure 16. BMD and CVS Contamination Check Comparison on PZEV Test Cell

	Average THC Bag Concentration (ppmC)	Standard Deviation of THC Bag Concentration (ppmC)	Estimated THC Impact on FTP (mg/mi)
BMD	0.015	0.0017	0.087 ± 0.012
CVS Net	0.016	0.0077	0.086 ± 0.087

Table 4. Results from Three Contamination Checks on PZEV Test Cell

The results of three contamination checks on the PZEV Test Cell, which were interspersed with vehicle tests, are presented in Figure 16. Both the BMD and CVS demonstrated significant improvements when compared to results from the prototype equipment in Reference [7]. Table 4 includes a summary of the average concentration, standard deviation, and estimated impact of the contamination checks on the emissions measurements from the BMD and CVS. We currently perform contamination checks on a weekly basis.

Dilution Ratio Checks - The nominal dilution ratio for the critical flow venturis used in the BMD was determined to be 5.51:1 (i.e. one part of exhaust gas is diluted by 5.51 parts of zero air) using laminar flow elements as a flow standard and dry synthetic air as the calibration gas. Weekly dilution ratio checks, using the technique outlined in Reference [6], are performed using 4% CO_2 in nitrogen calibration gas to ensure that the BMD dilution control system is working properly. The current requirement for this check is that the dilution ratio measured using the 4% CO_2 calibration gas must agree with the nominal dilution ratio, as determined by LFEs, to within ±1.0%. Venturi adjustment factors are used to account for the small gas composition effects on the flow through the "exhaust" venturi when exposed to the different gases.

TEST SITE COMPARISON

A test program was performed using the VEES to compare emissions measurements from different test cells in our laboratory, including the PZEV Test Cell. The VEES was programmed to simulate the phase-by-phase FTP emissions profile from a representative PZEV test vehicle (3 phases were completed for each "composite FTP" simulation). Table 5 below compares the performance of various test cells with different vintages of equipment. At least four simulated FTP tests were conducted on each test cell.

The results from the PZEV Test Cell are unique in that the BMD and CVS are sampling in parallel with bag measurements taken by the same analyzer bench. The use of "E-CVS" in Table 5 below indicates that the CVS in the PZEV Test Cell has been enhanced by numerous improvements as compared to the older vintage CVS systems, including purge strategies, multiple venturi flow rates, and electropolished stainless steel plumbing. The data indicates that for low-level emissions measurement, modern analyzers are extremely important and further accuracy is achieved with a BMD sampling system.

	PZEV Cell	PZEV Cell	Site A	Site B	Site C
Sampling System	BMD	E-CVS	CVS	CVS	CVS
Analyzers	7000-LE	7000-LE	Analog	Analog	Analog
THC target (mg/mi)	7.10	7.10	7.10	7.10	7.10
THC average (mg/mi)	7.07	6.01	6.45	8.02	5.88
THC COV (%)	0.24%	1.07%	10.29%	15.19%	25.50%
CH_4 target (mg/mi)	1.20	1.20	1.20	1.20	1.20
CH_4 average (mg/mi)	1.21	0.90	-0.39	2.36	1.25
CH_4 COV (%)	0.42%	6.10%	>100%	27.59%	57.61%
NOx target (mg/mi)	2.40	2.40	2.40	2.40	2.40
NOx average (mg/mi)	2.47	2.44	3.04	2.99	3.03
NOx COV (%)	0.55%	2.85%	67.57%	8.09%	8.75%

Table 5. Composite FTP Results from PZEV Experiments with VEES (COV = Coefficient of Variation)

	Sampling System	THC (mgs)	CH_4 (mgs)	NO_x (mgs)	CO (mgs)	CO_2 (gms)
Phase 1	BMD	70.9	16.6	29.8	1227.8	1346.3
	CVS	69.1	16.6	28.4	1228.5	1338.9
Phase 2	BMD	2.1	0.0	0.3	268.1	1314.8
	CVS	-1.7	0.1	1.6	310.5	1330.4
Phase 3	BMD	1.7	0.9	0.5	165.8	1140.2
	CVS	-1.3	1.0	0.8	191.5	1138.2
Highway	BMD	1.3	0.6	1.1	358.0	2310.4
	CVS	-3.8	1.2	-1.7	403.0	2314.2

Table 6. Paired FTP and Highway Data Obtained From PZEV Test Vehicle

The data from the VEES experiment can be analyzed further to evaluate the measurement limits of the sampling and analytical equipment. During the 2nd Phase of the simulated FTP test, the VEES was programmed to inject just 2.0 mgs of THC (1 mg of propane plus 1 mg of methane), 1.0 mg of CH_4, and 1.0 mg of NOx into the sampling systems in the different test cells. On the PZEV Test Cell, the BMD recovered an average of 1.7 mgs of THC, 1.0 mg of CH_4, and 1.1 mgs of NOx. The Coefficient of Variation (COV) from these measurements was 0.08% for THC, 2.76% for CH4, and 2.97% for NOx. The 2nd Phase CVS measurements of THC and CH_4, regardless of the test cell, were typically negative. The COV during the 2nd Phase for all three emissions constituents (THC, CH_4, and NOx) on the CVS systems varied extensively, but was consistently much larger than 10%.

VEHICLE TESTING

Although the PZEV Test Cell is still in the commissioning stage, a variety of vehicles have been tested in the facility to ensure proper operation and to demonstrate correlation with other certification test cells. Table 6 includes an example of paired (i.e. both BMD and CVS measurements are taken in parallel) emissions results on a PZEV-compliant vehicle during the EPA City (FTP) and Highway test cycles. The composite FTP emissions from this test vehicle, as measured by the BMD, were 3.6 mgs/mile of NMHC and 1.8 mgs/mile of NOx. The PZEV test vehicle had a 2.0-liter, four-cylinder engine and a test weight of 3000 pounds (1361 kgs). It is important to note that only BMD results will be reported on the PZEV Test Cell in the future. The CVS results are only necessary to demonstrate correlation, and will be analyzed on a periodic basis by laboratory-support engineers.

CONCLUSION

The technical evolution of the Bag Mini-Diluter sampling system continues at a rapid pace. As discussed in various papers in the past, the BMD has several inherent advantages over the traditional Constant Volume Sampling (CVS) System.

- Dilution is at a fixed ratio (the CVS is a variable dilution process)
- Dilution with dry zero air obviates the ambient bag and corrects dilution factor assumptions
- Gaseous composition of the bag is consistent for water, Oxygen and CO2 (at stoichiometry)
- Considerations for heating, purging, reducing outgassing levels, etc. are simpler with the inherently smaller BMD sampling system.

Concerns with past prototype systems have been with measuring the gas transport delay time (from the probe to the bag), providing a proportional bag fill with respect to the exhaust flow, and achieving sufficient accuracy and/or response for the exhaust flow measurement system. We believe these concerns have been addressed and that appropriate tools exist to ensure traceable operation of the BMD system. We look forward to working with the new equipment, driving further improvement, and sharing further experiences as data is collected and analyzed.

ACKNOWLEDGMENTS

The authors wish to thank the following people for their contributions to this paper:

- Noelle Baker, Richard E. Chase, Darius Harrison, Travis Henney, Steve Hunter, Kim Isbrecht, Ed Kulik, Terry Laskowski, Mark Polster, Andre Welch from Ford Motor Company.
- Dave Balaka, Al Dageforde, Bob Gierada, Neal Harvey, Karl Oestergaard, Dan Whelan, Denny Wu from Horiba Instruments.
- Mike Wusterbarth from Flow Technologies, Inc.

REFERENCES

1. California Air Resources Board, "California Exhaust Emission Standards and Test Procedures for 2003 and Subsequent Model Zero-Emission Vehicles, and 2001 and Subsequent Model Hybrid Electric

Vehicles, in the Passenger Car, Light-Duty Truck and Medium-Duty Vehicle Classes", May 30, 2001
2. California Air Resources Board, "Dynamometer Performance Evaluation and Quality Assurance Procedures", March 17, 2000
3. M. Sherman, R. Chase, A. Mauti, Z. Rauker, W. Silvis, "Evaluation of Horiba MEXA-7000 Bag Bench Analyzers for Single Range Operation", SAE paper 1999-01-0147
4. A. Tayama, K. Kanetoshi, H. Tsuchida, H. Morita, "A Study of a Gasoline-Fueled Near-Zero-Emission Vehicle Using an Improved Emission Measurement System", SAE Paper 982555
5. K. Inoue, M. Ishihara, K. Akashi, M. Adachi, K. Ishida, "Numerical Analysis of Mass Emission Measurement Systems for Low Emission Vehicles", SAE Paper 1999-01-0150
6. M. Guenther, T. Henney, W. Silvis, S. Nakatani, D. Wu, "Improved Bag Mini-diluter Sampling System for Ultra-low-level Vehicle Exhaust Emissions", SAE Paper 2000-01-0792
7. M. Guenther, K. Brown, M. Landry, M. Sherman, D. Wu, "Refinement of a Bag Mini-Diluter System", SAE 2001-01-0212
8. M. Landry, M. Guenther, K. Isbrecht, G. Stevens, "Simulation of Low Level Exhaust Emissions for Evaluation of Sampling and Analytical Systems", SAE Paper 2001-01-0211

CONTACT

Author Information:

1. Mark Guenther, Senior Technical Specialist, Correlation Engineering, Vehicle Environmental Engineering, Ford Motor Company, Allen Park Test Laboratory, Suite 3W-100, 1500 Enterprise Drive, Allen Park, MI, 48101-2053, phone: (313) 594-2054, fax: (313) 594-2044, e-mail address: mguenthe@ford.com
2. Michael T. Sherman, Senior Technical Specialist, Correlation Engineering, Vehicle Environmental Engineering, Ford Motor Company, Allen Park Test Laboratory, Suite 3W-100, 1500 Enterprise Drive, Allen Park, MI, 48101-2053, phone: (313) 594-2056, fax: (313) 594-2044, e-mail address: msherma1@ford.com
3. Mike Vaillancourt, Principal Engineer, Correlation Engineering, Vehicle Environmental Engineering, Ford Motor Company, Allen Park Test Laboratory, Suite 3W-100, 1500 Enterprise Drive, Allen Park, MI, 48101-2053, phone: (313) 322-5228, fax: (313) 594-2044, e-mail address: mvaillan@ford.com
4. Daniel Carpenter, Product Development Engineer, Product Analysis and Verification, Ford Motor Company, Allen Park Test Laboratory, 1500 Enterprise Drive, Allen Park, MI, 48101-205 phone: (313) 621-2396, fax: (313) 390-3089, e-ma address: dcarpent@ford.com
5. Rick Rooney, System Engineer, Engin Measurement Division, Horiba Instruments, Inc. 5900 Hines Drive, Ann Arbor, MI, 48108, phone (734) 213-6555 x596, fax (734) 213-6525, e-ma address: rick.rooney@horiba.com
6. Scott Porter, System Engineer, Engin Measurement Division, Horiba Instruments, Inc. 5900 Hines Drive, Ann Arbor, MI, 48108, phone (734) 213-6555 x869, fax (734) 213-6525, e-mai address: scott.porter@horiba.com

2002-01-0047

Evaluation of Improved Bag Mini-Diluter System for Low Level Emissions Measurements

David J. Luzenski, Kyle T. Bedsole, Jim Hill, Donald B. Nagy and Steven S. Decarteret
General Motors Corporation

ABSTRACT

General Motors Corporation has installed its first production Bag Mini-Diluter Emission Test Site. This site is capable of conducting chassis dynamometer tests in either traditional CFV mode, or BMD mode, and a combination mode in which both technologies sample simultaneously. This paper discusses ambient interferent (Hydrocarbon and humidity) contamination issues which impact data resolution and measurement accuracy as emission standards are lowered. Solutions are recommended which compensate for these effects. This document examines new technologies required for accurate BMD measurements, and presents data obtained from running the site in the dual sampling mode utilizing a four-cylinder PZEV vehicle.

INTRODUCTION

With the adoption of the new California (LEV II) and Federal EPA (Tier II) vehicle standards, it has been widely recognized that new technologies are needed in emission test equipment to measure these very low emission levels. Significant research has been put into one of these new equipment technologies, namely the Bag Mini-diluter (BMD). It is believed that BMDs offer the best solution to achieve needed accuracy goals at such low levels. This paper describes advances which further improve the BMD system's performance over that of prototype units. In addition, this study looks at the pitfalls associated with conventional sampling techniques (CFV/CVS) and how BMDs overcome these obstacles

BACKGROUND-PROBLEM WITH TODAY'S CFV

A BRIEF REVIEW OF TODAY'S CFV MEASUREMENT TECHNIQUES - To measure vehicle exhaust emissions, most testing laboratories make use of a Critical Flow Venturi (CFV) exhaust sampling system. Briefly, with this device, ***all*** of the vehicle's exhaust gases are routed into a sample collection system and then diluted with enough ambient air to prevent water condensation. After dilution, a small amount of the gas samples are put into a set of sample bags for later chemical analysis. In addition to these diluted exhaust bags, a set of ambient bags are also simultaneously filled for later chemical analysis. This well-documented technique is typically referred to as a full dilution system[1]. Diagrammatically, this measurement process looks something like this:

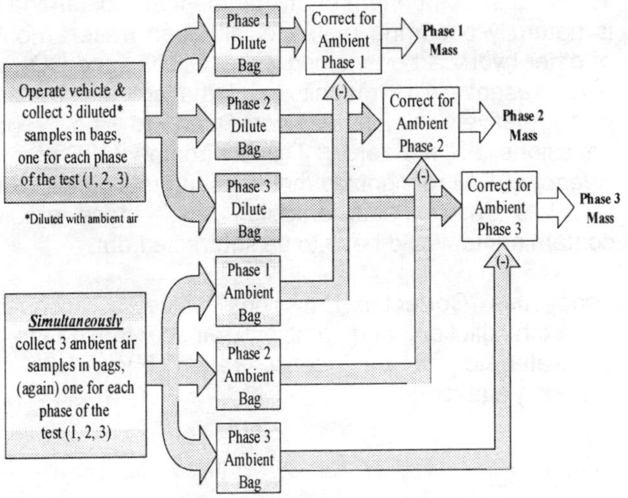

Figure 1: Full Dilution System

In this process, the test vehicle is operated over a defined driving schedule, which for a US FTP test is segmented into three different time phases (phases 1, 2 and 3). As previously mentioned, the entire exhaust gas is mixed with ambient air typically at a 15:1 dilution factor, to avoid water condensation in the bags. A 15:1 dilution means that one part of exhaust gas is mixed with 14 parts of ambient air, for a total of 15 parts. Separately, for each phase, after these two gases are thoroughly mixed, a small continuously proportional amount of this diluted exhaust gas is placed into sample bags. At the same time, samples of ambient air are also placed into separate sample bags, one for each phase.

Finally, these six sample bags are chemically analyzed for emissions, and the diluted sample bags are then "compensated" for ambient air contamination. Many people erroneously think that this ambient contamination is directly subtracted from the vehicle emissions. Actually, the ambient bag correction subtracts only that portion of the ambient air contamination which results from the dilution of the raw exhaust gases. Once corrected for ambient air contamination, the mass emissions can then be determined. This chemical analysis typically identifies the amount of total hydrocarbons (THC), methane (CH_4), carbon monoxide (CO), oxides of nitrogen (NO and NO_2) and carbon dioxide (CO_2) in the bags and hence, emissions from the vehicle. Note that the above measurement/analysis scenario is a highly abbreviated version, while the actual methodology requires many more steps to complete.

HISTORY OF AMBIENT AIR COMPENSATION - For the 1972 model year, the Federal Register Test Procedure was changed to specify a new type of vehicle emissions measurement technique, where ambient air is used to dilute the raw exhaust gases. At this time, the adopted standard for hydrocarbons was 3.4 weighted grams/mile. Even 30 years ago, this regulation recognized that a measurement error would occur with this full dilution technique due to (1) the dilution itself and (2) the mixing of ambient air contaminants with the raw exhaust gases. The largest component of this ambient air contamination is naturally occurring methane, although trace amounts of other hydrocarbons, along with CO, NO_X and CO_2, are also present. The magnitude of the ambient air error was relatively small as compared to the vehicle emissions, but not zero. The Environmental Protection Agency (EPA) recognized that to accurately measure the vehicle's true exhaust emission level, these ambient contaminants would have to be subtracted out.

Today this "Corrected Emissions" value, compensated for both dilution and ambient air contamination, is calculated on a bag concentration basis with the following equations:

$$Corrected\ Emissions_{conc} = Dilute\ Bag_{conc} - Ambient\ Bag_{conc} * (1 - \frac{1}{DF})$$

Equation 1

Where:

Dilute Bag_{conc} = The particular emission constituent (HC, CH_4, CO, NOx,CO_2) concentration as measured in the diluted exhaust sample bag.

Ambient Bag_{conc} = The particular emission constituent (HC, CH_4, CO, NOx,CO_2) concentration as measured in the ambient sample bag.

$$DF = Dilution\ Factor = \frac{13.4}{(3 * HC_{conc} + CO_{conc}) * 0.0001 + CO_{2conc}}$$

Equation 2

Where:

HC_{conc} = HC concentration (ppmC) found in the diluted exhaust bag
CO_{conc} = CO concentration (ppm) found in the diluted exhaust bag
CO_{2conc} = CO_2 concentration (%) found in the diluted exhaust bag
13.4 = Molar fraction of carbon in Indolene fuel

Again, the above equations are in a simplified form and can be different depending on the test type and fuel. Note that only a portion (1-1/DF) of the ambient is subtracted from the diluted exhaust bag, in keeping with the philosophy that only dilution air may be subtracted from the diluted exhaust sample bag.

AMBIENT AIR INTERFERENT

AMBIENT AIR IS AN INTERFERENT - The effect of ambient air contamination is a topic of much controversy, especially at the lower emission levels, i.e. SULEV. One way to look at the potential interference is to look at the measurement from a classical signal-to-noise perspective.

In any measurement system, the primary phenomenon one is trying to measure (temperature, pressure, emissions, etc.) is the ***desired response.*** The instrumentation used to make this primary measurement will almost always have some other phenomenon which will interfere with it, yielding an **undesired response**. Undesired responses will cause measurement errors. Ideally, one would like this undesired response (or interference) to account for only a small error to the primary measurement, maybe less than 10%. By actively measuring the interferent with another sensor, and compensating the primary measurement for this error, one can usually compensate for this interference. Although active measurement of the interferent is the most accurate way of compensation, other techniques could be used. For example, if the interferent is constant, then a simple mathematical subtraction could be used. Sometimes if the interferent is very small, one could ignore it altogether. Either way, the ratio of the interferent error to the desired primary measurement depends on (1) how much the interferent affects the primary measurement, and (2) the accuracy of the compensation.

EFFECTS OF AMBIENT AIR HYDROCARBON INTERFERENT - Applying this to the measurement of vehicle emissions, and specifically for hydrocarbons, how does ambient hydrocarbon ***interference*** potentially effect the measured emissions?

Ambient interferent has historically been described in units of parts per million (ppm). Using the amount of hydrocarbons present in ppm really didn't quantify the potential effect. A more useful way to look at this error is by converting from ppm to the regulatory standards of weighted grams/mile (wtd g/mi).

To convert ambient concentrations to the equivalent weighted grams/mile requires a few equations and assumptions. First, the Federal Register defines how to calculate the normal weighted grams per mile for each emission constituent as follows:

$$Y_{WM} = 0.43 * \left\{ \frac{Mass_{CT} + Mass_S}{Dist_{CT} + Dist_S} \right\} + 0.57 * \left\{ \frac{Mass_{HT} + Mass_S}{Dist_{HT} + Dist_S} \right\} \quad Equation \quad 3$$

where: Y_{WM} = Weighted mass (in grams / mile) of the particular emission constituent
$Mass_{CT}$ = Mass emissions for the cold transient phase of the test, or phase I
$Mass_S$ = Mass emissions for the stabilized phase of the test, or phase II
$Mass_{HT}$ = Mass emissions for the hot transient phase of the test, or phase III
$Dist_{CT}$ = Distance traveled (in miles) for the cold transient phase of the test
$Dist_S$ = Distance traveled (in miles) for the stabilized phase of the test
$Dist_{HT}$ = Distance traveled (in miles) for the hot transient phase of the test

For the purposes of estimating the magnitude of this ambient air interference, Equation 3 can be simplified by making the assumption that $Dist_{CT} = Dist_{HT}$. This can be done since both distances are derived by following identical driving schedules (time versus speed matrix). Also assume nominal distances for each phase of the test (i.e., $Dist_{CT} = Dist_{HT}$ = 3.59 miles and $Dist_S$ = 3.86 miles). With these assumptions Equation 3 can be broken down into individual phase components as follows:

$$Phase\ 1\ Mass = 0.43 * \left\{ \frac{Mass_{CT}}{7.45} \right\} \quad Equation \quad 4$$

$$Phase\ 2\ Mass = 1.0 * \left\{ \frac{Mass_S}{7.45} \right\} \quad Equation \quad 5$$

$$Phase\ 3\ Mass = 0.57 * \left\{ \frac{Mass_{HT}}{7.45} \right\} \quad Equation \quad 6$$

These Phase equations yield weighted grams/mile.

To calculate the simulated ambient air interference a final step is needed: to configure the CFV as a "blank test". A "blank test" is a simulation of a conventional FTP test schedule on a CFV system without a vehicle. For this test, the exhaust gas collection pipe is capped. Under these conditions the 3 diluted sample bags (phases 1,2 and 3) would be filled with the same dilution air as the 3 ambient bags (phases 1, 2 and 3), i.e., ambient air. Typically, this ambient air contains about 1 ppm C_3 of total hydrocarbons as read by a flame ionization detector, and 0.8 ppm C_3 of methane as read by a gas chromatograph.

Making these assumptions, and using equations 4, 5 and 6, the interference masses can be estimated by bag, phase and constituent. These calculated phase weighted masses and the total weighted mass are shown below:

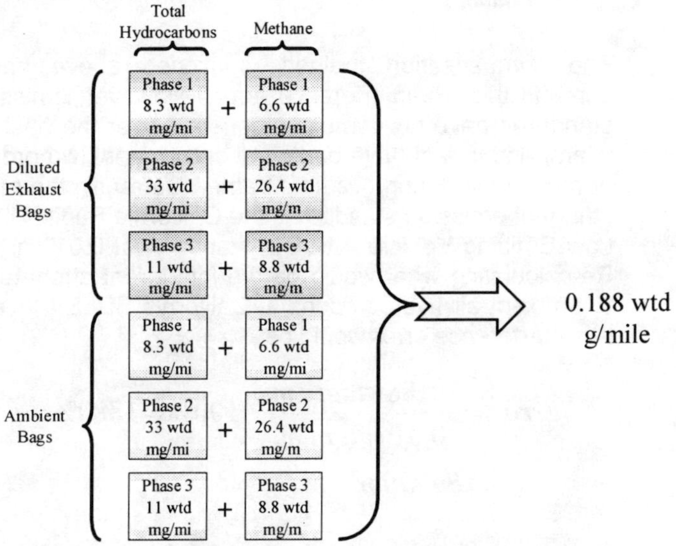

Figure 2: Ambient Interferent Contribution to FTP

Note in the above scenario, the masses from each phase bag are being added together to yield a total, measured mass of 0.188 wtd g/mile. This was done to demonstrate an important point: what is the total effect of the ambient air interferents? To compensate for the total effect of these interferents, and achieve zero emissions for this "blank" case, the CFV technique will have to compensate for the total, measured mass of 0.188 wtd g/mi. What the CFV technology actually does is to try to estimate what these ambient interferent errors are, then try to compensate for them. This is done by taking a representative gas sample of dilution air for each phase of the test (which is an estimate), then through a series of 6 bag fills, 6 total hydrocarbon analyses, 6 methane analyses, and 9 subtractions (3 total hydrocarbons minus methane; 6 diluted sample minus ambient), the **estimated** contribution of ambient air interference would be as shown below:

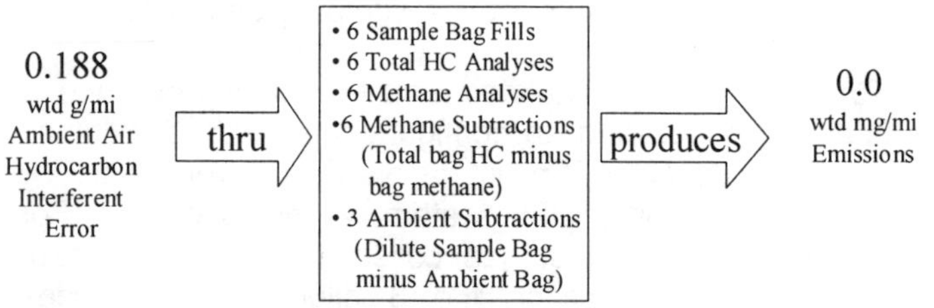

Figure 3: CFV Corrections for Ambient Interferent

In 1972 when the hydrocarbon emission standard was 3.4 wtd grams/mi, this ambient air interference was fairly small as follows:

$$\%Interference(or\ Error) = \frac{0.188wtdgr/mi}{3.4wtdgr/mi} * 100\% = 5.5\%$$

Equation 7

The compensation method mentioned above easily corrects this 5% interference error. However, emission standards have not remained constant over the past 30 years, and in fact have been reduced at least two orders of magnitude during this time frame. Currently, the most stringent emission standard is the California Super-Ultra-Low Emitting Vehicle (SULEV) standard, at 0.010 gr/mi. Re-calculating what would be the interferent attributable to ambient air hydrocarbons at California SULEV levels, the interference error would be:

$$\%Error = \frac{0.188wtdgr/mi}{0.01wtdgr/mi} * 100\% = 1880\%$$

Equation 8

That is, ambient air intereference is effectively ***19*** times higher than the SULEV standard**.**

AMBIENT AIR INTERFERENCE CHANGES WITH TEST SITE OPERATING CONDITIONS - Ambient interferent levels of **19 times** the SULEV standard is not a worst case scenario – it actually consists of fairly ideal measurement conditions. Our previous analysis assumed the CFV flowrate was 9.9 m^3/min (typical) and the ambient total hydrocarbon levels were about 1 ppm C_3. However, many of today's CFV's are multi-venturi units that can increase the flowrate up to 20 m^3/min for high speed tests like the US06, or even a higher flowrate (34 m^3/min) for alternative fueled vehicles. Similarly, ambient hydrocarbon levels can be as high as 3 ppm C_3 under certain circumstances. Calculating the effect these variants of CFV flow or ambient HC levels have on the ambient hydrocarbon interferent yields:

Ambient Total Hydrocarbon Concentration in ppm C_3	CFV Flowrate in M^3/Min: 9.9	20	34
1	**19 x** SULEV std	**38 x** SULEV std	**65 x** SULEV std
2	**29 x** SULEV std	**58 x** SULEV std	**99 x** SULEV std
3	**40 x** SULEV std	**80 x** SULEV std	**136 x** SULEV std

Table 1: Ambient Interferent as a Function of Site Flowrate and Ambient HC Levels (assumes Methane portion of ambient is constant at 0.8 ppm C_3)

While the 20 m^3/min flowrate is routinely used for the US06 test and alcohol-fueled vehicles, under unusual circumstances, the 34 m^3/min flowrate will be used. Using this flowrate and an ambient contamination of 2 ppm C_3, the ambient hydrocarbon interferent could be as high as 99 times the SULEV standard

AMBIENT AIR INTERFERENCE SUMMARY - The previous discussion illustrates that ambient air hydrocarbons are an interferent to the measurement of vehicle exhaust emissions. In addition, this interferent has dramatically increased its influence by orders of magnitude over the past 30 years, as emission standards have become more stringent by orders of magnitude.

This is not to say that the full dilution CFV sampling system has not done a remarkable job thus far...it has. However, at very low emissions levels, the previously described physics starts to inflict more variability into the measurement process. Clearly, one would not like to have interferences that are orders of magnitude above the desired measurement, and one would quickly move toward some other form of measurement system to eliminate the problem.

As emissions standards are not fixed, they could go even lower over time. Obviously, this would further exacerbate the previously mentioned problems. For accurate measurements, it is clear that ambient air must be eliminated as the primary diluent.

BMD ELIMINATES AMBIENT DILUTION AIR INTERFERENCE ERRORS - The Bag Mini-Diluter solution which eliminates the ambient air has been very well documented in SAE literature (see references). The BMD eliminates ambient air all together through partial dilution of the raw exhaust gases with zero air. By definition, zero air has nearly zero HC contaminants – perhaps 5 ppbC_3 or less. Note that compared to ambient air contaminants, zero air contaminants are but a small fraction as shown below:

$$\frac{ZeroAirHCContamination}{AmbientAirHCContamination} = \frac{0.005\,ppmC_3}{1\,ppmC_3} = 0.005 = 0.5\%$$

Equation 9

This small zero air contaminant can be automatically compensated for by zeroing the HC analyzer on the same gas, and thereby eliminating its contribution to the measurement error altogether.

SAMPLING MATERIAL INTERFERENCE - There is one other source of measurement interference that hasn't been discussed yet, and that is the materials used to collect the vehicle exhaust emission samples. Typically, this interferent, again, effects the hydrocarbon readings. The source of this new error is typically in polymer compounds used to make flexible connections like tubing, sealing gaskets, pump diaphragms and, most importantly, for the Tedlar® sample storage bags. These materials outgas multiple chemicals, which effect the hydrocarbon analysis. One major source of material contamination is a material called N,N-dimethylacetamide (DMA). DMA is added to Tedlar® to improve the elastic characteristics and eliminate problems associated with the bag manufacturing process. Much cleaner bag materials are available (e.g KYNAR®). Several recent SAE papers explore this sample bag contaminant in more detail.

One of the benefits of using a bag mini-diluter sampler over a conventional CFV sampling system is a constant and lower dilution ratio. On a CFV system the dilution ratio can range from 4:1 to 60:1, depending on the system's upper flow limit and on the vehicle's engine displacement. As a result, there will be varying amounts of moisture throughout the sample system. It has been determined that there is a correlation between humidity and hydrocarbon (HC) contamination (see below). Actually, these sample material contaminants effect both the CFV and the BMD by approximately (but not exactly) the same amount. Because the BMD provides a constant dilution it will be shown that the contamination levels are constant and predictable. However, because the CFV sample system does not produce constant humidity, the resultant contamination **can** become a significant portion of the standard.

Theory says that the CFV will compensate for such bag material interference because of the ambient air sample bag "subtraction," but data indicates that this is **not true**. Studies have shown that the magnitude of this interference is a function of the humidity in the sample. In the CFV case, the diluted exhaust sample bag will, by definition, have more humidity in it than the ambient sample bag. Therefore, even with perfect subtraction, one will only partially compensate for this interferent. Because there is no ambient bag to subtract, the BMD will experience the full magnitude of this material interferent. However, it turns out that the effect of this material interferent is similar to that of the CFV, just due to the difference in dilution ratios (15:1 vs. 6:1), which acts as a multiplier.

Experiment –One can estimate how much this material interferent will effect the BMD hydrocarbon readings. A study was run on both the CFV and BMD system where exhaust samples were collected simultaneously by the two methods. Note that this site utilizes clean KYNAR® sample bags. To determine this error, and thus validate the theory that only a small portion of the contamination effect is compensated by the ambient bag subtraction, humidified zero air was directly injected into the CFV & BMD sampling system.

The experiment consisted of filling bags with differing amounts of humidified, zero air according to the 3-phase FTP method. The CFV ambient bags were filled with dry zero air and the CFV sample and BMD bags were filled with varying amounts of humidified zero air, thus simulating the conditions as if a vehicle were under test. Humidified air was provided by flowing zero air through a bubbler. The saturated zero air was then injected into the BMD and CFV sample lines. The diluted sample bags were then analyzed for trace hydrocarbon levels.

Results – The results of the tests are shown in Graph 1. This plot indicates a relationship between HC

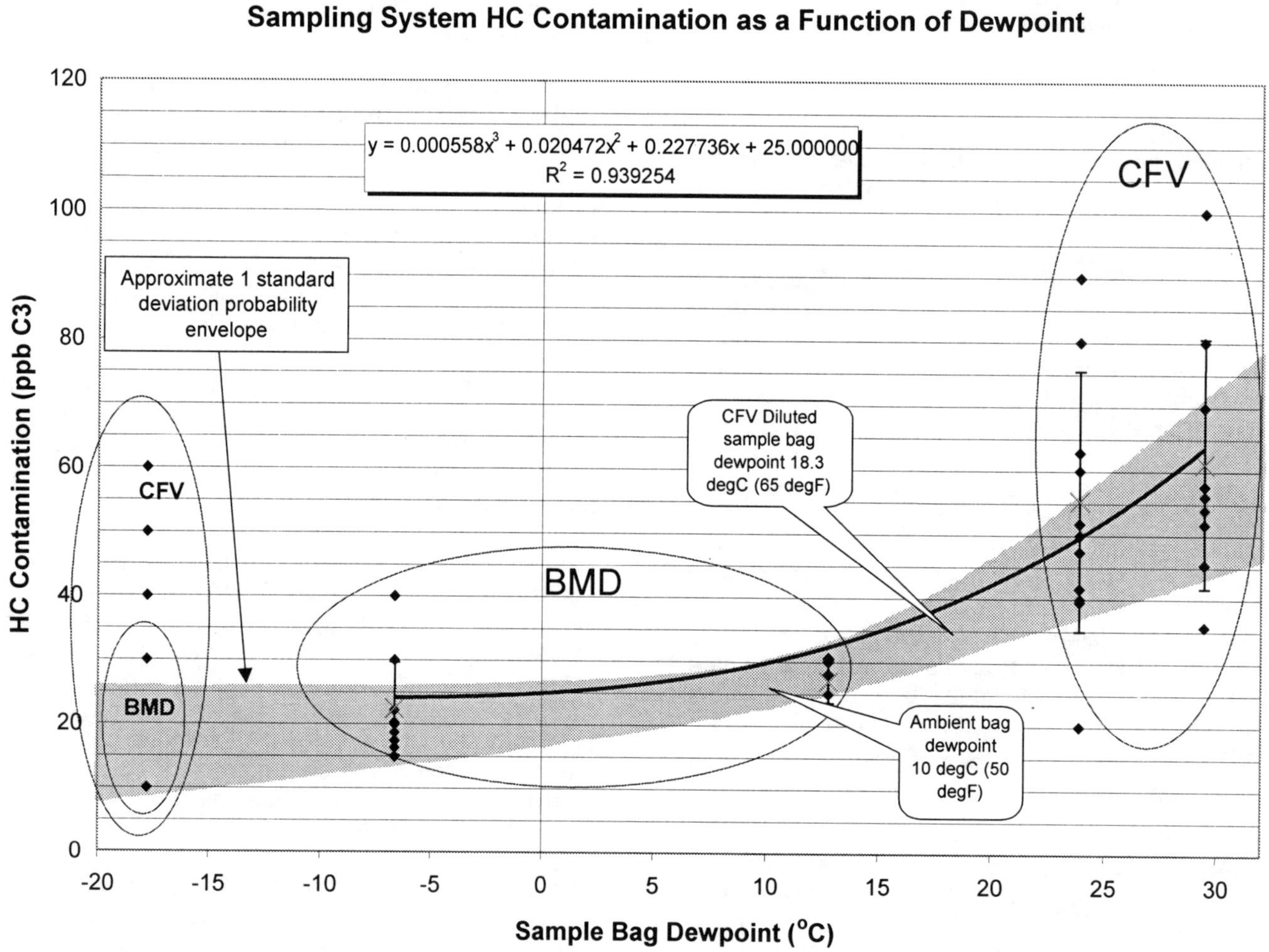

Graph 1: HC Contamination Dependence on Moisture for the CFV & BMD Sampling Systems

contamination in the sample system and sample bag dewpoint. A regression line was fitted through these data and a probability envelope (1 standard deviation) is also indicated.

Several key points can be ascertained from the plot shown in Graph 1. Note that there is a specific trend between humidity and HC contamination. This trend indicates as humidity increases so does the relative contamination level. However, this is mainly true for the CFV method. In comparing the readings obtained from the BMD method one can see that there is no significant contamination change as the humidity is increased in the samples. As indicated in this plot, the CFV bags have much higher humidity levels and consequently a higher resultant contamination level. Because the BMD operates at a constant dilution ratio (7:1 in this case), one can predict, with some confidence that the HC contamination is constant. This is not the case in the CFV method. The impact that this phenomenon has on vehicle mass data is discussed in following sections.

Also indicated from the plot shown in Graph 1 is that there is greater spread in the data obtained in the CFV sampling method as compared to the BMD method. This is true for both low and high humidity.

A mathematical simulation was performed to show the relationship between the bag mini-diluter method and the CFV method in terms of total hydrocarbon (THC) mass. Graph 2 summarizes these data. Note that in the CFV method, two engine displacements were studied in terms of mass. Recall that the humidity in the sample bag increases as displacement increases and subsequently the HC contamination also increases. This plot shows THC mass (weighted mg/mile) as a function of phase number and also as a composite average.

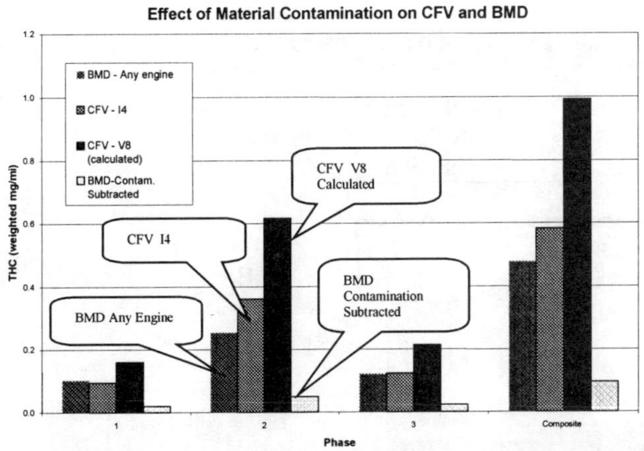

Graph 2: Contrast of Contamination of CFV and BMD

The data shown in Graph 2 was generated from conclusions obtained from the model shown in Graph 1. The calculated mass shown on the bar plot in Graph 2 for the BMD method is based on a contamination level of 25 ppbC_3. It should be noted that the BMD method is independent of engine size and consequently the contamination level is constant (as indicated earlier). The weighted composite mass for the BMD method is approximately 0.45 mg/mile.

In terms of contributory response from HC contamination, estimated comparisons are also made between an 8-cylinder engine and a 4-cylinder operated on a 9 m^3/min (325 ft^3/min) CFV test site. For these calculated mass values, the bag HC concentrations for the ambient and the sample were extracted from the model shown in Graph 1. For each CFV ambient bag, a HC concentration of 25 ppbC_3 was used. This value corresponds to an ambient sample bag dewpoint of 10^{0}C. For the sample bag in the 4-cylinder case a contamination level of 35 ppbC_3 was used. This value corresponds to a dewpoint of 18^{0}C. For the 8-cylinder simulation a HC contamination of 42 ppbC_3 was used. This level corresponds to a sample bag dewpoint of 24^{0}C. It should be noted that these are conservative estimates and that more dramatic results would have been observed had more extreme values been used. It can be seen from the plot in Graph 1 that the regression line is a cubic function; consequently, the contamination levels increase at a much faster rate as the bag humidity is increased above 13^{0}C.

In comparing the results shown in Graph 2, it is clear that with the CFV method there is a direct relationship, in terms of hydrocarbons, between engine size and sample bag humidity. It is worth noting again that these values are the result of zero air samples. Table 2 shows a comparison between the three representations in terms of the composite average as percentage of the SULEV standard.

	Percent of SULEV STD
BMD	4.5%
CFV – 4 CYL	5.4%
CFV – 8 CYL	9.2%
BMD – With Subtraction of dry zero air contamination	0.9%

Table 2: Comparisons as a % of SULEV Standard

Note that in this model the % error due to sampling materials is small, around 5% of the SULEV standard. Note also that previous data (not presented here) taken on the CFV system has shown much higher errors dependent on initial conditions.

BMD technology, by eliminating the ambient air as a dilution medium, has basically returned measurement interference levels back to those seen in 1972 (see Equation 7). With a modest compensation for the material contaminants, this error could be dramatically reduced as listed below:

1. Looking for better materials that outgas less hydrocarbons, thereby minimizing the interferent.
2. Implementing an existing CFR compensation technique used in diesel testing. In diesel hydrocarbon emissions analysis, the heated FID analyzer compensates for contamination offsets by zeroing through the end of the heated sample probe via a probe overflow technique. Likewise with the BMD, if one were to zero the bag bench FID analyzer through the BMD sampling system, it is anticipated that this material contamination error would be cut in half (2% of the SULEV standard).
3. Measuring the amount of this interferent on some periodic basis then subtract this amount from the bag readings. This is not the most accurate approach, but because the correction is so small (5% of the SULEV standard), its compensation accuracy could actually be quite liberal, maybe as much as ± 20%. This ± 20% compensation error would result in a total system error of ±1% of the SULEV standard for vehicle emissions measurements.

To reiterate, the experimental data indicates that if dry zero air were injected into the BMD sample system, the system background will be relatively constant at 20 ppbC_3. If this contribution is then subtracted from the total (as contributed from a wet sample, 25 ppbC_3), then the overall system HC contamination contribution would be 0.09 mg/mile. This represents **less than 1 percent** of the SULEV standard.

Schematic of General Motors Bag Mini-Diluter

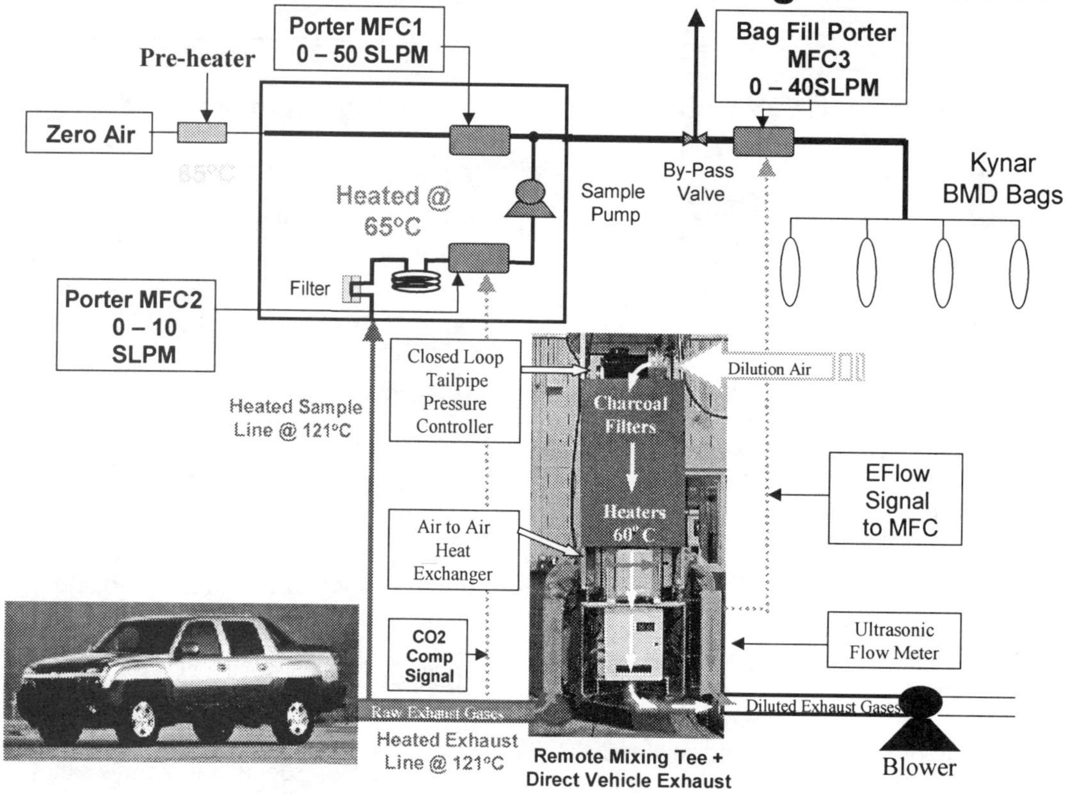

Schematic 1: Overview Of Bag Mini-Diluter Illustrating Heated Diluent Injection, MFC2 CO_2 Compensation Feedback, And MFC3 Proportional Control Via Direct Vehicle Exhaust Measurement.

BMD-EQUIPPED EMISSION TEST SITE

Utilizing several of the previously discussed technologies, the bag mini-diluter system is depicted above in Schematic 1. This system utilizes mass flow controllers (other researchers utilize other techniques, i.e. CFV[2]) to meter the exhaust sample and the zero air diluent. The mass flow controllers are located in an electrically heated oven maintained at 65°C, necessary to provide a stable thermal environment. MFC3 meters the diluted sample into the bags--continuously proportional to the vehicle's exhaust flow and delayed for the transport time lag. This proportionality is guaranteed by the accurate measurement of the exhaust flow via the DVE, a device utilizing non-invasive ultrasonic technology.

As this site is also capable of running traditional CFV tests, the diluted exhaust gases are collected via the site blower which maintains a constant flow rate via the main site venturi which for this site is 9 m^3/min (325 SCFM). Also note that the exhaust sample is heated from the extraction point to the dilution point.

This is the first BMD-based production emission test site commissioned at the Milford Emission Test Laboratory. The installed equipment takes advantage of some methodology advances gained from participation with the AIGER consortium. The following sections will detail improvements in DVE and BMD hardware.

Hardware Improvements

A. Vehicle Exhaust Measurement

BACKGROUND - Several researchers have noted that the largest source of variability for mini-diluter carbon dioxide calculations resides with the difficulty in accurate measurement of exhaust flow.[3] This enhanced direct vehicle exhaust (DVE) device decreases the measurement variability of CO_2 along with other constituents of HC, CO, and NO_X by providing a repeatable measurement of exhaust volume. This unit utilizes a Kaijo Transient-Time Ultrasonic (TTU) gas flow meter packaged into a remote mixing tee with exhaust backpressure control. This meter differs from devices used in earlier research in that the TTU is vertical as opposed to horizontal with respect to the exhaust flow path. This vertical orientation avoids any effects due to buoyancy theory[4], that is, the tendency of air currents to draft upward and create non-uniform flow when the meter is in a horizontal orientation. This direct measurement of the exhaust flow provides the control signal for the continuously proportional filling of the sample collection bags for the Bag Mini-Diluter.

FLOW METER DESIGN – Previous DVE designs are larger and are difficult to locate in test cells built to house less sampling equipment. One of the design goals was to minimize intrusion into the test cell and capital investment.[5] The unit is a standard General Motors Remote Mixing Tee (RMT) with backpressure control and dilution air heat. Flow Technology Inc of Phoenix, AZ modified the standard unit to include an air-to-air heat exchanger, the Kaijo TTU and associated electronics. As a constant flow rate is maintained by the site main venturi, the RMT controls the tailpipe backpressure by varying the amount of dilution air added downstream of the BMD sample probe and ultrasonic transducer.

The DVE uses an air-cooled version of the ultrasonic transducer that allows the use of an air-to-air heat exchanger to reduce the exhaust temperature to below 350 °C (660 °F), the maximum allowable for the ultrasonic transducer. The exhaust flows through the tube side of the heat exchanger and is conditioned by the dilution air flowing through the shell side of the heat exchanger. The ultra sonic transducer is cooled by shop-grade compressed air which has passed through a vortex tube.

During pre-test, the dilution air heater is used to keep the system above 60°C (140°F) to prevent condensation. The heater is also required for testing small displacement engines as these generate neither the heat nor flow to keep the exhaust water content in suspension. The choice of an air-to-air heat exchanger significantly reduced the suspended weight of the unit and eliminated the components associated with a water-to-air heat exchanger namely the pump, tubing, radiator, etc. The exhaust sample is extracted just prior to entering the heat exchanger. Via heated blankets and heated lines, the sample is kept above it's dew point from vehicle exit to the BMD where zero air dilutes the sample above it's dew point.

Picture 1: View of DVE with Backpressure Control

The addition of the heat exchanger and the Kaijo transducer increased the flow path's pressure drop above that of a standard RMT. Consequently, a proportionally controlled butterfly valve located in the dilution air path is required to force site vacuum to draw on the tailpipe during WOT. Without the valve, vacuum follows the path of least resistance and draws from the dilution air hopper. This condition leads to positive pressure being developed at the tailpipe during WOT. The backpressure controller is set so that the valve closes during positive pressure events, thus closing off the dilution air path. The Dilution air blowers develop enough flow to supply ambient air up to the rating of the test cell's exhaust blower for traditional CFV testing.

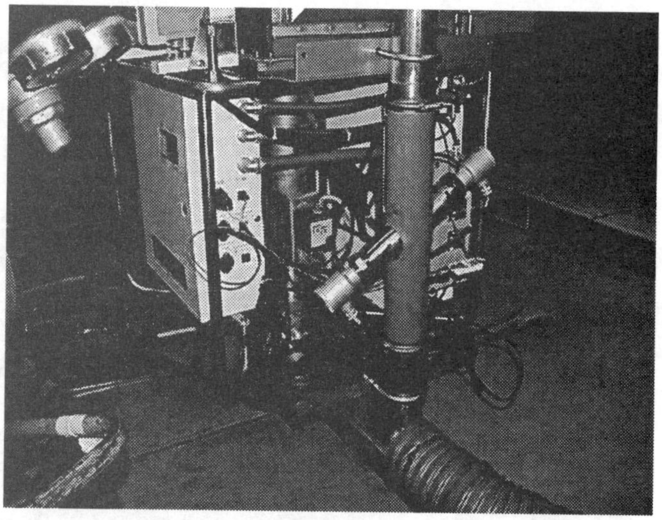

Picture 2: View of Ultrasonic Transducer

Lambda Meter– The DVE is modified with a meter which measures lambda,λ, in the raw exhaust flow stream. This data is used by the BMD system to adjust the raw exhaust flow controller in response to compositional changes of CO_2 and H_2O in the exhaust.

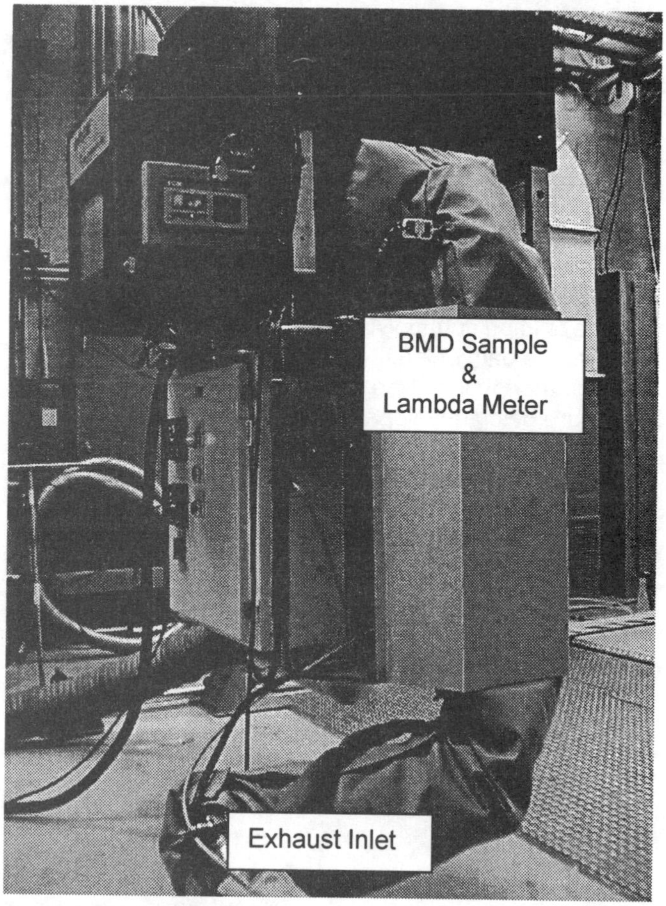

Picture 3: The BMD sample probe is located directly behind the shield in the raw exhaust stream.

CALIBRATION – The DVE is calibrated via a multiple SAO, smooth approach orifice, a device with ±1 % accuracy. Contrary to the manufacture's recommendation, this particular DVE is calibrated at its operational temperature of 68°C (155°F) instead of ambient. Early in the commissioning of this unit, it was found that the ambient calibration was not valid at operational temperatures by a factor of approximately 4.8% as seen in Graph 3. This effect is under investigation.

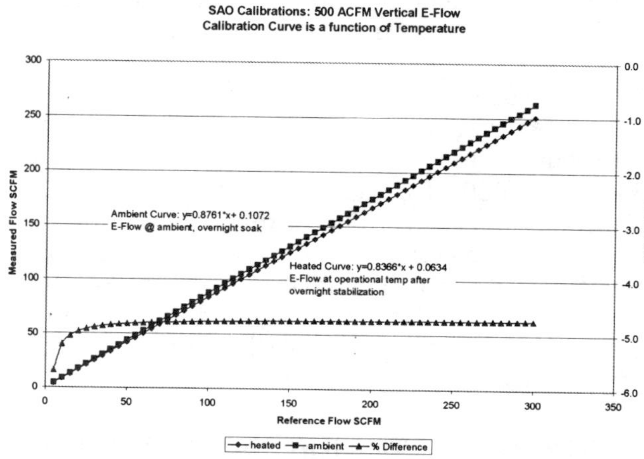

Graph 3: SAO Calibration: Heated vs. Ambient

A relatively stable exhaust temperature in the raw exhaust measurement zone appears to be a criterion for accurate exhaust volume measurement with this generation of DVE. The range of stability is currently under investigation by the manufacturer. However, using the heated calibration, the DVE exhibits excellent measurement linearity over the range of interest and excellent test-to-test repeatability, as seen in Graph 4.

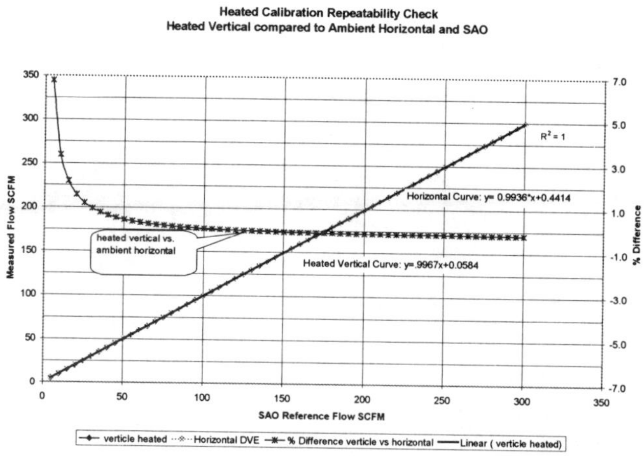

Graph 4: DVE Calibration Linearity

DVE OPERATION – As seen from Graph 5 below, the unit maintains a consistent temperature in the flow measurement zone while maintaining tailpipe backpressure to an average of zero relative to ambient. Some of the variation seen in the pressure data is due to a corrugated tailpipe used to connect the vehicle to the DVE. This tailpipe was chosen as it has proven to be more leak-proof than conventional metal bendable sampling pipes; unfortunately, the corrugations that allow the pipe to bend also increase the pipe's roughness and this, consequently, increases the pressure restriction. In contrast to the CFV, it is more imperative to be leak-free in the BMD sampling system when performing a direct measurement of the raw exhaust as opposed to the CFV's diluted exhaust. Table 3 shows the consistency of the exhaust volume measurement over six three-phase cold start emission tests.

DVE FURTHER WORK – As this technology is relatively new to exhaust emissions testing, it continues to require refinement and experimentation. The calibration standards are one such area. It has been expected that the unit perform within 1% of reading while in the fully developed turbulent flow segment and 1% of range below the turbulent region. More data needs to be gathered to statistically support calibration limits.

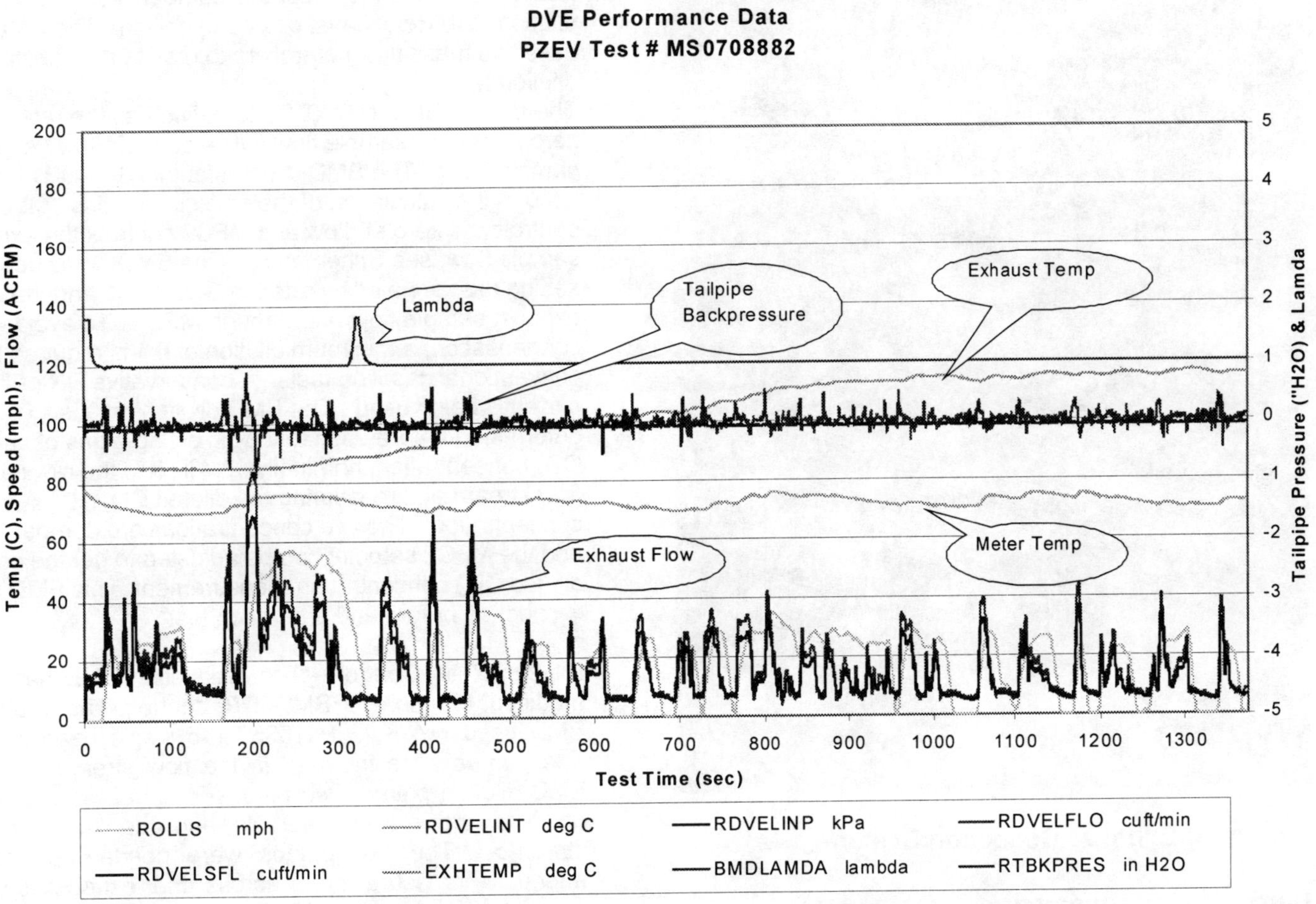

Graph 5: Performance Data obtained from a 4-cylinder PZEV test; first two phases of FTP

Table 3: Exhaust Volume Repeatability from six tests of a 4-cylinder PZEV; by Phase of FTP

	Test Number	Phase 1 Exhaust Volume		Phase 2 Exhaust Volume		Phase 3 Exhaust Volume	
		Cubic Feet	Cubic Meter	Cubic Feet	Cubic Meter	Cubic Feet	Cubic Meter
	1	165.5	4.687	168.3	4.765	140.8	3.989
	2	165.2	4.677	169.9	4.811	140.4	3.977
	3	166.7	4.722	169.0	4.785	141.9	4.018
	4	166.0	4.702	169.4	4.797	141.1	3.995
	5	165.2	4.677	168.8	4.779	143.2	4.057
	6	165.8	4.696	169.6	4.803	141.1	3.996
Average		165.7	4.694	169.1	4.790	141.4	4.005
Stand. Dev.		0.6	0.017	0.6	0.017	1.0	0.029
COV		0.36 %	0.36 %	0.35 %	0.35 %	0.71 %	0.71 %

Picture 4: Exhaust Collection System

Hardware Improvements

B. Bag Mini-Diluter

PRODUCTION BAG MINI-DILUTER - The Production Bag Mini-Diluter is part of a combined Exhaust Collection System (ECS), see Picture 4, that allows exhaust sampling in either BMD or CFV mode. ECSs are installed on all new GM emission test sites. This allows current testing using the CFV mode and future testing using the BMD mode, when DVEs are installed on the sites. The production BMD has been modified in two ways to improve the BMD dilution ratio. The first is a new sample extraction pump for improved BMD dilution ratio control, and the second is compensation of BMD dilution ratio for exhaust gas composition. The BMD uses two mass flow controllers to maintain a constant dilution ratio.

The BMD dilution ratio (DR), is defined as the ratio of zero air flow to sample flow (DR = zero air flow / exhaust sample flow). The BMD dilution factor (DF), DR+1, is used in the calculation of the emissions mass. MFC1 controls the zero air flow and MFC2 controls the exhaust sample flow, see Schematic 1. The BMD DR is set by setting the zero air flow rate through MFC1 and the exhaust sample flow rate through MFC2. To avoid water condensation, a minimum dilution of 6:1 is required for conventional gasoline fuels. A conservative DR of 7:1 is currently being used. The BMD dilution ratio is determined before each test by injecting a gas of known CO_2 concentration, nominally 13.4%, into the raw sample flow stream and measuring the diluted CO_2 concentration. The two concentrations are compared and the MFC2 setpoint can be adjusted to get the correct diluted CO_2 concentration measurement. This BMD DR is checked before and after each test[6].

Dilution Ratio Control – The production BMD has been modified to improve BMD DR control. The prototype BMD used two KNF PJ10357-145 single head pumps. The pumps were installed in the flow stream after the BMD dilution point. Two pumps in series were drawing on both the exhaust sample and the zero air flow streams. The two pumps were needed to provide adequate pressure drop across the exhaust sample mass flow controller (MFC2) to assure proper flow control. Proper flow control is essential to maintain a constant BMD DR. The production BMD has been modified to use one KNF PU830-N035.0 heated dual-head pump. The pump is installed in the flow stream after exhaust sample MFC2 and before the BMD dilution point. In this improved design, the pump draws directly on the exhaust sample MFC2 instead of both MFC1 and MFC2. This simplified system, using only one pump, providing improved BMD DR control, is illustrated in Schematic 1.

Dilution Ratio Compensation – During stoichiometric vehicle operation, CO_2 and H_2O levels remain nearly constant. For non-stoichiometric operation (i.e. the initial start of a test, deceleration fuel shutoff, and lean burn), CO_2 and H_2O concentration levels change by a maximum offset of 4% and 0.9% respectively. These changes in CO_2 and H_2O concentration levels change the flow through MFC2, thus changing the BMD DR. The improved BMD uses a real-time feedback adjustment to MFC2's flow rate to compensate for this change in flow. An ECM Lambda Pro meter was added to the DVE to measure lambda in the raw exhaust at the BMD sample point. The meter produces a 0 to 5 volt, linearized output for lambda in the range of 0.55 to 1.75. An algorithm was developed to correct the exhaust sample MFC2 setpoint for CO_2 and H_2O concentration in the raw exhaust based on the measured value of lambda. This small correction for exhaust gas composition should lead to more accurate BMD DR during non-stoichiometric operation.

Graph 6 shows dynamometer roll speed, lambda measured at the DVE, and MFC2 flow rate for phase one of an FTP test. The BMD DR is 7:1; MFC1 flow rate is 35 slpm and MFC2 flow rate is 5 slpm. The BMD DR is set using a gas of 13.4% CO_2, representing stoichiometric CO_2 for conventional gasoline fuels. MFC2 requires a setpoint of 5.0 slpm of 13.4% CO_2. Because MFC2 is calibrated in zero air, a setpoint of 5.2 slpm is required for an equivalent 5.0 slpm of 13.4% CO_2. As there is no CO_2 at the beginning of the test, lambda reads 1.85 and the MFC2 flow rate is 5.1 slpm. As CO_2 concentration increases to stoichiometric levels, and lambda goes to 1, the MFC2 flow rate is increased to 5.2 slpm N_2 (5.0 slpm 13.4% CO_2). This keeps the BMD DR constant. Deceleration fuel shutoff can be seen on cycle 2. Here, as lambda increases from 1 to 1.75 and back, the exhaust MFC2 flow rate sample is decreased from 5.2 slpm to 5.1 slpm and back. This compensation again keeps the BMD DR constant.

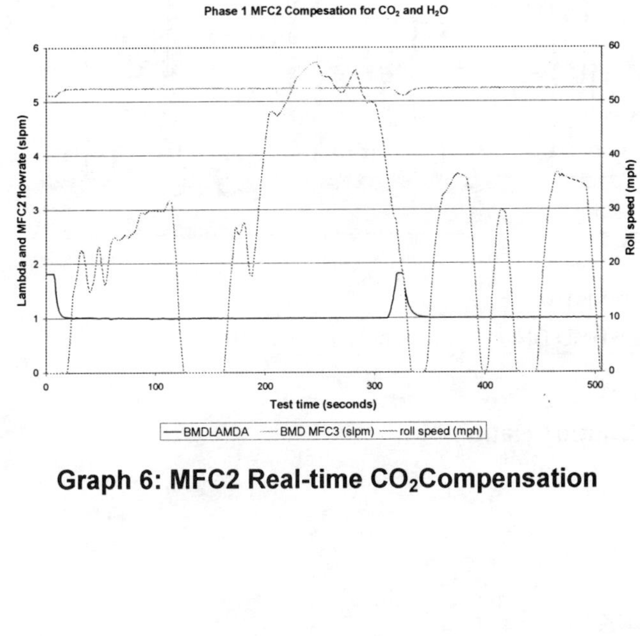

Graph 6: MFC2 Real-time CO_2Compensation

PZEV TEST RESULTS

PZEV Test Results – A PZEV vehicle was tested on Milford Site 7S using an ECS with a production BMD modified to allow sampling in CFV mode and BMD mode simultaneously. The test data is from five cold start Federal Test Procedures (FTP).

Graph 7 is a comparison of BMD and CFV CO_2 concentration by phase for the five tests. The BMD CO_2 concentrations are consistent from phase to phase and test to test showing the constant BMD DR. The lines for the 3 phases lie on top of each other. Monitoring BMD CO_2 concentrations, for given BMD DR, can be used as a quality assurance check on a test to test basis.[7]

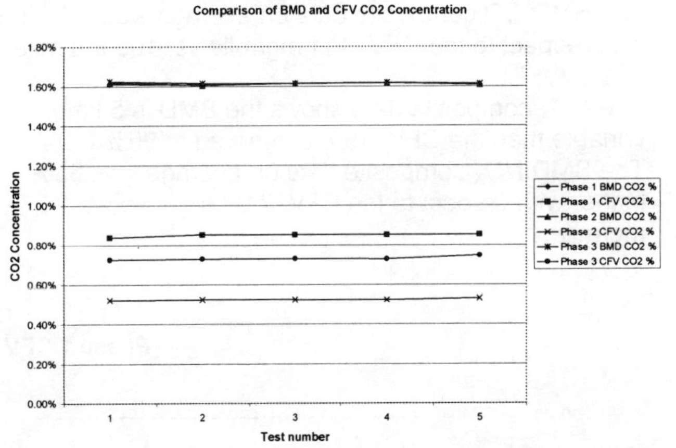

Graph 7: CO_2Concentration by Test Phase

The CFV CO_2 concentrations, on the other hand, differ by phase, but are fairly constant for a given phase from test to test on the same vehicle. Graph 8 compares one second BMD DR to CFV DR over a phase one of a single test. This shows the BMD DR is constant through out the phase at 7:1. The CFV DR varies inversely to tailpipe exhaust flow, as measured by the DVE, from as low as 4:1 during high accelerations to as high as 75:1 during idles. This data further illustrates today's problem with the CFV: during any given test, the system may experience a dilution ratio that is both too low to avoid water condensation and too high for low level concentration measurement.

Graphs 9~14 show the test data in triangle plots. The triangle plots show the individual data points on the base of the triangle and the mean as the point of the triangle. Note that negative phase masses have not been set to zero in the calculation of composite data. The NO_X Phase 2 data had several negative values and this resulted in the calculation of one negative composite value.

The THC composite data shows the BMD is 2% less variable than the CFV, 21% compared to 23% COV. The BMD THC composite data on average was 4%

higher with respect to the CFV, 7.49 mg/mile vs. 7.21 mg/mile.

The CH_4 composite data shows the BMD is 9% more variable than the CFV, 29% compared to 20% COV. The BMD CH_4 composite data on average was 30% lower with respect to the CFV, 0.9 mg/mile vs. 1.28 mg/mile.

The NMHC composite data shows the BMD is 4% less variable than the CFV, 20% compared to 24% COV. The BMD NMHC composite data on average was 10% higher with respect to the CFV, 6.72 mg/mile vs. 6.10 mg/mile.

The CO composite data shows the BMD is 1% more variable than the CFV, 35% compared to 34% COV. The BMD CO composite data on average was 5% higher with respect to the CFV, 111 mg/mile vs. 106 mg/mile.

The NO_X composite data shows the BMD is 5 times less variable than the CFV, 19% compared to 95% COV. The BMD NO_X composite data on average was 65% higher with respect to the CFV, 2.09 mg/mile vs. 1.27 mg/mile.

The CO_2 composite data shows the BMD is 3 times less variable than the CFV, 0.2% compared to 0.6% COV. The BMD CO_2 composite data on average was 1% lower with respect to the CFV, 289 G/mile vs. 292 G/mile.

This data shows that the CFV continues to under-report THC and NMHC and over-report CH4 as compared to the BMD. Previous data shows the CFV under-reporting THC by 2.7% for all tests and 5.9% for tests where THC was under 50 mg/mile[8]. For these PZEV tests of an average composite THC of 7.5 mg/mile, the CFV under-reported the BMD by 4%. Previous data shows the CFV under-reporting NMHC by 3.2% for all tests, and 11.3% for tests where NMHC was under 50 mg/mile. For these PZEV tests, of an average composite NMHC of 6.7 mg/mile, the CFV under-reported the BMD by 10%. For CH_4, previous data shows the CFV to over-reporting CH_4 by 2.9% for all tests and 42% for tests where CH_4 was under 5 mg/mile. For the PZEV tests, at average composite CH_4 of 0.9 mg/mile, the CFV over-reported the BMD by 30%.

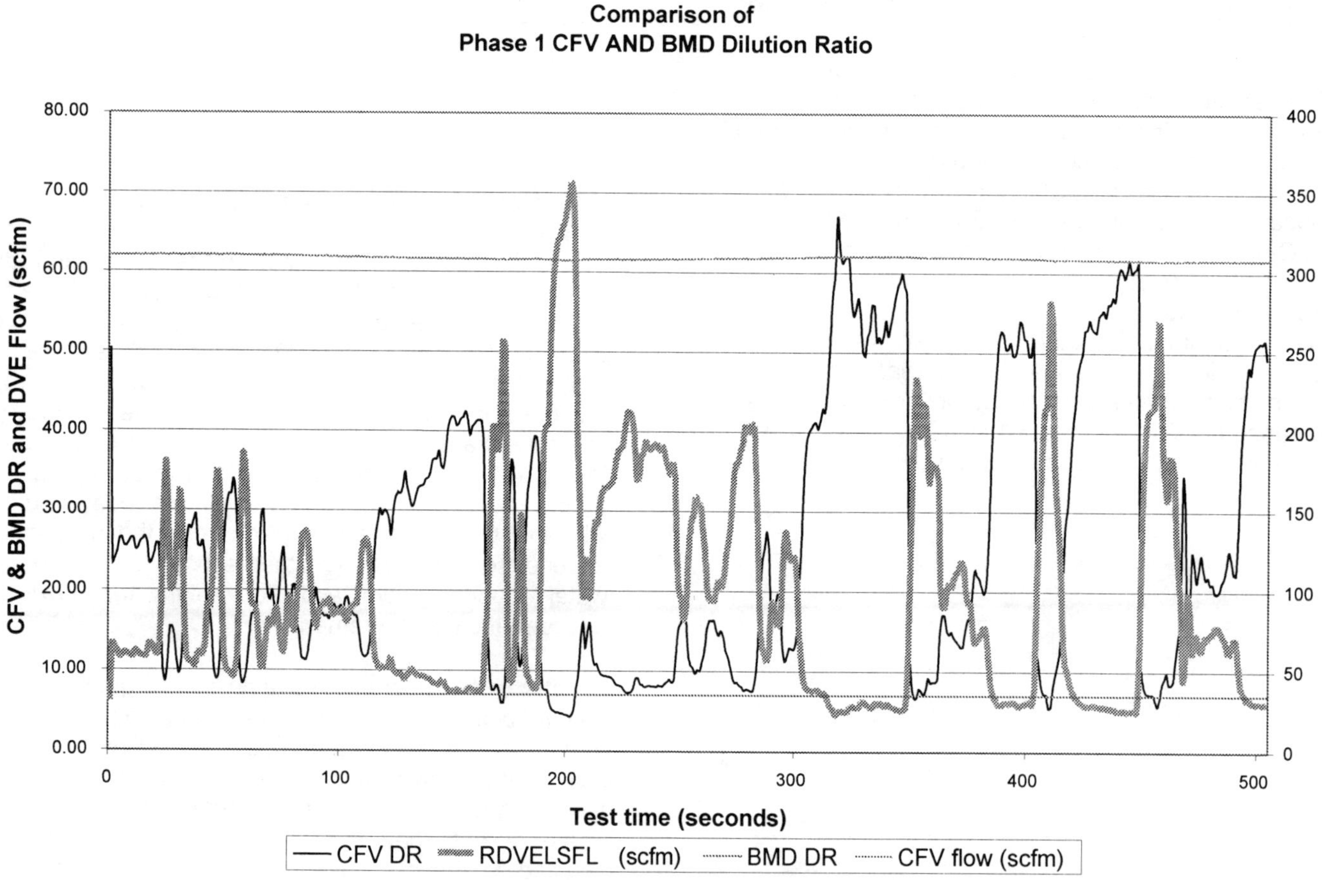

Graph 8: BMD & CFV Dilution Ratio

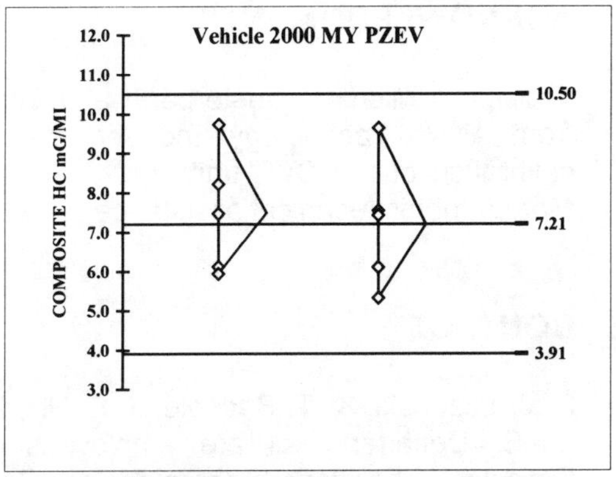

	BMD	CFV
N	5	5
MEAN	7.49	7.21
STD DEV	1.57	1.65
COV	21%	23%

Graph 9: Composite HC

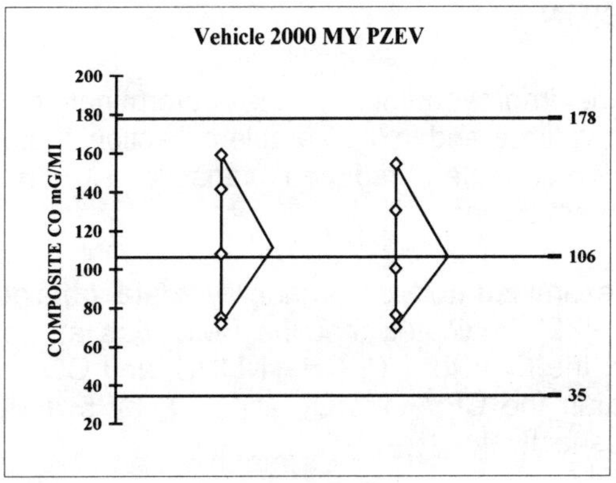

	BMD	CFV
N	5	5
MEAN	111	106
STD DEV	39	36
COV	35%	34%

Graph 12: Composite CO

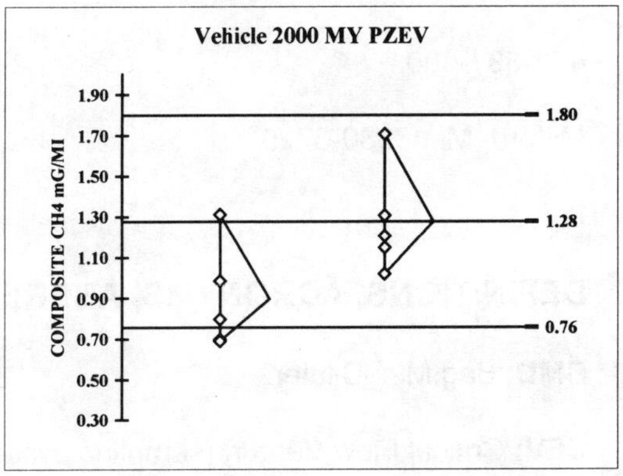

	BMD	CFV
N	5	5
MEAN	0.90	1.28
STD DEV	0.26	0.26
COV	29%	20%

Graph 10: Composite CH4

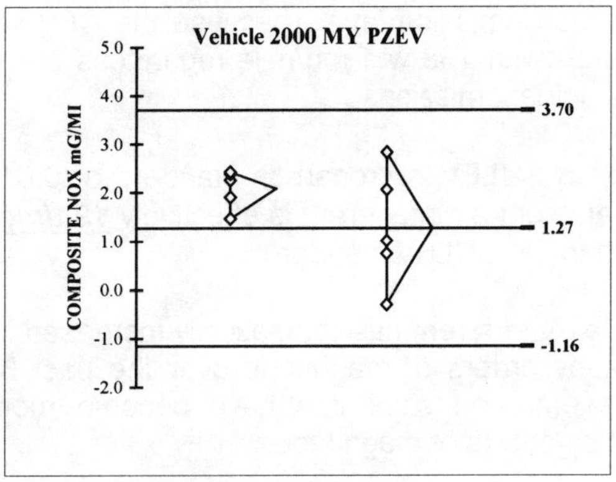

	BMD	CFV
N	5	5
MEAN	2.09	1.27
STD DEV	0.40	1.21
COV	19%	95%

Graph 13: Composite NOx

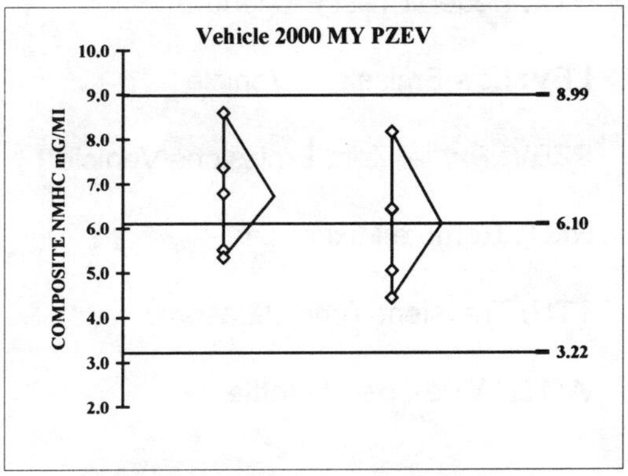

	BMD	CFV
N	5	5
MEAN	6.72	6.10
STD DEV	1.35	1.44
COV	20%	24%

Graph 11: Composite NMHC

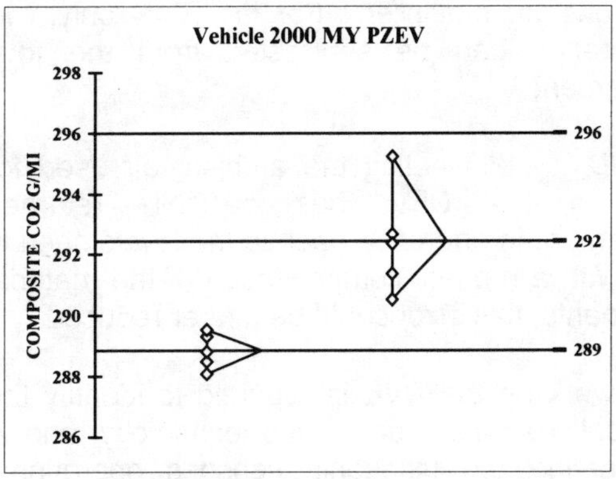

	BMD	CFV
N	5	5
MEAN	289	292
STD DEV	0.6	1.8
COV	0.2%	0.6%

Graph 14: Composite CO2

Limits = ±2 Standard Deviation of the CFV Mean

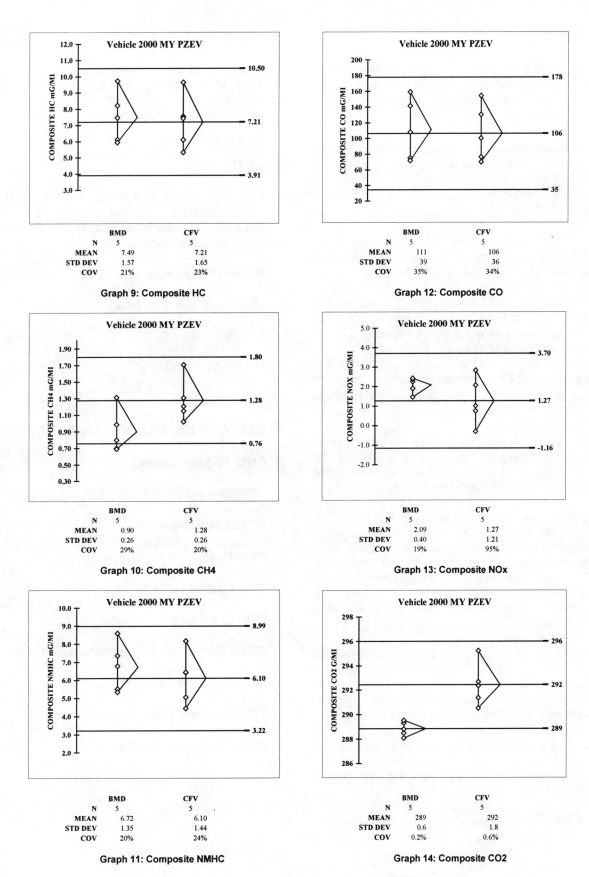

Graphs 9 ~ 14 : Comparison of BMD and CFV by Composite Constituent

CONCLUSION

- Hardware improvements in the determination of exhaust volume and in the sample collection system have demonstrated reduced variability in most reported emissions.
- When examined for low variability, data obtained from a PZEV indicated that the BMD was at least equal to the CFV for HC, CH_4, NMHC, and CO and better than the CFV for NO_X and CO_2 by 5 and 3 times respectively.
- The CFV continues to show that it under-reports HC and NMHC and over-reports CH4 when compared to the BMD by 4%, 10%, and 30% respectively.
- Ambient air contaminants will have an effect on low concentration emission levels because the CFV will always start with 188 wtd mg/mile regardless of the vehicle's actual emissions.
- Based on a SULEV hydrocarbon standard of 0.010 mg/mi., ambient air interferent is effectively ***19 times higher*** than the SULEV standard.
- Ambient air interferent has dramatically increased its influence by orders of magnitude over the past 30 years as emission standards have become more stringent by orders of magnitude.
- Sample material contaminants effect both the CFV and the BMD by approximately (but not exactly) the same amount. Humidity and hydrocarbon (HC) contamination can be correlated experimentally. The CFV has an exponential relationship while the BMD has a constant multiplier. For the BMD only, this contamination can be subtracted from the total measurement.
- The BMD, by eliminating the ambient air used for dilution in the CFV, has basically returned measurement interference back to the levels seen in 1972. With a modest compensation of the material contaminants, this error could be further reduced.
- Further work on the DVE is required to identify the correct calibration standards and technology, and to further define the ultrasonic sensor's operational temperature limits.
- Correction of non-stoichiometric exhaust flow allows for more robust dilution ratio control.

ACKNOWLEDGMENTS

Thanks to Michael Wusterbarth and Thomas McKenz from Flow Technology Inc for their assistance calibration of the DVE unit; and Jeff Loo from Gener Motors for his technical assistance.

CONTACT

D. J. Luzenski, K. T. Bedsole, J. F. Hill, D. B. Nagy, ar S. S. Decarteret all are employees of the Vehic Emission Laboratory, General Motors Powertrain Grou at the Milford Proving Grounds.

General Motors Proving Ground – Powertrain

3300 General Motors Road

483-331-000

Milford, MI 48380-3726

DEFINITIONS, ACRONYMS, ABBREVIATIONS

BMD: Bag Mini-Diluter

CFV: Critical Flow Venturi (sampling system)

DR: Dilution Ratio

DVE: Direct Vehicle Exhaust

FTP: Federal Test Procedure

LEV: Low Emissions Vehicle

PZEV: Partial-Zero Emissions Vehicle

RMT: Remote Mixing Tee

TTU: Transient-Time Ultrasonic

WOT: Wide-open Throttle

REFERENCES

[1] 40 CFR, § 86.101. Subpart B-Emission Regulations for 1977 and Later Model Year New Light-Duty Vehicles and New Light-Duty Trucks; Test Procedures

[2] William M. Sikvis et al. A CFV Type Mini-dilution Sampling System for Vehicle Exhaust Emissions Measurement, SAE 1999-01-0151

[3] Mark Gunther et al. Improved Bag Mini-Diluter Sampling System for Ultra-Low Level Vehicle Exhaust Emissions, SAE 2000-01-0792

[4] John D. Wright. The Performance of Transit Time Flowmeters in Heated Gas Mixtures, FEDSM98-5290. 1998 ASME Fluids Engineering Division Summer Meeting

[5] Donald B. Nagy et al. Evaluation of the Bag Mini-Diluter and Direct Vehicle Exhaust Volume System for Low Level Emissions Measurement. SAE 2000-01-0793

[6] ERC Technical Report, "The Bag Mini-Diluter Alternative to CVS Sampling" May 1, 2001

[7] See reference 3

[8] See reference 5

2002-01-0048

Studies on Enhanced CVS Technology to Achieve SULEV Certification

H. Behrendt and O. Mörsch
DaimlerChrysler AG

C. T. Seiferth
BMW AG

G. E. Seifert
Porsche AG

J. W. Wiebrecht
Audi AG

ABSTRACT

For the measurement of exhaust emissions, Constant Volume Sampling (CVS) technology is recommended by legislation and has proven its practical capability in the past. However, the introduction of new low emission standards has raised questions regarding the accuracy and variability of the CVS system when measuring very low emission levels.

This paper will show that CVS has the potential to achieve sufficient precision for certification of SULEV concepts. Thus, there is no need for the introduction of new test methods involving high cost.

An analysis of the CVS basic equations indicates the importance of the Dilution Factor (DF) for calculating true mass emissions. A test series will demonstrate that, by adjusting the dilution and using state of the art analyzers, the consistency of exhaust results is comparable with those of LEV concepts, measured with conventional CVS systems and former standard analyzers.

Blank tests and vehicle emission tests demonstrate that emissions at SULEV level can be measured with high accuracy.

Finally, this paper discusses the influence of individual components of the CVS system on exhaust results. It suggests an enhanced CVS system that detects SULEV limits with high precision.

INTRODUCTION

Since CVS technology was recommended in 1972 by legislation of the USA /1/ including California for the determination of exhaust emissions, this technology has established itself as a standard worldwide. All emission limits defined since then are checked by using the CVS technique. All measures introduced to tighten the regulations were taken on this basis, and consequently SULEV, as the most stringent exhaust emission limit currently in the USA, is also based on this definition.

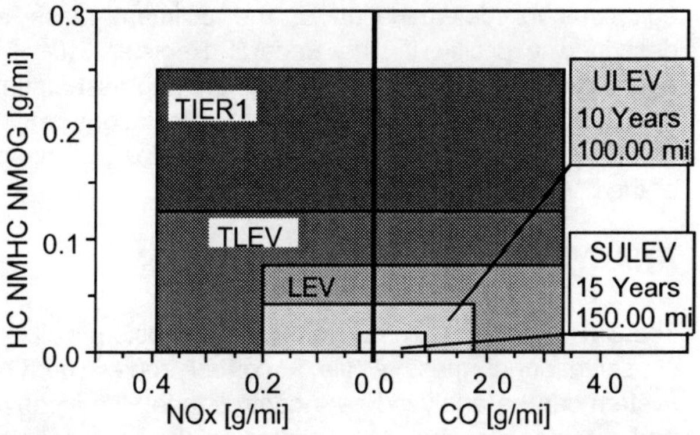

Figure 1: Pollutant limits

The limit for hydrocarbons HC/NMHC/NMOG, i.e. 0.01 g/mi, corresponds to 0.28% of the original value from 1972 (3.6 g/mi). While the exhaust emission limits have been reduced to a fraction as can be seen in Fig. 1, nothing has been changed in the basic formulas or in the measuring technology principle since introduction of CVS technology in the seventies. Requesting the measurement of the extremely low emission limits, CVS technology is now the subject of discussion. On the one hand, there is doubt about the attainable measuring accuracy since vehicles in some cases produce pollutant concentrations in the exhaust gas below those in the ambient air. On the other hand, the calculation of the emitted pollutant mass may lead to negative results. A question that arises here is whether these are measurement errors or results based on the principle of CVS technology. For this reason alternative methods are currently discussed as replacements for CVS technology /2,3,4/. However, new methods not only have to display the necessary measuring accuracy, but also verification that their results are equivalent to those of CVS technology. This applies both to determination of the mass of the pollutant emissions and to fuel economy.

To clarify the potential of CVS technology and the calculation of negative mass emissions a theoretical analysis will be conducted here with a derivation of the basic equations. An error analysis and resulting enhancement options will be presented. Improvements of CVS technology have already proven effective in practice for ULEV concepts /5/. Further studies for enhancement of CVS technology for SULEV were conducted and are reported on here. The enhancement capabilities determined in this process must be examined with an eye to technical feasibility, suitability in practice as well as cost aspects in order to maintain worldwide acceptance.

CVS-PRINCIPLES

The US Federal Register defines the pollutant masses indicated in the limits as the "true mass" (absolute emitted mass) at the tailpipe of the vehicle /6,7/. CVS technology is specifically described in the US /7/, European /8/ and Japanese /9/ test regulations for determination of pollutant emissions. The calculation equations for determination of the pollutant mass are described explicitly in the Federal Register /10/. The equations indicated there contain simplifying assumptions that may be noticeable in mass calculations at the SULEV level. This will be shown in the following analysis of the CVS basic equation.

ANALYSIS OF BASIC CVS EQUATION

The CVS equations are derived from the basic principle of conservation of mass within a control zone. The CVS system can be analyzed as a control zone with incoming and outgoing masses, differentiated for the components i, Fig. 2.

For the CVS system control zone the following balance can be drawn up for the exhaust gas component i at the ***entry***. The inflowing pollutant mass $m_{i\ entry}$ of the component i is determined on the basis of the dilution air volume V_{air}, the pollutant concentration $c_{i\ air}$ of the dilution air, as well as its density ρ_i and the incoming undiluted exhaust gas volume V_e with the respective concentration $c_{i\ e}$ and density ρ_i:

$$m_{i\ entry} = V_{air} \cdot c_{i\ air} \cdot \rho_i + V_e \cdot c_{i\ e} \cdot \rho_i \qquad \textbf{Eq. 1}$$

The significance of the designations will be pointed out in the following balance analyses.

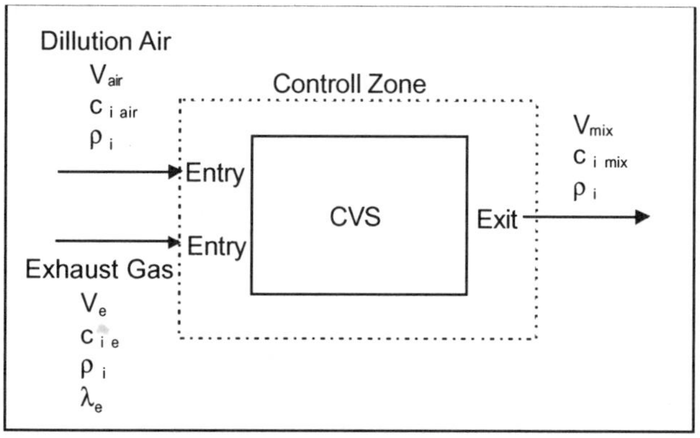

Figure 2: Control zone: CVS system

At the ***exit*** of the CVS system the pollutant mass $m_{i\ exit}$ is determined through the CVS volume V_{mix}, the respective pollutant concentration $c_{i\ mix}$ and the density ρ_i, based on:

$$m_{i\ exit} = V_{mix} \cdot c_{i\ mix} \cdot \rho_i \qquad \textbf{Eq. 2}$$

Under the condition that a change in mass for the component i in the CVS system due to further chemical reactions is ruled out, the following applies:

$$\boldsymbol{m_{i\ entry} = m_{i\ exit}} \qquad \textbf{Eq. 3}$$

i.e. the incoming component mass $\mathbf{m_{i\ entry}}$ and the outflowing component mass $\mathbf{m_{i\ exit}}$ are equal. Furthermore, it must be ensured in the CVS system that the condition $\mathbf{V_e \leq V_{mix}}$ is met at all times.

The following variables are measured by means of current CVS technology:

- CVS volume $\mathbf{V_{mix}}$,
- Diluted pollutant concentration $\mathbf{c_{i\ mix}}$,
- Dilution air concentration $\mathbf{c_{i\ air}}$.

The volume of the dilution air $\mathbf{V_{air}}$ or of the undiluted exhaust gas $\mathbf{V_e}$ is not measured. CVS technology circumvents this problem as follows:

By using **Eq. 1 – 3**, the following results:

$$V_e \cdot c_{i\ e} \cdot \rho_i = V_{mix} \cdot c_{i\ mix} \cdot \rho_i - V_{air} \cdot c_{i\ air} \cdot \rho_i \qquad \textbf{Eq. 4}$$

According to the CVS principle, the following applies:

$$V_{air} = V_{mix} - V_e \qquad \textbf{Eq. 5}$$

For the absolute pollutant mass follows:

$$m_{i\,e} = V_{mix} \cdot c_{i\,mix} \cdot \rho_i - (V_{mix} - V_e) \cdot c_{i\,air} \cdot \rho_i \qquad \textbf{Eq. 6}$$

The unknown exhaust gas volume V_e is shown with the dilution ratio DR as follows:

$$DR = \frac{V_{mix}}{V_e} = \frac{V_{mix}}{V_{mix} - V_{air}} \qquad \textbf{Eq. 7}$$

Accordingly, the absolute pollutant mass $m_{i\,e}$ is determined as calculated "true mass" by using Eq. 6 and Eq. 7 as follows,

$$m_{i\,e} = V_{mix} \cdot \rho_i \cdot (c_{i\,mix} - c_{i\,air} \cdot (1 - 1/DR)) \qquad \textbf{Eq. 8}$$

as it also continues to be used for ULEV-SULEV limits in /10/ for determination of the mass.

For the dilution ratio DR, Eq. 7, the Federal Register for gasoline fueled vehicles (H/C ratio of the fuel = 1.85) gives the following definition DF as an approximation of DR /10/:

$$DF = \frac{134000}{CO_{2\,mix} + CO_{mix} + HC_{mix}} \qquad \textbf{Eq. 9}$$

with HC_{mix} in ppm C1, CO_2 and CO in ppm

A derivation for this definition is not indicated in the CFR, it has not changed since the introduction of DF in the early 70th.

ANALYSIS REGARDING DILUTION FACTOR DF

DF is calculated according to Eq. 9 from the relationship between the theoretically formed CO_2 concentration in the humid, undiluted exhaust gas of an ideal combustion and the measured diluted concentrations of the hydrocarbon products (CO_2, CO, HC).

The basic assumption behind this DF calculation is that for actual combustion, only the stochiometric amount of air is used.

$$C_8H_{14,8} + 11{,}7 \cdot (O_2 + 3{,}77 \cdot N_2) \longrightarrow 8\,CO_2 + 7{,}4\,H_2O + 44{,}14 N_2$$

(H/C=1,85; m=111g) 13,4Vol% + 12,4Vol% + 74,2Vol%

Eq. 10

For other fuels, such as methanol, LPG and CNG, other DF definitions are given in /10/. They are also based on stochiometric equations.

In the case of secondary air feed, fuel cutoff in the overrun, etc. the exhaust gas volume actually emitted by the engine, V_e, is not exactly determined by Eq. 9. It is reduced to stochiometric conditions.

Another simplification in Eq. 9 is that the measured diluted concentrations used without correcting the influence of dilution air background level of 0.04 Vol%.

To determine the dilution ratio, Eq. 7, V_e must be measured directly or indirectly via $V_e = V_{mix} - V_{air}$ through measurement of V_{air}.

The calculation of masses according to Eq. 8 with DF, Eq. 9, and DR, Eq. 7, have to lead to different results even with identical CVS measured values. Using Eq. 8 and Fig. 3 the mass difference

$$\Delta m_{ieDF\text{-}DR} = m_{ieDF} - m_{ieDR}$$

is expressed by:

$$\Delta m_{ieDF\text{-}DR} = \rho_i \cdot V_{mix} \cdot c_{i\,air} \cdot (1/DF - 1/DR) \qquad \textbf{Eq. 11}$$

with

$DF = V_{mix}/V_{eDF}$, $DR = V_{mix}/V_e$

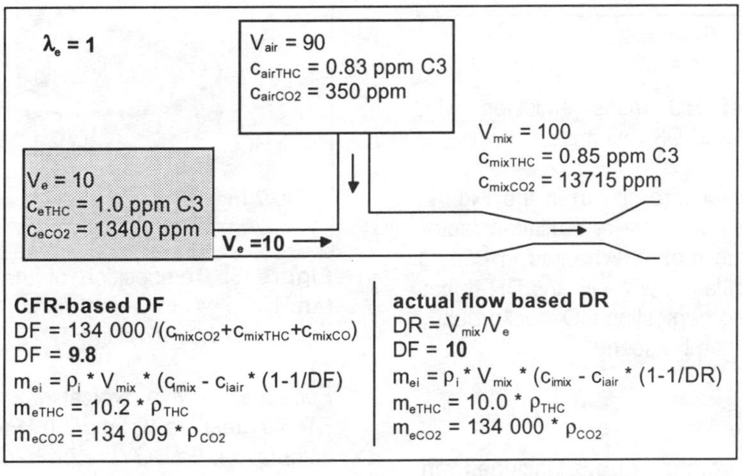

Figure 3: Comparison of the mass calculation according to DF and DR, example 1

and the approximation, appendix 1

$V_{eDF} \approx V_e/\lambda_e$

follows:

$$\Delta m_{ieDF\text{-}DR} \approx \rho_i \cdot V_e \cdot c_{i\,air} \cdot (1-\lambda_e) \qquad \textbf{Eq. 12}$$

Eq.12 shows that only for $\lambda_e \neq 1$ a mass difference proportional to V_e, $c_{i\,air}$ and $1 - \lambda_e$ appears. There is no influence of $c_{i\,mix}$ in mass difference $\Delta m_{ieDF\text{-}DR}$. That means mass difference is in good approximation independent on emission level.

To illustrate this, examples 1 – 3 are calculated with DF and DR, Fig. 3 and Appendix 2.

1. Exhaust gas values of a spark ignition engine operated on an average $\lambda_e = 1$
2. Exhaust gas values of a spark ignition engine with an average $\lambda_e = 1.2$
3. Exhaust gas values of a spark ignition engine with an average $\lambda_e = 1.2$ and complete exhaust gas purification

Mass calculations by DF and DR, for examples 1 - 3 are carried out for component HC (low emission limit, dilution air and diluted exhaust gas concentrations within same range) and will be examined against true mass in undiluted exhaust.

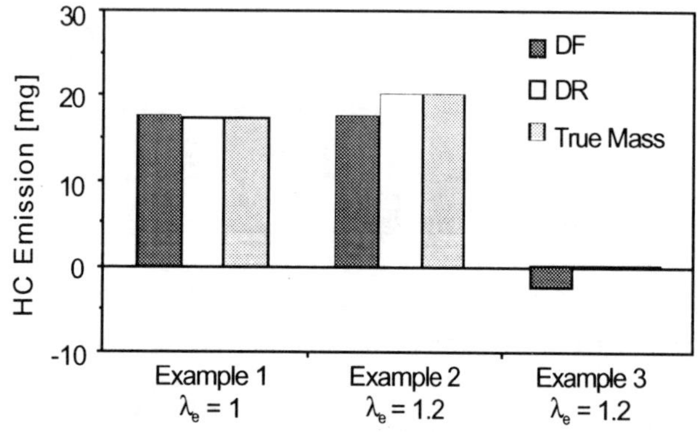

Figure 4: Comparison of HC mass emission with calculation according to DF and DR (example 1-3)

The results of this comparison are shown in the Fig. 4. The calculation of the mass with the real dilution factor DR corresponds to the "true mass". Whereas in case 1 the difference between calculation with DF and DR is not significant and is only due to neglecting CO_2 background, in case 2 and 3 differences can be seen.

To summarize the result:

- CVS-Mass calculation formula, Eq.8, includes an approximation for Dilution Ratio DR by Dilution Factor DF. The calculation of DF by diluted exhaust concentration, Eq. 9, is a basic assumption and has not changed since introduction of CVS- Technology for emission testing. /1/.
- Eq. 11 and 12 explain a difference in mass-calculation by Eq. 8 if DF is not equal DR. In the case of extremely low-emission vehicles like SULEV (example 3) the simplifying assumptions in mass-calculation, Eq.8 , will lead to negative emissions.

ERROR ANALYSIS OF MASS CALCULATION

With CVS Technology true mass is calculated from CVS-measurement values. It must be expected in general for all measured values that they are not free of measurement-errors. To examine the error in the calculation of mass, an error propagation calculation based on Eq. 8 is carried out.

Although SULEV represents an NMOG limit value, a more far-reaching analysis is conducted only for HC and not for NMOG. The reason for this is that NMHC as the basis for NMOG can be determined in different ways and one single standardized mathematical error analysis is thus not possible. Instead, typical HC values for SULEV vehicles are examined here. This also maintains comparability to previously conducted error analyses regarding Tier1, LEV I and LEV II /5,11 /.

Based on Eq. 8, the error is made up of $dm_{i\,e} = f(dV_{mix}\ (V_{mix} = f(Q,t)),\ dc_{i\,air},\ dc_{i\,mix},\ dDF,\ dd)$. For the distance d, time t and flow rate Q the legally permissible deviations were taken as the basis and for the concentration measurement the manufacturer's specifications for the analyzers (Appendix 3).

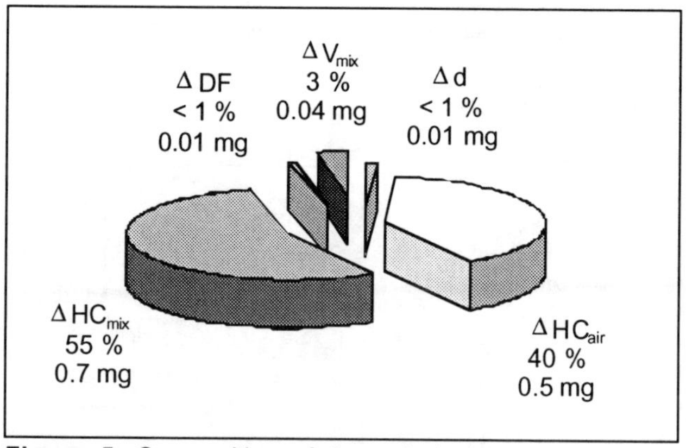

Figure 5: Composition of the calculated probable error (ΔmHC e = $\pm$1.3 mg/mi $\equiv$ 100%) at weighted mass emission of 6 mg/mi in FTP75

For a SULEV a typical weighted HC emission in an FTP75 test is 6 mg/mi. From the theoretical error calculation, Eq. A7, the probable error for HC in this case is $\pm$1.3 mg/mi (= 22%). The decisive sources of errors in HC-mass calculation are given by concentration measurement of air and exhaust sample bags, Fig. 5.

Further error variables are negligible. The error for DF shown here is merely the random error that results through uncertainty in the measurement of the concentration of CO_2 (and CO, HC). It has nothing to do with the systematic error due to the simplifications of DF that were discussed in the previous section, Eq. 11 and Eq. 12.

ENHANCING CVS- TECHNOLOGY

To reduce the statistical error in the CVS based mass calculation the following possibilities result from the above considerations:

- Reduce diluted exhaust gas volume V_{mix} for a higher pollutant concentration in diluted exhaust gas and a reduced mass calculation error by an optimized dilution (Eq. A3 – A5),
- Improve accuracy of measurement of $c_{i\,mix}$ and $c_{i\,air}$ (Fig. 5, Eq. A3 – A5).
- Reduce background concentration $c_{i\,air}$ (Eq. A3 – A5)

OPTIMIZED DILUTION - As shown before, a reduction of V_{mix} means an increase of the share of exhaust gas volume V_e in the diluted exhaust gas V_{mix} and thus in the exhaust sample bag. As the error analysis shows before, Eq. A3 - A5, a reduction of V_{mix} proportionally reduces the mass calculation error.

In the following section, the theoretical minimum dilution for an FTP75 will be derived. The CVS-principle requires:

- $V_{mix} \geq V_e$ (DF ≥ 1) at all times,
- no condensation of water in the sampling system at any time,
- no condensation of water in the sample bags at any time.

As DF-calculations based on CO_2-measurements show, Fig. 6, the lowest possible average dilution factor that ensures DF ≥ 1 at all times during phase 1 of an FTP 75 is typically DF ≈ 4. The same applies for phase 2 and 3. Under these extreme conditions the minimum DF at t ≈ 200 seconds of cold start is around 1.2. This is almost the theoretical limit of the constant volume sampling with respect to the optimized dilution. No measurements were done under these extreme conditions.

As described in Eq. 10, at stochiometric combustion of gasoline with air the raw exhaust gas includes approx. 12.4 volume percent water vapor. The conditions in the CVS system are to be designed such that this water vapor cannot condense at any time during emission measurement. This means that for a given average DF the temperature of the sampling system has to be kept above the dew point at the point of maximum water vapor content in V_{mix}. Time-resolved measurements show that, apart from the cold start phase, the necessary information on the water content in the diluted exhaust gas is obtained through the CO_2 curve over time. In Fig. 7 the maximum humidity is reached at t ≈ 200 s with an absolute humidity of 17 $g_{H2O}/kg_{dry\ air}$.

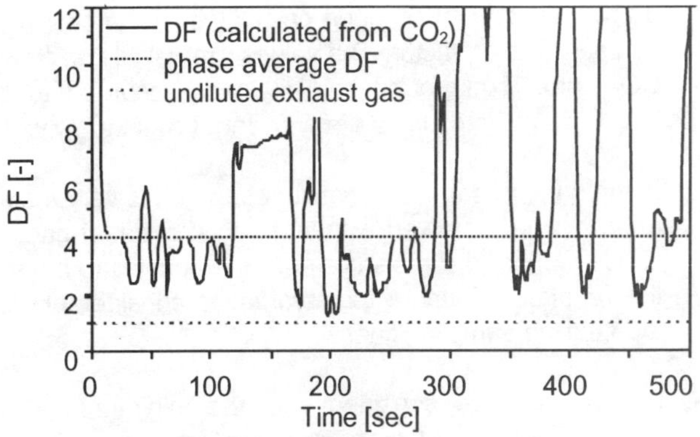

Figure 6: Time resolved dilution factor for an average DF of 4 (calculated from CO_2), Phase 1 FTP75

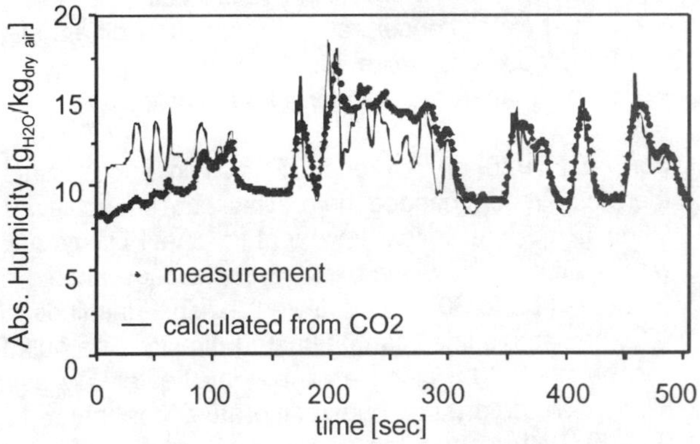

Figure 7: Abs. humidity in mixed exhaust gas at sample venturi in phase 1 of the FTP-75, comparison of a measurement (capacitive humidity sensor) and values calculated from CO_2 (DF = 13)

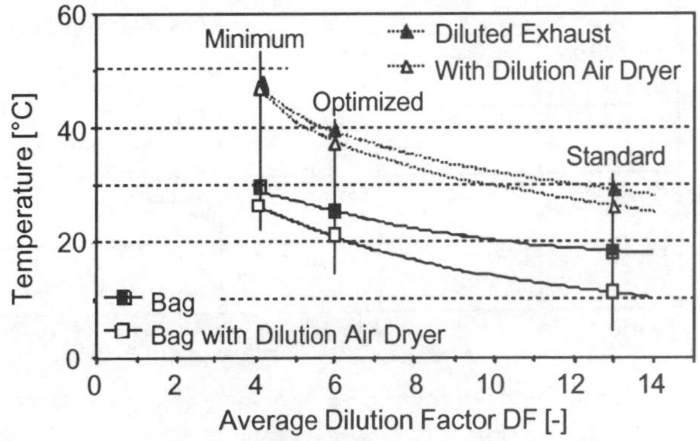

Figure 8: Temperature to avoid water vapor condensation in diluted exhaust gas lines and bags as function of average dilution factor DF

The corresponding minimum temperature of the sample lines and the exhaust gas bags for an FTP75 with a DF = 13 can be taken from Fig. 8. It also shows the required minimum temperature in the diluted exhaust gas lines and in the exhaust gas bags with and without application of dilution air dryer has been calculated as function of the average dilution factor. The values are calculated from the CO_2 data. Background humidity was assumed to be 2 $g_{H2O}/kg_{dry\ air}$ with dilution air dryer and 8 $g_{H2O}/kg_{dry\ air}$ without.

In the sample bag the exhaust gas, which is very moist for a short time, is mixed with less moist exhaust gas. As a consequence, the moisture in the exhaust gas bags and the required minimum temperature is considerably lower than for the sampling lines.

Based on these considerations the minimum DF of 4 requires heating of the sampling system to T ≈ 50 °C and the sample bags to T ≈ 30 °C. The theoretical DF of 4 is not feasible in practice for several reasons, such as:

- unknown vehicle emissions before test,
- actual water concentration in cold start differs slightly from theoretical value, Fig 7,
- safety margin required for a sound process.

For an FTP75 an average DF = 6 in each phase is, therefore, recommended here. This can be achieved by using a multiple critical flow venturi or a PDP-type CVS with multiple speed operation. Sample bags should then be heated up to 30 °C and diluted exhaust gas lines to 50 °C. The sample lines can be heated directly. The bulk flow can be heated by using a dilution air heater. By using a dilution air dryer, the bag temperature can be reduced (Fig. 8). With respect to standard CVS dilution of DF = 13 a DF of 6 reduces V_{mix} to about 50 per cent in phase 1 and 75 per cent in phase 2, respectively, table 1. As stated above this decreases the statistical error of mass calculation accordingly.

FTP-75 Test cycle	Standard CVS sampling	DF-optimized CVS sampling	Increase in exhaust gas volume in exhaust gas bag
Phase 1	DF = 13	DF = 6	2–fold
Phase 2	DF = 22	DF = 6	4–fold
Phase 3	DF = 16	DF = 6	3–fold

Table 1: DF of a standard and optimized CVS setting

Equivalent considerations for special tests, such as SC03 or US06, are presented in /5/.

ACCURACY OF ANALYZERS - For the exhaust gas measurement of extremely low emission vehicles besides standard analyzers, nowadays more expensive analyzers that have been specially optimized for very low measurement ranges (low-emission analyzers) are also available.

In Fig. 9 the zero and span stability of modern standard analyzers based on zero scatter and full-scale scatter is compared to low-emission analyzers. Zero and full-scale scatter are defined via mean and standard deviation of the difference between zero and span calibration before and zero and span check after the bag analysis. The data were obtained through evaluation of exhaust gas tests over a period of approx. 4 months. The measurement ranges 10 ppm C3 for HC and 10 ppm for NOx were used. HC and NO_x LE analyzers are slightly improved by lower zero-scatter against standard analyzers in practice though a systematic upward deviation exists with the HC full-scale value. The full-scale scatter is significantly less than the manufacturer's specification (± 0.1 ppm C3) with all analyzers. However, the standard analyzers lie slightly outside the specification (± 0.004 ppm) for the zero scatter, though the above described approach displays the real scatter in the measurement and does not correspond to the determination of the verifiable limit according to specification.

Since particularly the full-scale deviation of state of the art standard analyzers provides considerably better values in practice than the specification, there is no decisive advantage in using the significantly more expensive LE analyzers examined here for extremely low measurement ranges. In addition, with the HC analyzer concentrations of less than 0.7 ppm C3 do not occur anyhow because of the background.

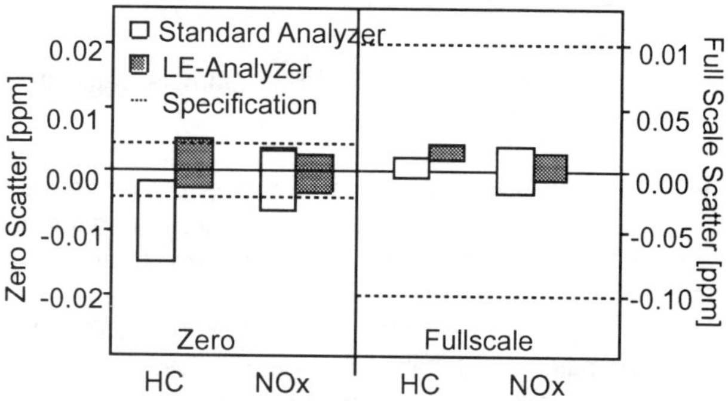

Figure 9: Re-zero and re-span scatter of state of the art standard and LE analyzers

SAMPLE BAG OUT-GASSING - Emissions of hydrocarbons (mostly DMA = NN-dimethylacetamid and phenole /12/) from the bag material (e.g. Tedlar) are a known phenomenon /13/ and certainly of special interest for measurement of HC/NMHC/NMOG - emissions at SULEV level. To quantify the out-gassing, the bags were filled with synthetic air and in some tests additionally with CO2 and water vapor. The bag temperature was 35°C. The HC concentration in the bag was determined with the standard FID of the analyzer bench after 10, 25 and 40 minutes. With this procedure contamination from the samplng system is included.

Fig. 10 shows the results. In the case of the filling with synthetic air, a HC concentration in the bag of approx. 0.06 ppm C3 results after 10 minutes. The differences in the HC concentrations between synthetic air and CO2 in synthetic air are not significant here. For CO2 and H2O in synthetic air, however, considerably higher values result, i.e. approx. 0.1 ppm C3, which means the out-gassing of HC is reinforced by moisture (DMA is soluble in water).

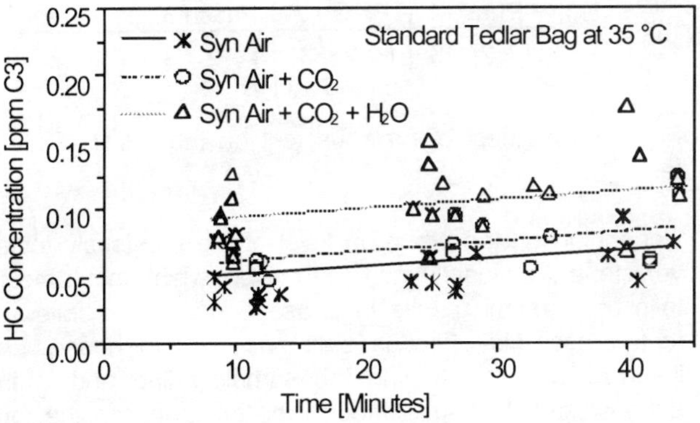

Figure 10: Out-gassing from Tedlar bag at 35 °C for different gas compositions

Through pretreatment of the tedlar bags /14/ out-gassing can be reduced from 0.06 to 0.02 ppm C3, Fig. 11. As discussed above, 30 °C is sufficient for an FTP 75 test. Out-gassing is than further reduced to approx. 0.005 ppm C3 which is almost identical to the out-gassing from an unheated bag (24 °C).

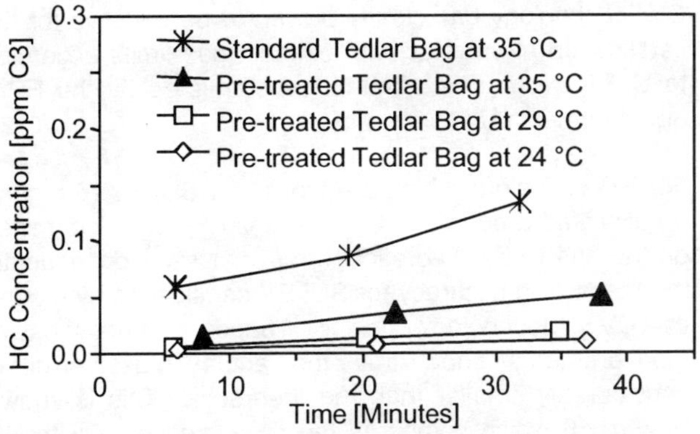

Figure 11: Out-gassing from Tedlar bags at different temperatures (filled with synthetic air)

CVS SYSTEM TEST (BLANK TESTS) - Repeatability, zero and mass-noise on zero-emission level of the entire CVS system (CVS unit, analyzer system, measuring procedures) can be estimated through so-called blank tests. A blank test is defined such that an emission test (FTP75) is carried out without vehicle. The exhaust inlet is closed and exhaust and ambient sample bags are filled with dilution air only. By Eq. 9 and 8 mass emission is calculated using CVS - measurement values and a DF = 8. Because no emitting vehicle is operated in blank test, for all pollutants mass-emission zero is expected.

Blank tests were carried out on a SULEV test cell with heated bags (35°C). NMHC as in all following figures was determined measuring CH_4 with a gas chromatograph. For HC and NMHC almost identical values were found with NMHC scatter only slightly larger. Both components showed a scatter of ± 0.1 mg/mi with an offset of –0.2 mg/mi. For NO_x the scatter is ± 0.05 mg/mi, an offset was not observed, Fig. 12.

For more realistic blank tests, sample flow to diluted exhaust bag was moistened in an additional series of measurements to a absolute humidity of 18 $g_{H2O}/kg_{dry\ air}$. Compared to the blank test without moistening, a detectable increase in scatter to ± 0.35 mg/mi results for HC and NMHC, which corresponds to 3.5% of the SULEV emission limit. Together with the offset of 0.2 mg/mi the maximum deviation, i.e. 0.6 mg/mi, is still only 6% of the SULEV emission limit. The observed shift in the mean value is due to increased out-gassing in the exhaust gas bag due to moisture. Out-gassing in sample and ambient bag in this case is not completely compensated. Scatter of NO_x in blank tests is not influenced by increased humidity in diluted exhaust, Fig. 12.

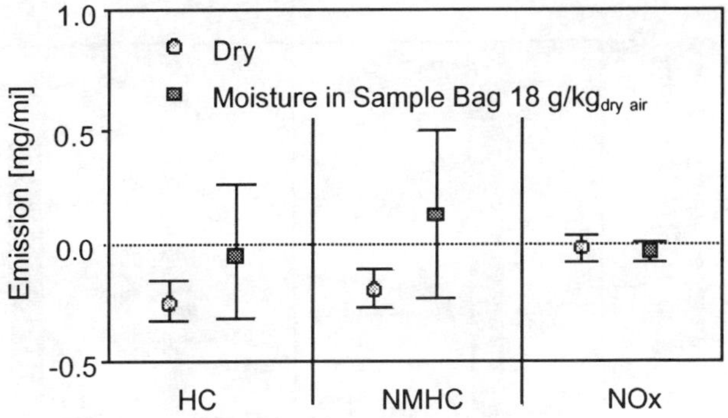

Figure 12: Mean value and standard deviation of 10 FTP75 blank tests with and without moisture in the exhaust gas bag (bag heated to 35°C)

SCATTER OF SULEV EMISSION TESTS - Most important for suitability of CVS-technology is scattering in vehicle tests at SULEV level. Fig. 13 shows emission results of FTP75 exhaust gas tests with 4 different SULEV development cars in an emission test cell configured as shown in Fig. 17. Each vehicle was tested 4-times on a FTP75-cold test. For NMHC an average scatter of approx. ± 1 mg/mi or approx. 10% of the SULEV emission limit was detected. The scatter of the NO_x emission results is around ± 2 mg/mi, which also corresponds to 10% of the SULEV emission limit.

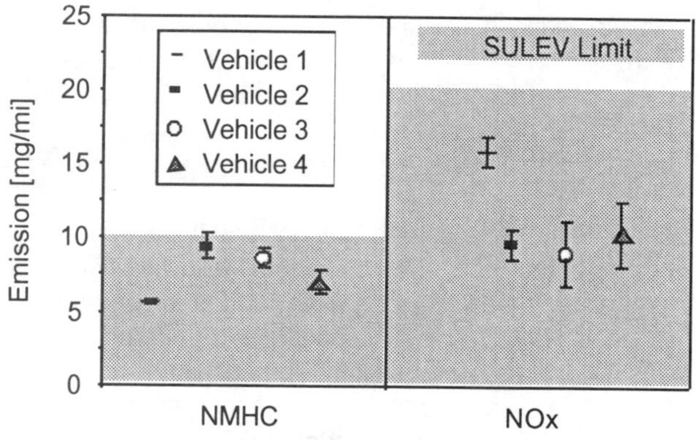

Figure 13: Mean value and standard deviation NMHC and NOx of 4 successive FTP75 cold tests on 4 SULEV development cars in each case (NMHC shown instead of NMOG)

If one compares the scatter occurring in these vehicle tests to the scatter from the above described blank test, Fig. 14, it can be seen that the total scatter of the FTP75 tests in phase 1 is one order of magnitude above the scatter in the blank test. In phase 2, in which the vehicle produces practically no HC or NO_x emissions, the scatter of the blank test is close to that of the FTP75 exhaust gas test. This means that the scatter in phase 1 is not significantly caused by the measuring technique. A significant influence on the emission result therefore stems from the vehicle itself or from not known influences in the test procedure.

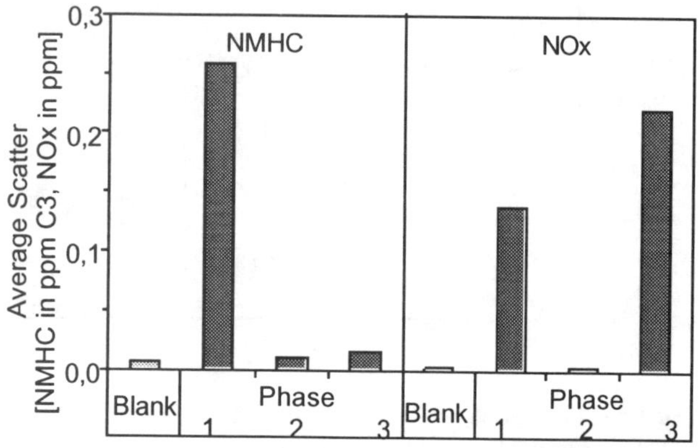

Figure 14: Comparison of the scatter in the blank test from Fig. 12 to the average scatter of the FTP75 from Fig. 13

HANG-UPS - An example of how the test handling influences the emission result concerns so-called hang-ups. These represent contamination of the measuring system partially released in the following exhaust gas test and thus falsifying the emission result. To study the extent to which these hang-ups in the exhaust gas test have an impact on the following test, three blank tests were conducted subsequent to the exhaust gas tests, Fig. 15.

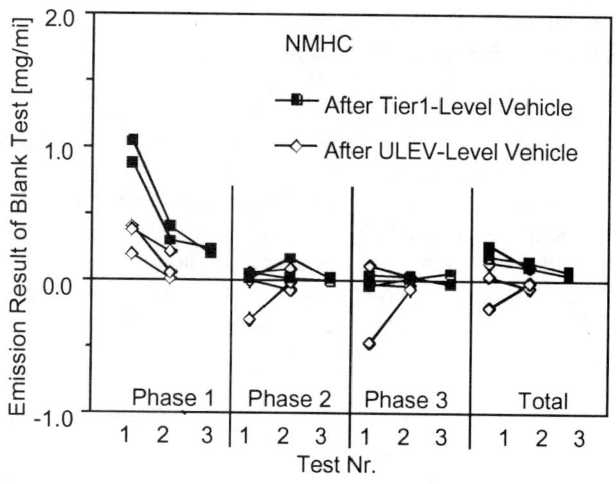

Figure 15: Effect of previous test on the result of blank tests (hang-ups)

Depending on the emission level of the previously tested vehicle, a significantly increased value, which may amount to over 1 mg/mi results in phase 1 during the following test. Since this effect is only observed in phase 1, it involves contamination in the sampling line and in the exhaust sample bag. Although the influence on the total emission result, i.e. a maximum of about 0.3 mg/mi, is still acceptable in the cases shown, a SULEV emission test after a high-emission vehicle may be strongly influenced by hang-ups. To minimize the effects of the previous test, it is useful to equip the test cell with an additional sampling line for higher emissions (dirty line and dirty bag). Additional improvements can be achieved by purge procedures (e.g. blank test).

DISCUSSION

CVS technology has clearly proven to be effective for the certification of vehicles with exhaust gas limits according to ULEV in the past. This is substantiated by the EPA, among others /12/.

For HC emissions which are typical for SULEV (6 mg/mi) a calculated probable error of over 20 % was determined on the basis of a theoretical error analysis. Consequently the measuring accuracy for SULEV certification by means of CVS technology initially appears questionable. Measurements show that the actual mass error is considerably smaller than the theoretical. This is shown by Fig. 16 which compares the calculated error with HC scatter in the blank test (± 0.1 mg/mi HC), and in FTP75 phase 2 vehicle tests (± 0,2 - 0,5 mg/mi). Both are significantly below the scatter (probable error) from the error calculation, which is of ± 2.4 mg/mi for phase 2, appendix 2. Reasons for this:

- The error in the measurement of $c_{i\ mix}$ and $c_{i\ air}$ are not independent as assumed by the error calculation. By using the same analyzer (identical calibration, zero/span adjustment, etc.) the influence of errors on the mass calculation will be compensated such as:

- Zero and full-scale drift
- Non linearity (especially at very low emission levels since the concentrations of sample and ambient bag are almost identical)

• State of the art analyzers measure significantly more reproducible than the manufacturer's specifications state.

In addition to the theoretical errors discussed above, the CVS principle mostly compensates out-gassing from bag material (same bag material, aging and temperature).

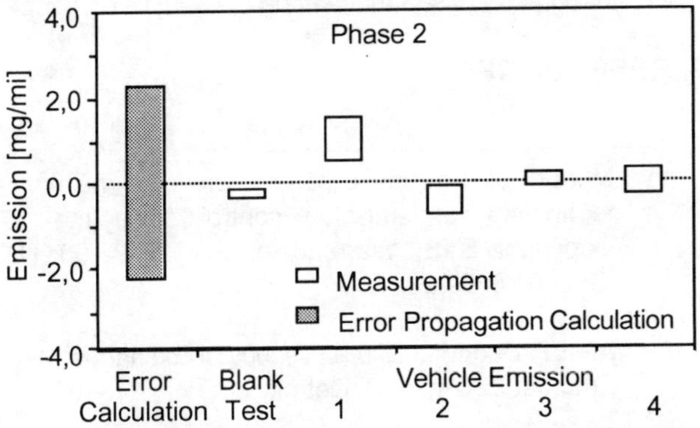

Figure 16: FTP 75, Phase 2, comparison of calculated error (appendix 2), blank test scatter and vehicle measurement scatter, Fig. 14.

Since out-gassing is not completely compensated due to different moisture in exhaust and air bag during emission tests, excessive heating above 30 °C has to be avoided.

In test series with SULEV development cars a HC scatter of approx. 10% of the SULEV emission limit was verified. The same applied to NO_x, Fig. 13. This shows that the SULEV emission limit can be verified with a high degree of certainty with CVS technology.

For vehicles with internal combustion engines with $\lambda_e > 1$, the existing definition of the DF in the legislation may lead to the calculation of negative total emissions, especially in phase 2 and 3. This can be avoided by determining DR by measurement of the dilution air volume V_{air} or by dilution air refinement. However, the typical error is negligible even at SULEV level. It is further reduced when negative emission results are set to zero as stipulated by the EPA /15/.

Fig. 17 shows a SULEV test cell with components that are meaningful also from a cost/benefit point of view.

In particular, heating and drying of the dilution air, heating of gas lines up to 50°C and bags up to 30°C is recommended to avoid condensation with optimized dilution.

Application of catalytic conditioning of the dilution air can be a possibility to further increase accuracy and to eliminate the systematic difference due to Eq. 12. This studies show, that SULEV certification without DAR can be done with high accuracy.

The commonly applied charcoal filter predominantly serves to smooth the background /16/.

A separate sampling line for high emissions (e.g. the cold start phase) and the use of pretreated bags reduce the problem of hang-ups and out-gassing.

The heat exchanger in the CVS provides for a nearly constant volume flow, which makes its determination considerably more accurate. Through the adjustable flow rate (multi-venturi or PDP) the dilution can be optimized for each test phase.

State of the art analyzers with a high degree of

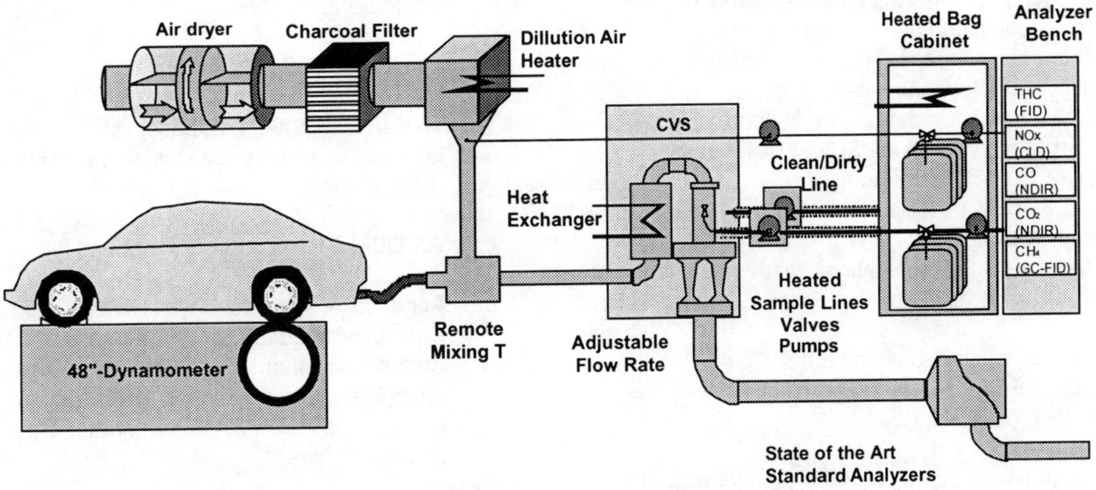

Figure 17: Recommended equipment for a test cell for low-emission vehicles

reproducibility are required, but more expensive analyzers specially optimized for very low measuring ranges (HC < 3 ppm C3; NOx < 10 ppm) are not necessary. On the one hand, because appropriate calibration gases are not available and, on the other hand, because for HC, for example, no concentrations less than 0.7 ppm C3 occur.

Furthermore, a precisely defined test procedure is decisive for less scatter in vehicle measurement, i.e.:

- Use SULEV test cell only for extremely low-emission vehicles
- Purge procedure
- Defined vehicle preconditioning
- Defined vehicle state (e.g. fuel composition, etc.)
- Precisely defined ambient conditions

CONCLUSIONS

The CVS technology is explicitly proposed by the legislature. A departure from the basic calculation formulas in the CFR would lead to a change in the mass emissions in comparison to the LEV I and LEV II definition. A discussion of the fundamental assumptions of CVS technology for the calculation of pollutant mass emission showed:

- The basic calculation formulas in the CFR can lead in some cases to negative mass emission results.
- For vehicle concepts with $\lambda_e \sim 1$ these are negligible even at SULEV level.

The results of a theoretical error analysis were compared to blank tests and SULEV emission tests. This showed that the real measuring uncertainty is considerably less than the uncertainty estimated via the error analysis.

Practical studies show:

- Due to the differential measurement principle of CVS technology, some errors are almost compensated.
- Through reduction of the dilution to DF=6, the concentration difference between ambient air and exhaust sample bag and, hence, the mass calculation accuracy can be nearly doubled.
- Vehicle measurements at SULEV level resulted in a scatter of ± 10% of the emission limit which shows that the SULEV emission limit can be verified with a high degree of certainty.
- Meticulous test preparation and execution is absolutely imperative for reproducible measurement results.
- SULEV test cells need enhanced CVS:
  - Charcoal filter
  - Heating dilution air
  - Remote mixing T
  - Separation of sampling lines (clean/dirty line)
  - Heat exchanger in the CVS
  - Heated sampling lines
  - pre-treated bags
  - State of the art analyzers

A reduction of DF additionally requires:
- Drying of dilution air
- Optimized dilution
- Pre-treated bags heated up to 30 °C
- Sample Lines heated up to 50 °C

CVS technology has been tested for many years and studies show that an enhanced CVS test cell has the required accuracy for certification of SULEV concepts. As a consequence, the worldwide comparability of the exhaust gas results is maintained.

REFERENCES

1 Mondt, J. Robert: Cleaner cars: the history and technology of emission control since the 1960; Copyright SAE, Warrendale; PA 15096-001 USA; ISBN 0-7680-0222-2

2 AIGER Bag Mini Diluter Technical Exchange Meeting , August 22-24,2000, Detroit.

3 W. M. Silvis, R. N.I Harvey, and A. F. Dageforde, A CFV Type Mini-dilution Sampling System for Vehicle Exhaust Emissions Measurement, SAE 1999-01-0151

4 K. Guenther, T. Henney, W.M. Silvis, S. Nakatani, and Dien-Yeh Wu, Improved Bag Mini-diluter Sampling System for Ultra-low Level Vehicle Exhaust Emissions, SAE 2000-01-0792

5 A. Gifhorn, H. Tieber, J. W. Wiebrecht, R. Ballik, Basic Investigations into the use of the CVS System for Ultra- low-emission Vehicles with Gasoline Engines, MTZ 61(2000)2, Pg. 106, MTZ Worldwide 61 (2000)2, Pg. 17

6 40 CFR 86.082-2 Definitions

7 40 CFR 86.109-94 Exhaust gas sampling system; Otto-cycle vehicles not requiring particulate emission measurement

8 70/220/EEC;1999/102 EC; Appendix 8

9 Automobile Type Approval Handbook For Japanese Certification; Edited as of December 31,1993; Japan Automobile Standards Internationalization Center; 5 Technical Standard for 10-Mode and 11-Mode Exhaust Emission Measurement for Gasoline-Fueled Motor Vehicles.

10 40 CFR 86.144-90 Calculations; exhaust emissions.

11 M. T. Sherman, R. E. Chase, and K. M. Lennon, Error Analysis of Various Sampling Systems, SAE 2001-01-0209

12 E.I. Sun, and W. N. McMahon, Evaluation of Fluorocarbon Polymer Bag Material for Near Zero Exhaust Emission Measurement, SAE 2001-01-3535

13 Sawano, Yoda, and Uchida, A Study of Low-Emission Measurement Techniques, JSAE 9631010, 1996

14 HORIBA Europe procedure, Sulzbach, Germany

15 Correcting Negative Calculated Emission Levels to Zero¸ United States Environmental Protection Agency; February 8, 2001

16 J. Heckelmann, Analyse und Verification verfahrenstechnischer Apparate zur Bereitstellung von stabiler und getrockneter Verdünnungsluft bei der CVS Abgasmessung, Diplom Thesis Fachhochschule Heilbronn, Institut für Verfahrenstechnik, 2001

CONTACT

Harald Behrendt and Dr. Oliver Mörsch
DaimlerChrysler AG
HPC D606
D-70546 Stuttgart
Germany
email: Harald.Behrendt@daimlerchrysler.com
email: Oliver.Moersch@daimlerchrysler.com

Christa T. Seiferth
BMW AG
Petuelring 130
D-80788 München
Germany
email: Christa.Seiferth@bmw.de

Dr. Gerd E. Seifert
Dr. Ing. h.c. F. Porsche AG
Porschestraße
D-71287 Weissach
Germany
email: Seiferge@porsche.de

Dr. Jörg W. Wiebrecht
AUDI AG
N/EA-521
D-74172 Neckarsulm
Germany
email: Joerg.Wiebrecht@audi.de

SYMBOLS

c	concentration
c_{iair}	dilution air pollutant concentration.
c_{ie}	exhaust gas pollutant concentration
c_{imix}	diluted pollutant concentration
CO	carbon monoxide concentration
CO_2	carbon dioxide concentration
HC	total hydrocarbon concentration
d	distance
DR	dilution ratio
DF	dilution factor
k	k-factor
m	mass
$m_{i\,eDF}$	Mass calculated using DF
$m_{i\,e}$	Mass calculated using DR
$m_{i\,entry}$	incoming component mass
$m_{i\,exit}$	outflowing component mass
M	weighted mass
p	pressure
Q	CVS flow rate
T	temperature
t	time
V	volume
V_{air}	dilution air volume
V_e	undiluted exhaust gas volume
$V_{e\,DF}$	Ve calculated using DF
V_{mix}	CVS volume
δ	differential
Δ	difference
λ	air/fuel ratio
λ_e	average air/fuel ratio over test
ρ_i	density

ABBREVIATIONS

CFR	Code of Federal Register
CVS	Constant Volume Sampler
CFV	Critical Flow Venturi
DAR	Dilution Air Refinement
D. L.	Detection Limit
FTP	Federal Test Procedure
LEV	Low Emission Vehicle
MR	measuring range
MV	measured value
NMHC	non-methane hydrocarbon
NMOG	non-methane organic gas
PDP	Positive Displacement Pump
SULEV	Super Ultra Low Emission Vehicle
ULEV	Ultra Low Emission Vehicle

Appendix 1:

The stochiometric combustion of one mole of the theoretical molecule $C_8H_{14,8}$ representing gasoline fuel with H/C = 1,85 according the formula:

$$C_8H_{14,8} + 11,7 \cdot (O_2 + 3,77 \cdot N_2) \quad \rightarrow 8\,CO_2 + 7,4\,H_2O + 44,14N_2$$

yields 59.5 moles of exhaust gas. Therefore, V_{eDF} = 59.5. The table shows the exhaust volume in moles for different λ_e values. Assuming ideal gas behavior, the total is equal to the real exhaust volume V_e. It can be seen that the assumption $V_e = \lambda_e V_{eDF}$ is a good approximation.

λ [–]	1	1,2	1,5	2	3
CO_2 [mole]	8	8	8	8	8
H_2O [mole]	7.4	7.4	7.4	7.4	7.4
N_2 [mole]	44.1	52.9	66.15	88.2	132.3
O_2 [mole]	0	2.3	5.85	11.7	23.4
total $=V_e$ [mole]	59.5	70.7	87.4	115	171
$\lambda\, V_e$DF [mole]	59.5	71.4	89.3	119	179
Error [%]	0	1.0	2.1	3.1	4.2

Appendix 2: Comparison of calculation DF/DR

The examples 1- 3 are theoretical analysis on the CVS system as a control zone. Based on the specifications of the engine emissions V_e and $c_{e\,i}$ as well as the dilution air V_{air}, $c_{air\,i}$, the mass calculations can be carried out according to the CFR with the dilution ratio DF (§86.544-94) and according to the real dilution ratio DR. It is shown that the mass calculation with the real dilution ratio is identical to the "true mass".

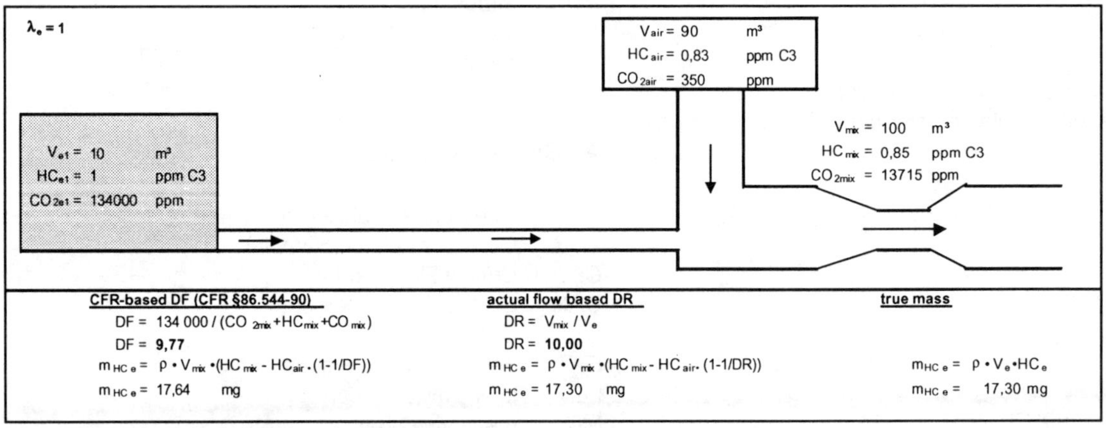

Example 1: λ_e = 1, SULEV emission level

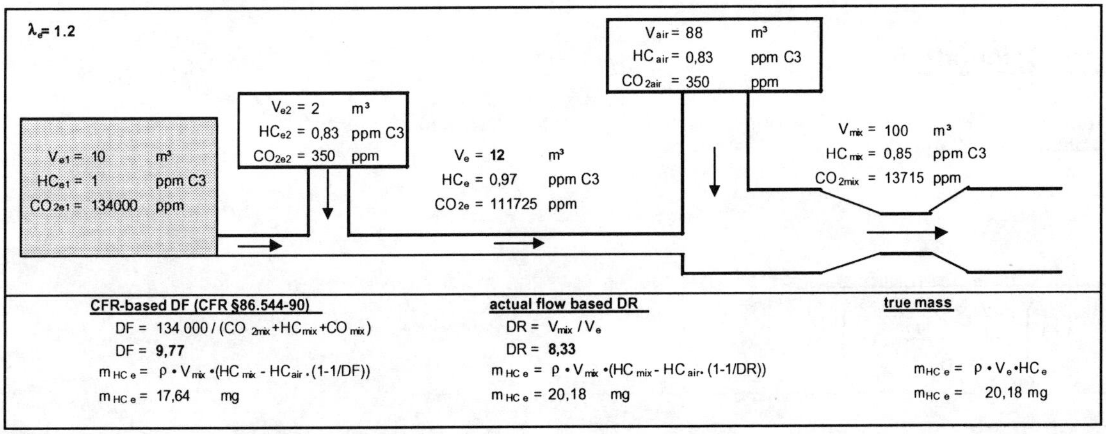

Example 2: λ_e = 1.2, SULEV emission level

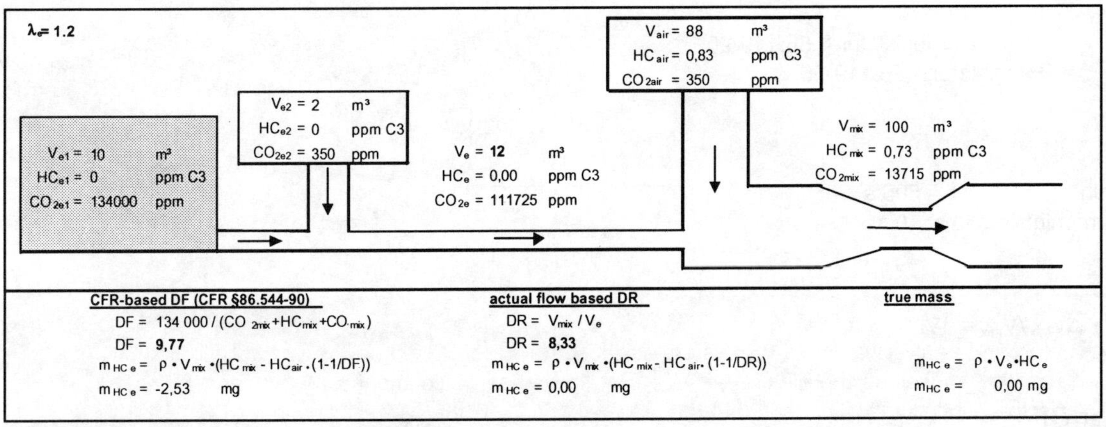

Example 3: λ_e = 1.2, complete catalytic clean up.

For λ_e = 1 there is very good consistency with the "true mass" in the HC determination of mass with DF and DR. The difference of 0.36 mg is due to the fact that the DF calculation is based only on the sample bag concentrations without correction for CO_2 background. The difference between calculation with DF and DR for λ_e = 1.2 is 2.53 mg. In the case of catalytic clean up, this leads to negative values. Since negative emission is set to zero, the final results for DF and DR in example 3 are zero in both cases.

Appendix 3: Error Calculation

1 CVS Volume V_{mix}

$V_{mix} = Q \bullet t$ **Eq. A1**

$$\Delta V_{mix} = \sqrt{\left(\frac{\delta V_{mix}}{\delta Q} \cdot \Delta Q\right)^2 + \left(\frac{\delta V_{mix}}{\delta t} \cdot \Delta t\right)^2}$$

$$\Delta V_{mix} = \sqrt{(t \cdot \Delta Q)^2 + (Q \cdot \Delta t)^2}$$

$\Delta Q/Q = \pm 0.5$ % $\Rightarrow$ legislator § 86.119-90
$\Delta t = \pm 0{,}05$s $\Rightarrow$legislator: §86.119-90

example:

for $Q = 9\ m^3/min$; $t = 505$ s
$\Delta Q = 0.09\ m^3/min$; $\Delta t = 0.1$s

$V_{mix} = 75.75\ m^3$
$\Delta V_{mix} = 0{,}757\ m^3$ => $\Delta V_{mix}/V_{mix} = 1\%$

2 Dilution Factor DF

$$DF = \frac{134000}{CO_{2mix} + CO_{mix} + HC_{mix}}$$ **Eq. A2**

$$\Delta DF = \sqrt{\left(\frac{\delta DF}{\delta CO_{2mix}} \cdot \Delta CO_{2mix}\right)^2 + \left(\frac{\delta DF}{\delta CO_{mix}} \cdot \Delta CO_{mic}\right)^2 + \left(\frac{\partial DF}{\partial HC_{mic}} \cdot \Delta HC_{mix}\right)^2}$$

$$\frac{\delta DF}{\delta CO_{2mix}} \cdot \Delta CO_{2mix} = -\frac{134000}{(CO_{2mix} + CO_{mix} + HC_{mix})^2} \cdot \Delta CO_{2mix}$$

$\frac{\delta DF}{\delta CO_{mix}} \cdot \Delta CO_{mix}$; $\frac{\delta DF}{\delta HC_{mix}} \cdot \Delta HC_{mix}$: analogous

$\Delta CO_{2mix} / CO_{2mix} = 1.49\%$ (see table under 7.)

errors of CO and HC measurement are negigible, because the absolut values are small in comparision to CO_2

$\Rightarrow \Delta DF / DF \approx 1{,}5\%$

3 Mass Emission per Phase

$$m_i = \rho_i \cdot V_{mix} \cdot \left[c_{i\,mix} - c_{i\,air} \left(1 - \frac{1}{DF} \right) \right]$$

Eq. A3

ρ: constant

$$\Delta m_i = \sqrt{\left(\frac{\delta\, m_i}{\delta\, V_{mix}} \cdot \Delta V_{mix} \right)^2 + \left(\frac{\delta\, m_i}{\delta\, DF} \cdot \Delta DF \right)^2 + \left(\frac{\delta\, m_i}{\delta\, c_{i\,mix}} \cdot \Delta c_{i\,mix} \right)^2 + \left(\frac{\delta\, m_i}{\delta\, c_{i\,air}} \cdot \Delta c_{i\,air} \right)^2}$$

$$\frac{\delta\, m_i}{\delta\, V_{mix}} \cdot \Delta V_{mix} = \rho_i \cdot \left[c_{i\,mix} - \left(1 - \frac{1}{DF} \right) \cdot c_{i\,air} \right] \cdot \Delta V_{mix}$$

$$\frac{\delta\, m_i}{\delta\, DF} \cdot \Delta DF = -\rho_i \cdot V_{mix} \cdot c_{i\,air} \frac{1}{DF^2} \cdot \Delta DF$$

$$\frac{\delta\, m_i}{\delta\, c_{i\,mix}} \cdot \Delta c_{i\,mix} = \rho_i \cdot V_{mix} \cdot \Delta c_{i\,mix}$$

Eq. A4

$$\frac{\delta\, m_i}{\delta c_{i\,air}} \cdot \Delta c_{i\,air} = -\rho_i \cdot V_{mix} \cdot \left(1 - \frac{1}{DF} \right) \cdot \Delta c_{i\,air}$$

Eq. A5

$\Delta V_{mix} / V_{mix} = 1\%$; $\Delta DF / DF = 1{,}5\%$
$\Delta HC_e / HC_e$, $\Delta HC_{air} / HC_{air} \approx 3{,}5\%$ (see table under 6)

4 Weighted Mass M (g/mi)

$$M = 0{,}43 \frac{m_1 + m_2}{d_1 + d_2} + 0{,}57 \frac{m_2 + m_3}{d_2 + d_3}$$

Eq. A6

d_1, d_2, d_3: distance in each phase
$d_1 + d_2 = d_2 + d_3 = 7{,}48$ mi
$\Delta d = 0{,}0003$ mi
(EPA:Specification for electric chassis dynamometers, attachment A RFD c 100081T1, 1991)

⇒ measurement uncertainty ΔM:

$$\Delta M = \sqrt{\left(\frac{\delta M}{\delta m_1} \cdot \Delta m_1 \right)^2 + \left(\frac{\delta M}{\delta m_2} \cdot \Delta m_2 \right)^2 + \left(\frac{\delta M}{\delta m_3} \cdot \Delta m_3 \right)^2 + \left(\frac{\delta M}{\delta d_1} \cdot \Delta d_1 \right)^2 + \left(\frac{\delta M}{\delta d_2} \cdot \Delta d_2 \right)^2 + \left(\frac{\delta M}{\delta d_3} \cdot \Delta d_3 \right)^2}$$

Eq. A7

$$\frac{\delta M}{\delta m_1}\cdot \Delta m_1 = \frac{0{,}43}{d_1 + d_2}\cdot \Delta m_1$$

$$\frac{\delta M}{\delta m_2}\cdot \Delta m_2 = \frac{1}{d_1 + d_2}\cdot \Delta m_2$$

$$\frac{\delta M}{\delta m_3}\cdot \Delta m_3 = \frac{0{,}57}{d_2 + d_3}\cdot \Delta m_3$$

$$\frac{\delta M}{\delta d_1}\cdot \Delta d_1 = -\frac{0{,}43(m_1 + m_2)}{(d_1 + d_2)^2}\cdot \Delta d_1$$

$$\frac{\delta M}{\delta d_2}\cdot \Delta d_2 = -\frac{0{,}43(m_1 + m_2) + 0{,}57(m_2 + m_3)}{(d_1 + d_2)^2}\cdot \Delta d_2$$

$$\frac{\delta M}{\delta d_3}\cdot \Delta d_3 = -\frac{0{,}57(m_2 + m_3)}{(d_2 + d_3)^2}\cdot \Delta d_3$$

5 Measuring error of modern Equipment

$$\text{total error} = \sqrt{(\text{linearity})^2 + (\text{reproducibility})^2 + \left(\frac{\text{drift}}{4}\right)^2}$$

	FID	**CO_{low}**	**CO_2**
detection limit	6 ppb für MR:10 ppm	50 ppb für MR: 50 ppm	15 ppm für MR: 20%
drift	< 0,5% MV + 2xD .L./h	<1,0% MV + 2xD .L./h	<1,0% MV + 2xD .L./h
linearity	< 1,0% MV + 2xD .L.	<1,0% MV + 2xD .L.	< 1,0% MV + 2xD .L.
reproducibility	< 0,5% MV + 2xD .L.	< 0,5% MV + 2xD .L.	< 0,5% MV + 2xD .L.

D. L. Detetion Limit
MV: measured value
MR: measuring range

6 Error Calculation in Bag Concentration Measurement

test cycle : FTP 75	HC ppm C3	CO ppm	NO_x ppm	CO_2 ppm
Sample bag phase 1	1.677	5.125	0.474	13600
phase 2	0.887	0.185	0.042	13287
phase 3	0.877	0.165	0.078	17040
Ambient bag phase 1,2,3	0.980	0.130	0.033	420

measuring uncertainty for concentrations

		single error			total. error	total. error.
Sample bag		lin.	repro.	drift	Δc ppm	Δc/c %
phase 1:HC_{mix}	ppm	0.029	0.020	0.014	0.038	2.26
phase 2:HC_{mix}	ppm	0.021	0.016	0.013	0.030	3.34
phase 3:HC_{mix}	ppm	0.021	0.016	0.013	0.030	3.37
phase 1:$CO_{2\,mix}$	ppm	166	98	64	203.12	1.49
Ambient bag						
HC_{air} (all phases)	ppm	0.022	0.017	0.013	0.031	3.12

7 Total Error in Calculation of Mass

ρ_{HC} = 576.8 g/m³
ρ_{CO2} = 1835 g/m³

		phase1	phase2	phase3
flow rate Q	m^3/s	0.15	0.15	0.15
time t	s	505	869	505
Volume V_{mix}	m^3	75.75	130.4	75.75
distance d	mi	3.59	3.89	3.59
Dilution Factor DF	-	9.85	10.08	7.86
emission				
$m_{(HC)}$	mg	98	0.9	2.8
$\Delta m_{(HC)}$	mg	6	9.1	5.2
$\Delta m_{(HC)}$ /$m_{(HC)}$	%	6.0	965	184
$\Delta m_{(HC)}$	mg/mi	1.7	2.4	1.5

Weighted Total HC Emission		
M	mg/mi	6.0
ΔM	mg/mi	1.3
ΔM/M	%	22

2002-01-0049

Improved Low-Emission Vehicle Simulator for Evaluation of Sampling and Analytical Systems

Daniel Whelan, Gary Lewis and Scott Kiyabu
Horiba Instruments, Inc.

Travis Henney, Darius Harrison, Jennifer Farrugia and Michael Gabbert
Ford Motor Co.

ABSTRACT

The Vehicle Exhaust Emissions Simulator was developed to evaluate the performance of vehicle emissions sampling and analytical systems. The simulator produces a representative tailpipe volume flow rate containing up to five emission constituents, injected via mass flow controllers (MFCs). Eliminating the variability of test results associated with the vehicle, driver, and dynamometer makes the simulator an ideal quality control tool for use in commissioning new test cells, checking data correlation between test cells, and evaluating overall system performance.

Earlier vehicle emissions simulators being used in the industry were primarily for checking Constant Volume Samplers (CVSs) and Bag Benches but they did not have the ability to properly simulate tailpipe volume. Therefore, these systems were unable to check the other components of the test cell sampling and analytical equipment, including modal benches, mini-diluters, ultrasonic flow-meters, and other exhaust flow measurement techniques, such as CO_2 Tracer and Remote Mix Tee (RMT) dilution air volume subtraction from total diluted exhaust volume. The simulator system described here contains a squirrel cage blower with a stepper valve and sub-sonic venturi, enabling variable tailpipe exhaust flow-rates in the range of 0.2 to 4.2 SCMM (standard cubic meters per minute).

Numerous improvements have been made in order to address other limitations that have plagued previous simulator systems. Accuracy of CO_2 mass injection has been a challenge because of MFC calibration and gas temperature stability. We have worked with Colorado Engineering Experiment Station Inc. (CEESI) to establish a primary flow standard for CO_2 MFCs, and have incorporated the use of a thermostatically controlled gas heater for reliable CO_2 temperature control over the flow range of 0-150 SLPM (standard liters per minute). We have evaluated MFC drift and differences in results when using different gas compositions and the system computer, a PC running Windows™ and LabVIEW™ software, has undergone several software enhancements, such as automatic MFC calibration, MFC zero correction, and flexible gas injection profiles.

This paper, besides describing all the functionality and features of the simulator system, also details the performance data, which has been collected and analyzed at several different test sites over the last year. In addition, recommended future investigations and enhancements are discussed.

INTRODUCTION

As vehicle tailpipe exhaust emissions continue to decrease, the sampling and analytical systems, used to measure these very low emission levels, have had to become more sophisticated in order to maintain the required levels of accuracy. Equally challenging is developing Quality Control tools, used to evaluate the performance of low level samplers and analyzers. The VEES-100 vehicle exhaust emission simulator was originally developed in a joint venture between Horiba Instruments Inc., and Ford Motor Co. It was introduced near the end of a 2001 SAE paper (Reference 1), by Michael Landry, et al. of the Ford Motor Co., which described the predecessor to the system described here. The VEES-100 has now evolved to the point that it is capable of checking all of the components of an emission test system.

Figure 1 shows the components that make up the VEES-100 system. The squirrel cage blower pulls air into the system to simulate exhaust volume. The flowrate is controlled by a motorized damper valve and is measured with a sub-sonic venturi. The flowrate range for the air is approximately 0.2 to 4.2 SCMM. Mass flow controllers are used to inject known amounts of propane, CO, CO_2, and NO_X into the duct to mix with the air prior to exiting the system. A 5th MFC is available either as a second MFC for a given constituent, enabling a

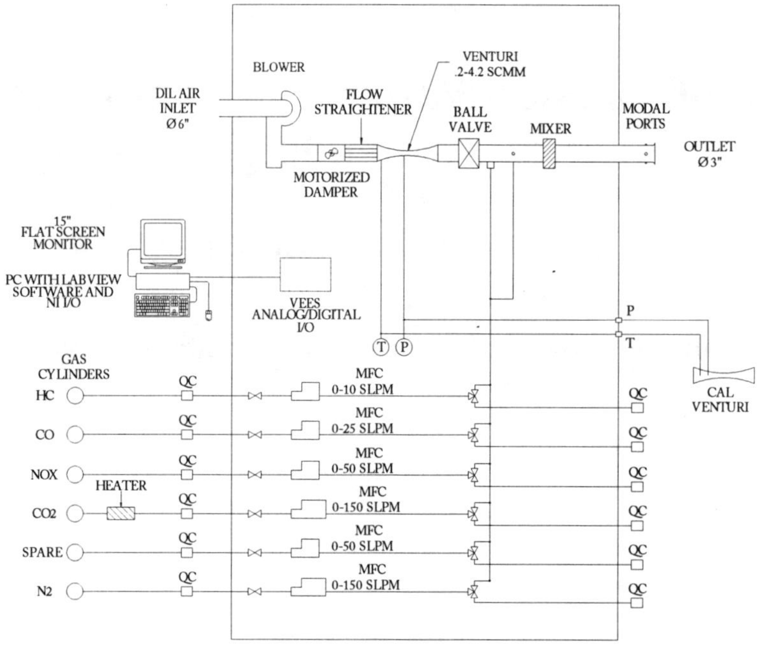

FIGURE 1: VEES-100 VEHICLE EXHAUST SIMULATOR FLOW SCHEMATIC

wider injection range, or for another gas. Also, Nitrogen can be used as the diluent, instead of the air from the blower, by closing a valve and injecting the N_2 via the 6th MFC. This mode of operation may be desirable in cases where very small emission levels are being simulated and the background contribution of the ambient air (particularly THC) would be a significant portion of the total mass injected. The MFC for CO_2 is from Brooks Instruments and the other five are from STEC-Horiba. Three-way valves and quick-connect ports on the unit allow for injected gases to bypass the mixing duct and are also used for MFC calibration purposes.

The VEES system is controlled by a PC, running Microsoft Windows™ and National Instruments LabVIEW™ with National Instruments analog and digital I/O. The customized program allows the user to select from four principle modes of operation:

- Fixed Mode – the most basic operation, with fixed MFC and exhaust flow-rates over a specified time interval.
- Ramp Mode – repetitive ramp up/down sequences for exhaust flow and MFCs.
- Phase Mode – build up to five timed phases, each containing an exhaust flow-rate and MFC set-points in either volume or mass units.
- Modal Mode – build a multi-phase, second by second profile of grams/second of each constituent and exhaust flow. Alternatively, a profile with target tailpipe concentration of each constituent and exhaust flow-rate can be used.

During testing, the system generates log files containing injected mass and volume data on a 10 times per second basis, and totals by phase. The injected mass calculations also can optionally account for any background contamination that exists in the air from the squirrel cage blower. Other software features in the system include automatic MFC calibration, using calibration venturis that plug into the quick-connect ports and MFC auto-zero offset correction, which is carried out each time a test sequence is initiated.

Figure 2 shows a typical test cell sampling and analytical system, and how the VEES system would be utilized. The air/gas mixture, exiting the VEES, simulates the tailpipe exhaust from a vehicle and can be connected to an ultrasonic flow meter and/or a remote mix tee (RMT) constant volume sampler (CVS) system. Also connected to the outlet of the VEES are two heated sample lines; tailpipe modal sample and bag mini-diluter (BMD) sample. This test cell has three methods by which tailpipe volume can be measured: 1) direct signal from the ultrasonic flow meter, 2) subtracting metered RMT dilution air volume from CVS volume, and 3) the CO_2 Tracer method in which the ratio of CVS diluted CO_2 to tailpipe CO_2 equals the ratio of tailpipe volume to CVS volume. The known tailpipe volume from the VEES can be used to check any or all three of these methods. Fundamental to the operation of the bag mini-diluter is that its bags be filled at a rate proportional to the tailpipe exhaust flowrate. Part of a system QC sequence would be to vary the exhaust flowrate out of the VEES and check how the other devices respond. The VEES can also be used to determine the time delays associated with the bag mini-diluter and the modal analyzer benches.

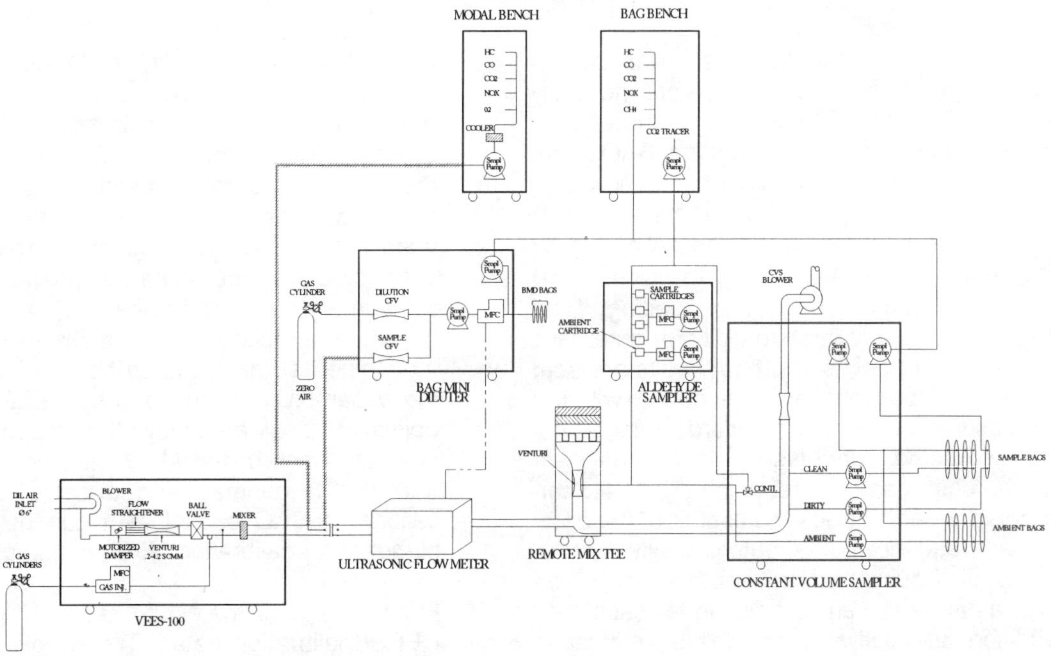

FIGURE 2: TYPICAL TEST CELL UTILIZING VEHICLE SIMULATOR

Much of the testing that was conducted for the purposes of validating the VEES system was targeted to simulate phase 1 of an EPA-75 test for a typical ULEV vehicle. Table 1 shows those emission levels. Looking at the average grams/minute and the CVS bag concentrations, these levels are easily achievable by the VEES. However, the tailpipe concentration data shows very large ratios of maximum to minimum concentrations for THC, CO and NO_X. If we build a modal mode profile with typical phase 1 second by second data, the VEES system will limit the MFC flow set-points to 2-100% of full-scale, so simulating the extremes is not feasible with a single gas bottle and MFC. As mentioned earlier, the spare MFC can be used as a 2nd MFC for a given constituent (usually THC), so that the injection range can be broadened. Typical phase 1 tailpipe exhaust flow-rates for ULEV vehicles range from approximate 0.1 to 2.3 SCMM. The VEES' exhaust flow from the on-board blower can be controlled to a flow range of approximately 0.2 to 4.2 SCMM, but after adding in the injected gases from the MFCs, the minimum exhaust flow-rate is approximately 0.3 SCMM.

Constituent	Grams/Mile	Total grams/phase	Average grams/min.	CVS Bag Conc.**	Max. TP Conc.	Min. TP Conc.
THC	0.11	0.396	0.047	9 ppmC	6,600 ppmC	<1 ppmC
CO	0.8553	3.08	0.366	35 ppm	53,500 ppm	<1 ppm
CO_2	343.3	1235.6	146.8	0.89 %	15.7 %	4.3 %
NO_X	0.1696	0.614	0.073	4.2 ppm	2125 ppm	<1 ppm

Table 1. ULEV Phase 1 Emission Levels

**CVS flowrate at 9 SCMM

FACTORS THAT AFFECT SIMULATOR TEST RESULTS

In order to achieve the desired results for measured vs. injected emissions, besides gas bottle concentration accuracy, there are considered to be three crucial factors that must be fully understood and resolved. They are MFC calibration, MFC drift, and analyzer response

MFC CALIBRATION – It would be reasonable to expect that the MFC manufacturer's factory calibration curve (voltage vs. flow-rate in slpm) would be suitable for use by the VEES. Generally, this is the case, except for the Brooks MFC used for pure CO_2, in which Brooks calibrates the MFC with nitrogen and applies a theoretical gas factor to account for the difference in gas composition. As we will show, the gas factor used is erroneous, and we are addressing this situation with Brooks. Regardless of the validity of the factory calibration data, VEES users often will re-calibrate the MFCs anyway, using their own calibration standards and procedures. Often, this is required for ISO purposes.

Using LFE System - As discussed in Reference 1, Ford evaluated (3) different methods for calibrating MFCs; Soap film Flow Meter, Molbloc System, and Laminar Flow Element (LFE). They found that all of the devices provided acceptable results within their respective flow ranges, but the LFE was the preferred device because of its ease of use and wider flow-range. When the Brooks CO_2 MFC is calibrated with the LFE system, pure CO_2 is flowed through both devices and the LFE's PC software applies a theoretical gas factor, since the LFE system was calibrated using air.

Using an SAO - Horiba personnel have evaluated the use of sub-sonic venturis for MFC calibration. These sub-sonic venturis will heretofore be referred to as SAOs or Smooth Approach Orifices. The calibration SAO can be connected to the quick-connect port at the outlet of the VEES (downstream of the MFC) and the built-in pressure and temperature sensors, that are normally used to monitor the main VEES exhaust volume venturi, can be utilized. There are two calibration SAOs that have been used; a 150 slpm SAO for the higher flow CO_2 and N_2 MFCs and a 50 slpm SAO for the other (4) MFCs. Both of these venturis have been calibrated by Colorado Engineering Experiment Station, Inc. (CEESI) and are accurate to within ± 0.5 % of reading. (The VEES software contains a calibration utility for automatic 10-point calibration. Set points of 0.5 to 5 volts are sent to the MFC and the output voltage is matched up with the flow-rate as measured by the flow standard. After all the data points are collected, curve fits (1st to 4th order) are generated for the output and the set-point. The user can select which curve order provides the best fit of the data. The system can read flow-rates automatically from a calibration SAO or flow-rates from another calibration standard, such as an LFE, can be typed in for each data point. In addition, the software includes a verification function, used to periodically check an MFC against the standard.

<u>CEESI Calibration of CO_2 MFC</u> – As previously mentioned, the validity of the Brooks 150 slpm CO_2 MFC factory calibration data was questionable, based on injected vs. measured CO_2 test results. In order to better understand these issues, a Brooks MFC (Model 5853S) and Horiba's 150 slpm calibration SAO were calibrated by CEESI with nitrogen and CO_2 gases. Calibration with the actual gas of interest eliminates the controversy that surrounds theoretical gas correction factors, commonly used by industry. The two measurement devices were connected in series with a Bell-prover, used by CEESI as the primary standard for the dual calibration. During the calibration procedure, the Brooks flow control valve was maintained at the full open position, and flow control for the assembly was provided by CEESI equipment. The objective was to study the response and linearity characteristics of the MFC with the two different gases and also to generate a dual set of coefficients for the calibration SAO such that it can be used as the flow standard for calibrating MFCs in the future.

Figure 3 shows the MFC response, plotted against the CEESI calibration data. The Brooks factory calibration curve is shown along with response curves using N_2 and using CO_2. The x-axis is MFC measured output voltage plotted against CEESI flow-rate on the y-axis; for purposes of discussion, the x-axis has been converted to equivalent CO_2 flow-rate using factory calibration data. Factory calibration of the MFC involves calibrating with nitrogen as the flowing gas, and applying a gas

Figure 3. Brooks CO2 MFC Dual Calibration by CEESI

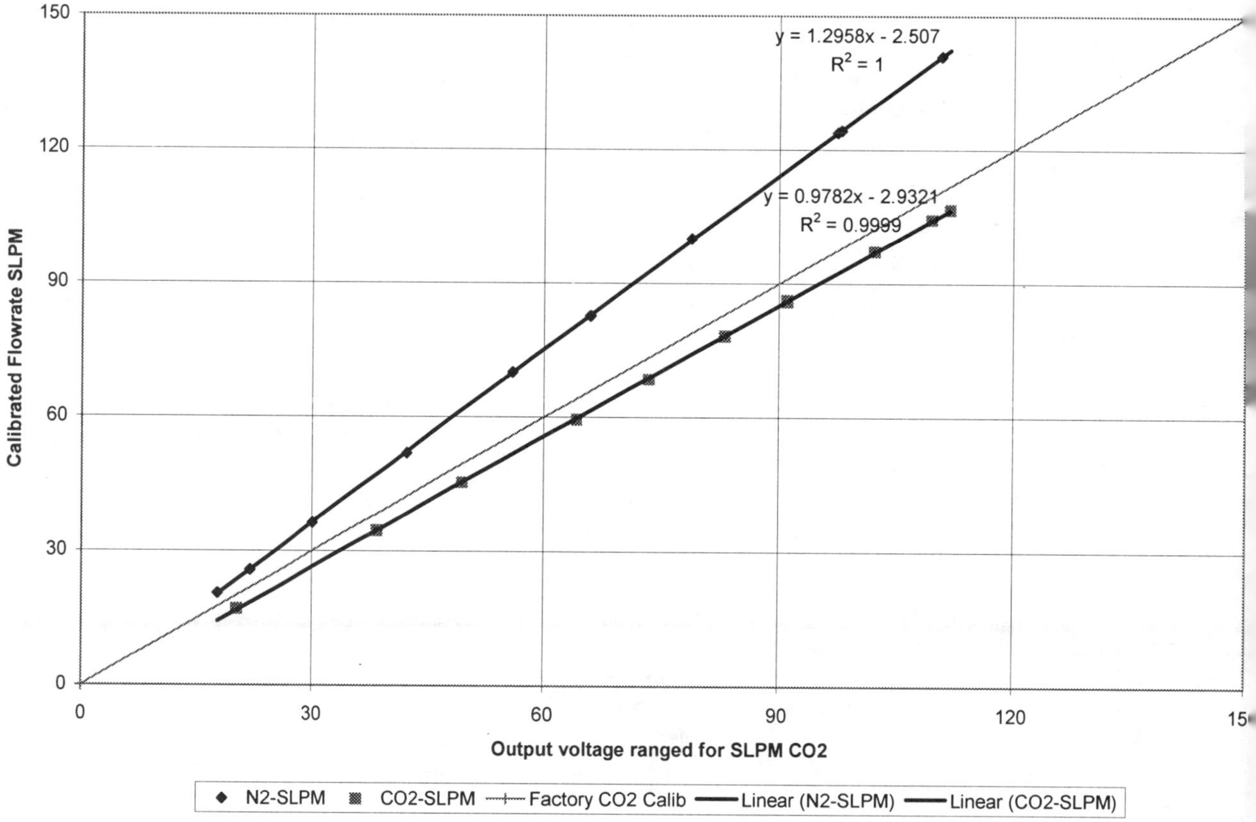

conversion factor "Gasfactor" in order to correct the electrical output for the gas of interest – CO_2 in our case. The "Gasfactor" for nitrogen is 1.0000 and 0.773 for CO_2. From the CEESI data, we are able to re-compute the "Gasfactor" for CO_2 to a more valid number. The following conclusions can be deduced from the graph:

1. Linear curve-fit is sufficient to characterize the plotted data; the "R-squared Factor" being very close to the ideal value of 1.

2. There appears to be a 2+% offset lower than factory calibration (-2.5 for nitrogen and –2.9 for CO_2).

3. Calculation of CO_2 "Gasfactor" using calibrated flow-rate at output voltage equivalent to 100 slpm

   - Using coefficients from equation of top curve (nitrogen), calibrated output Y is 127.073 slpm

   - Using coefficients from equation of bottom curve (CO_2), calibrated output Y is 94.888 slpm

   - New "Gasfactor" = 94.888/127.073 = 0.747

   - Brooks' "Gasfactor" is (.773-.747)/.747 = 3.5% high.

<u>Dual Calibration of SAO</u> – CEESI has calibrated the 150 slpm SAO venturi with N_2 and CO_2 gases, provided 4th order curve fit equation coefficients, and calculated percent differences from the calculated equations for each gas. These results have been plotted, illustrating the level of confidence in using this device for MFC calibration.

Figure 4 illustrates the calibration error, expressed as a percent of reading, over the N_2 and CO_2 calibration flow-rate ranges. The curves show that the SAO flow-rate calibration and curve-fit coefficients result in errors, which fall within an error band of ± 0.25% of reading, with the exception of one point at 0.4%. This information provides a confidence factor when discussing system accuracy, and illustrates that the SAO venturi can be used to calibrate MFC devices to better than 1% accuracy.

Figure 4. SAO Venturi Calibration Error, % of Point

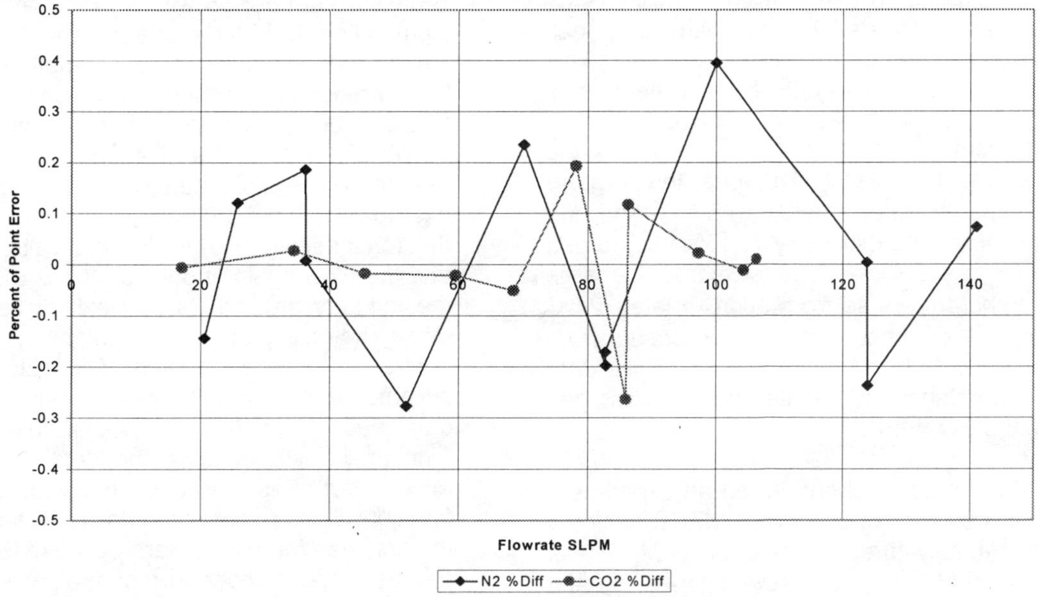

Figure 5 shows flow-rate of CO_2 and N_2 as a function of venturi throat pressure, and includes the calculated ratio of CO_2 / N_2 for throat pressures on the right axis. It illustrates that the ratio is well behaved without discontinuities; the ratio changes less than one percent over the calibrated flow-rate range. It also suggests that for calibration accuracy better than one percent, the SAO venturi should be calibrated with the actual gas (CO_2 in this case), and not rely on N_2 calibration with correction factors.

It is important to note that Horiba has been using SAO venturi devices for almost 20 years as the preferred method of flow-rate calibration. The combination of CEESI calibration data and the use of quality pressure transducers which incorporate zero and 80% span calibration features provides an accurate method for flow-rate measurement, typically between 0.5% to 1.0% of reading. We have collected test data of measured vs. injected mass that is presented later in this paper, using both LFE and SAO based MFC calibrations. The data (Table 4) shows that the results using the SAO MFC calibration are improved for CO_2 but for the other constituents, the differences are not statistically significant. We attribute the improved CO_2 recovery results, when using the SAO MFC calibration, to the dual calibration of the SAO at CEESI.

Figure 5. SAO Venturi Throat Pressure vs Flowrate for N2 and CO2

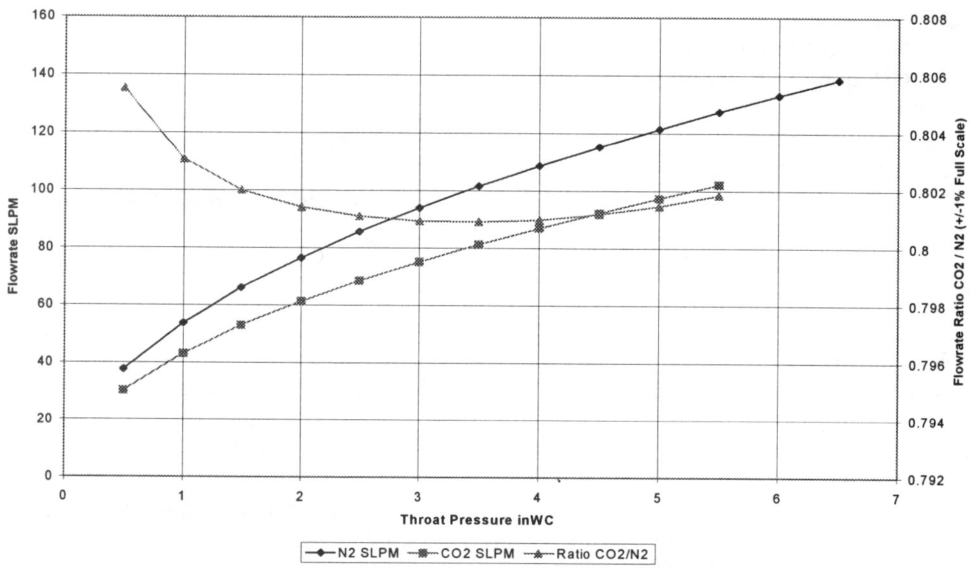

Treatment of CO_2 Gas – During calibration of the Brooks CO_2 MFC, drift has been observed, as seen on the calibration standard being used. It is hypothesized that the drift is caused by temperature drift of the CO_2 gas, mostly evident when changing flow-rates frequently. The pure (99.8%) bone dry CO_2 gas is very cold as it exits the cylinder and heating the gas is required to prevent freeze up and malfunctioning of the cylinder regulator. Typically, fixed rate heaters, on the order of 300 watts, have been utilized in between the cylinder and the regulator. In addition, expansion tanks have been periodically used in between the heater and the regulator in a further attempt to stabilize temperature and/or prevent any liquid droplets from making it through to the MFC. The expansion tanks have been shown to cause temporary pressure disturbances and flow-rate instability when the flow-rate is changed, so we do not recommend their use in cases where the CO_2 MFC is programmed for a transient cycle. Horiba has recently purchased a 1000-watt, thermostatically controlled heater. We conducted an experiment in which tailpipe CO_2 concentration was monitored along with CO_2 gas temperature during a 300 second fixed injection. This test was carried out using two different gas heaters: a 120-watt fixed heater, and the aforementioned 1000-watt thermostatically controlled heater. Figure 6 shows the results for the small heater. The gas temperature dropped from approximately 16°C to 8.5°C and the tailpipe CO_2 rose from approximately 11.15% to 11.5%, a 3.1% increase. Figure 7 shows the results from the large heater. The temperature rose from approximately 24°C to 28°C and the concentration dropped from approximately 11.15% to 11%, or 1.3%. This data illustrates the importance of maintaining constant gas temperature at the CO_2 MFC.

MFC DRIFT – Ford personnel conducted an experiment to study the drift behavior of the VEES MFCs. Over a period of approximately three months, each MFC was checked against the LFE standard several times, in each case at five different MFC set-points.

Figure 8 shows the results for the NO_X MFC, which is an STEC model SEC-4550 with a measurement range of 0-50 slpm. The MFC has a full scale voltage of 5 VDC so in the case of the NO_X MFC, 5 VDC corresponds to approximately 50 slpm and 1 VDC corresponds to approximately 10 slpm. The data is typical for the other STEC MFCs used for THC and CO injection. At a given set-point voltage, say 1 volt, the bars are in chronological order showing how the MFC output signal compared with the LFE on that particular day. The data shows that there is no evidence of a steady drift in one direction or another, just some up/down variability from one test to the next. It is curious that all of the data, except a few points during the 2/2/01 run, shows the MFC reading low compared to the LFE as evidenced by the average % difference values. This suggests that perhaps the original MFC calibration, which was conducted on 2/1/01, was invalid or that some early zero drift had occurred.

Figure 9 shows the results for the CO_2 MFC, which is a Brooks model 5853S with a measurement range of 150 slpm. This MFC also doesn't show any evidence of a steady drift in one direction or another and the average % differences are closer to 0% than was the case for the STEC model. However, there is a great deal more variability in how the direction for % difference with respect to the LFE changes with set-point voltage. For example, on the 3/1/01 test, the MFC was reading low on the 1 volt setting, high for 2 and 3 volts, right on at 4 volts and then low again at 5 volts. It is a widespread belief in the industry that MFCs can exhibit zero drift behavior. Often, adjusting the zero point brings the whole curve back to calibration condition. The VEES system now has an auto zeroing function that was not present when this data was collected. We can surmise that this added functionality will improve the drift results of the STEC MFCs, but it is not expected that the auto zeroing function will improve the data from the CO_2 MFC since the drift was more irregular. Some of the variability could be due to gas temperature control, which was discussed in the last section.

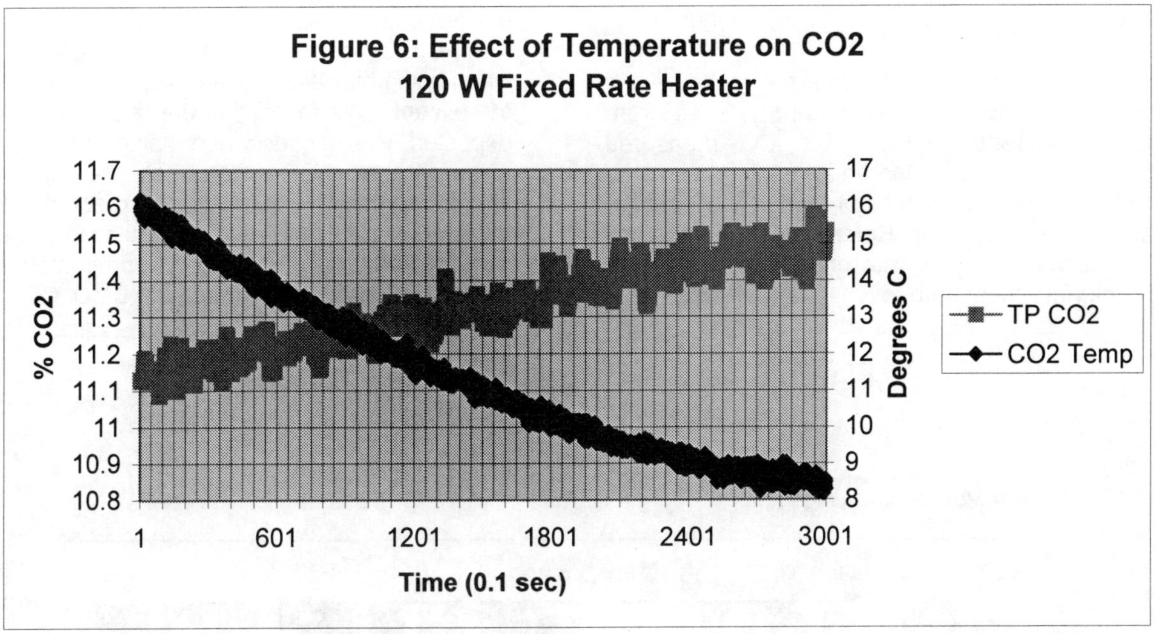
Figure 6: Effect of Temperature on CO2
120 W Fixed Rate Heater
% CO2
Degrees C
Time (0.1 sec)
TP CO2
CO2 Temp
11.7
11.6
11.5
11.4
11.3
11.2
11.1
11
10.9
10.8
17
16
15
14
13
12
11
10
9
8
1
601
1201
1801
2401
3001

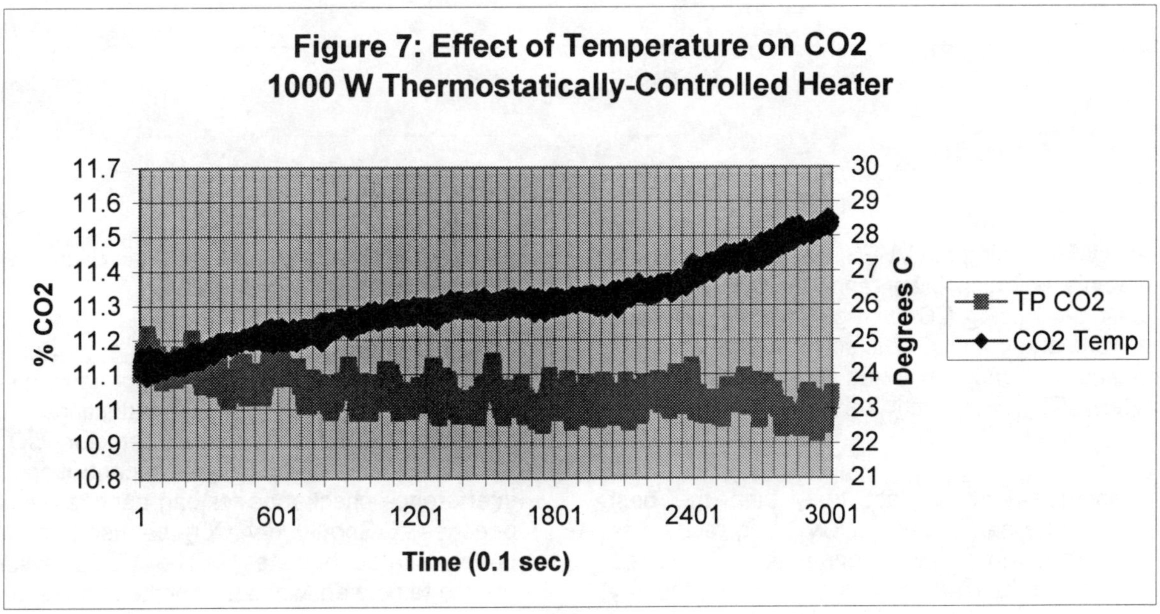
Figure 7: Effect of Temperature on CO2
1000 W Thermostatically-Controlled Heater
% CO2
Degrees C
Time (0.1 sec)
TP CO2
CO2 Temp
11.7
11.6
11.5
11.4
11.3
11.2
11.1
11
10.9
10.8
30
29
28
27
26
25
24
23
22
21
1
601
1201
1801
2401
3001

ANALYZER RESPONSE – We have studied two analyzer response issues, both dealing with Non-Dispersive Infrared analyzers (NDIR). The effects of oxygen and water on the CO_2 analyzer and the effects of water and CO_2 interference on CO measurements.

Oxygen/water effects on CO_2 Measurement – As discussed in Reference 1, Michael Landry, et al. from Ford Motor Co. asserted that CVS bag CO_2 measured vs. injected results are subject to an approximate error of –2% because of the low water content of the sample. This assertion was based on Reference 2, a paper by Horiba, Ltd. For ordinary vehicle emission bag testing, the NDIR interferences from oxygen and water offset each other. When using the VEES, the water content of the bag sample is much lower and the CO_2 analyzer, which is calibrated using CO_2/Nitrogen blends, will read low. The Ford study included injecting CO_2/Air blends into a CO_2 analyzer and they found the analyzer to read approximately 2% lower than the label value of the gases.

If these assertions are in fact true, then the best agreement that one can expect for CVS bag recoveries is –2%. We have run a few experiments to determine whether this issue has an impact on VEES recovery results. First, we looked at CVS bag recoveries at different CVS flow-rates. As the CVS dilution rati increases with the higher CVS flow-rates, the oxyge concentration in the bag will increase. We found that th CO_2 mass recovered actually increased slightly as th oxygen concentration was increased. Next, we checke tailpipe CO_2 concentration at the outlet of the VEE when using blower air as the diluent vs. using nitrogen a the diluent. We found that the tailpipe concentration vs expected was closer when using the air dilution an when nitrogen dilution was used, the concentration wa approximately 1% lower than expected. Finally, ba mini-diluter recovery tests were carried out using air vs nitrogen dilution. The results, presented later in thi paper (Table 5) show that the BMD CO_2 recovery i about 1% higher when using the air dilution and is closer to expected.

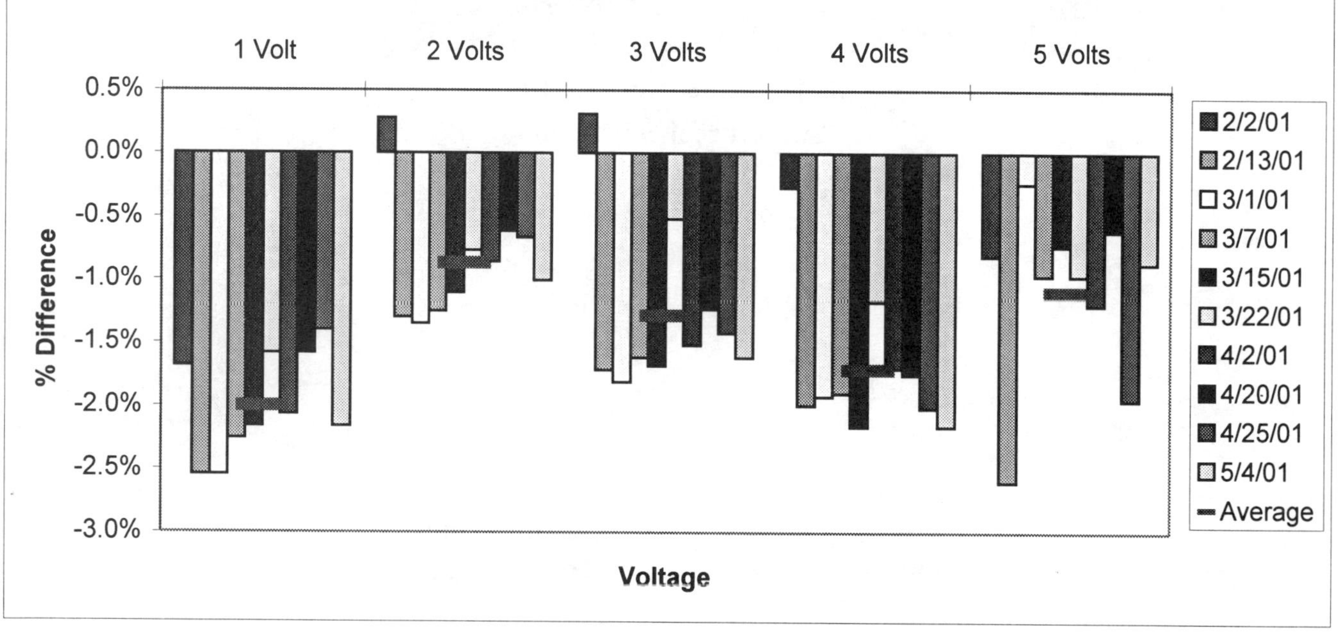

Figure 8. MFC Drift Check (% Diff. wrt LFE) NOx MFC (0-50 SLPM)

CO_2/H_2O effects on CO Measurement – Some frequently asked questions within the emission testing industry deal with procedures for conducting interference checks/adjustments on dual detector CO analyzers. What should the CO_2 concentration, used for interference checks, be for bag benches and for tailpipe benches? Should dry CO_2 be used or should it be bubbled through water? The VEES system can be utilized to help answer some of these questions.

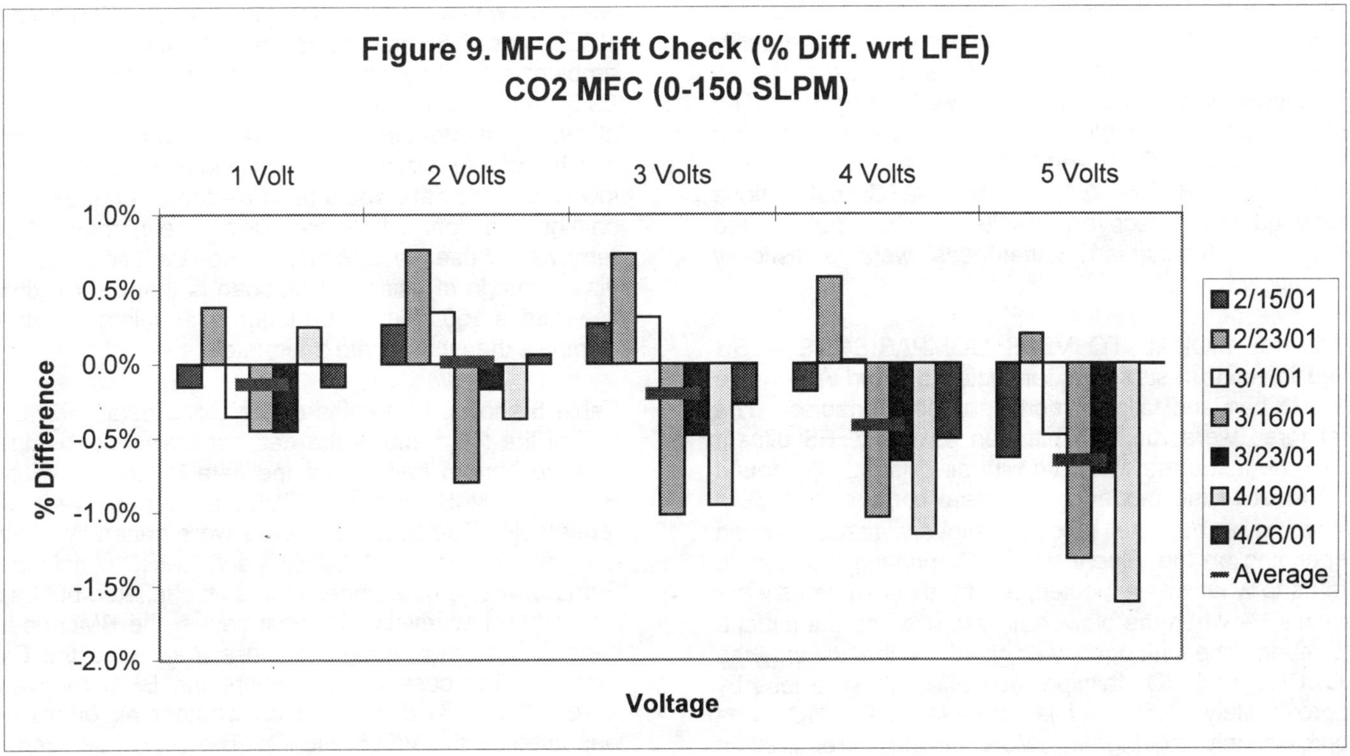

We have run an experiment to help understand these questions. Typically, while sampling vehicle exhaust, we have observed changes in tailpipe CO concentration, depending on what concentration of CO_2 is used for the interference adjustment. We conducted interference adjustment on the tailpipe CO(L) analyzer using dry (no water in bubbler) 20% CO_2, dry 4% CO_2, and wet 20% CO_2. In each configuration, we flowed a fixed CO injection level and varied the CO_2 concentration, with a fixed nitrogen dilution. We found that the closest accuracy for the tailpipe CO measurements occurred when the interference adjustment was performed using the wet 20% CO_2 gas. This was consistent across the range of tailpipe CO_2 concentrations. Tailpipe CO concentrations, when using the dry 20% CO_2 interference gas, were consistently measuring about 1% low. When using the dry 4% CO_2 interferent, the readings were about 0.5% low.

INJECTED VS. MEASURED TEST RESULTS

CVS BAG ONLY TESTS (FIXED AND TRANSIENT) – Ford personnel developed a three-505 second phase modal test profile for use in evaluating the VEES system performance. Phase 1 uses a fixed or steady-state injection profile, phase 2 uses a trapezoidal function and phase 3 uses a transient profile. Phase 3 contains some rapid and irregular shifts of the MFC flow-rates (eg., 0 to 100% of the full scale of the MFC in 3 seconds), and was designed to test the response capability of the MFCs. The CO_2 MFC injection was steady-state for all of the phases because for gasoline engines, the tailpipe CO_2 concentration is relatively constant. The other constituents were commanded to follow the trapezoidal and transient profiles. The total exhaust flow-rate (air plus injected gases) was maintained at a steady-state 0.57 SCMM for all phases of the test. Each of the three phases produces about the same mass of each constituent. The target mass levels were designed to be close to the ULEV bag 1 levels presented earlier in Table 1, although the THC levels were 3 times higher because of gas bottle availability. Our approach was based on the fact that we can use the higher mass levels to validate the VEES system and then once that is accomplished, switch to lower levels to check out SULEV systems. Seven of these tests were run with the MFCs calibrated with Ford's LFE system and another seven were run with the MFCs calibrated using Horiba's SAOs. All of the tests were run at the Ford Motor Company Allen Park Test Laboratory (APTL) Test Site 30.

Table 2 shows VEES vs. VEES results for seven of the tests, which illustrate how well the VEES was able to maintain the requested profile. The maximum error is approximately 0.5% of the requested mass level, occurring during phase 3 for the HC MFC. Table 3 shows the CVS / Bag bench recoveries for the same seven tests. These seven tests were conducted with the Horiba SAO based MFC calibrations. Most of the recovery data shows agreement to be within ±1% of the expected values except for the NO_X which is at –2 to –3%. This is thought to be a test cell problem, but as of the writing of this report, the issue is still under investigation.

Table 4 summarizes the recovery results for all fourteen tests. The top half of the table illustrates the repeatability of the data. For both sets of seven tests, the % deviations for all constituents are well under 1%. The bottom half of the table compares the recovery results for the SAO vs. LFE based MFC calibrations. As stated earlier, the Horiba SAO based MFC calibrations produced better recovery results for CO_2, but for the other constituents, any differences were statistically insignificant.

TAILPIPE MODAL TO VEES COMPARISONS – Six fixed injection tests were conducted at Ford APTL Site 31 with bag and tailpipe modal sampling enabled. The first three were run with nitrogen as the VEES diluent while the last three were run with air dilution. We found that most of the modal to bag gram comparisons were within ±2% but the tailpipe recovery results varied depending on the diluent used. Comparing the tailpipe grams to what was expected, all of the constituents were within ±2% when the blower air was used as the diluent. However, when nitrogen was used as the diluent, the CO, CO_2, and NO_X tailpipe recoveries all were less by approximately 1.5% while the tailpipe THC was approximately 1% higher. More testing is required in order to fully understand the source of these differences.

COMPARISON OF BMD AND CVS RECOVERIES – Eight 300 second fixed injection tests were conducted at Ford APTL Site 31 using CVS and Bag Mini-Diluter (BMD) samplers. The mass injected was near SULEV levels. Four of the tests were run using nitrogen as the VEES diluent and the other four used the VEES air blower for dilution. The perceived benefit for using nitrogen was that the background contamination of the ambient air from the blower would be a significant contribution of the total mass that is injected and if the concentrations are unstable, this could be a major source of variability in the results. In the four tests that were run with air dilution, the total mass of CH_4 from the VEES was 6.55mg and 3.6mg of this was from the ambient air. The data was post-processed using the CVS ambient bag concentrations. The VEES software allows for a pre-test entry of background concentration which it can factor into the calculations, however, at very low levels, the data would be impacted by changes in the background concentrations. Using the nitrogen diluent removes these factors from the equation, but the disadvantage of using the nitrogen is that the maximum flow-rate is 150 slpm, not enough to simulate a vehicle's tailpipe exhaust flow-rate adequately.

Table 5 shows the results of the eight tests. The top half of the chart shows the data for the nitrogen dilution and the bottom half shows the data for the air dilution. Both the BMD and the CVS recoveries were very repeatable. The BMD recoveries were generally within a milligram for the CO, NO_X, CH_4 and NMHC constituents. Some of the % differences look a little inflated but this is due to the low levels. In most cases, the BMD does a better job of recovering the mass than does the CVS, except in the case of CO_2 where the BMD recoveries were –3 to –4% depending on whether air or nitrogen was used as the VEES diluent. The BMD uses zero air as its diluent gas. We ran one test where we switched the BMD to use nitrogen as the VEES *and* the BMD diluent. The CO_2 recovery improved from –4% to –3% but other constituents such as NMHC got worse. Comparing the results of the tests with the nitrogen vs. air VEES dilution, the BMD recoveries were closer for all constituents when using the air dilution, indicating that the ambient background contamination is not such a serious issue, provided that the concentrations can be accurately quantified. In general, this data illustrates that the VEES can be used to demonstrate any differences in sampling techniques used to measure SULEV emission levels. The discrepancy for BMD CO_2 recovery needs to be investigated further.

TABLE 2. VEES vs. VEES PERFORMANCE FOR SEVEN TESTS

		Phase 1			Phase 2			Phase 3			Total		
Test Date & #	Constituent	Actual (g)	Profile (g)	% Diff	Actual (g)	Profile (g)	% Diff	Actual (g)	Profile (g)	% Diff	Actual (g)	Profile (g)	% Diff
Nov 2, 2001 - 3	C3H8	1.191136	1.190054	0.09%	1.193995	1.192608	0.12%	1.19265	1.186893	0.49%	3.577781	3.569554	0.23%
Overall Test #	CO	2.412056	2.40922	0.12%	2.419426	2.41439	0.21%	2.405998	2.402819	0.13%	7.23748	7.226429	0.15%
11	NOx	0.768713	0.7689	-0.02%	0.770481	0.77055	-0.01%	0.766411	0.766857	-0.06%	2.305604	2.306307	-0.03%
Calibration	CO2	1165.221	1165	0.02%	1168.799	1164.349	0.38%	1163.546	1162.5	0.09%	3497.566	3491.849	0.16%
Horiba	CH4												

		Phase 1			Phase 2			Phase 3			Total		
Test Date & #	Constituent	Actual (g)	Profile (g)	% Diff	Actual (g)	Profile (g)	% Diff	Actual (g)	Profile (g)	% Diff	Actual (g)	Profile (g)	% Diff
Nov 2, 2001 - 4	C3H8	1.191026	1.190054	0.08%	1.193987	1.192608	0.12%	1.192453	1.186893	0.47%	3.577465	3.569554	0.22%
Overall Test #	CO	2.41301	2.40922	0.16%	2.419767	2.41439	0.22%	2.4064	2.402819	0.15%	7.239177	7.226429	0.18%
12	NOx	0.768924	0.7689	0.00%	0.770561	0.77055	0.00%	0.766497	0.766857	-0.05%	2.305981	2.306307	-0.01%
Calibration	CO2	1165.356	1165	0.03%	1168.824	1164.349	0.38%	1163.586	1162.5	0.09%	3497.765	3491.849	0.17%
Horiba	CH4												

		Phase 1			Phase 2			Phase 3			Total		
Test Date & #	Constituent	Actual (g)	Profile (g)	% Diff	Actual (g)	Profile (g)	% Diff	Actual (g)	Profile (g)	% Diff	Actual (g)	Profile (g)	% Diff
Nov 5, 2001 - 1	C3H8	1.191377	1.190054	0.11%	1.193048	1.192608	0.04%	1.191053	1.186893	0.35%	3.575478	3.569554	0.17%
Overall Test #	CO	2.412307	2.40922	0.13%	2.419563	2.41439	0.21%	2.405957	2.402819	0.13%	7.237826	7.226429	0.16%
13	NOx	0.768402	0.7689	-0.06%	0.770301	0.77055	-0.03%	0.766413	0.766857	-0.06%	2.305116	2.306307	-0.05%
Calibration	CO2	1165.539	1165	0.05%	1168.918	1164.349	0.39%	1163.654	1162.5	0.10%	3498.111	3491.849	0.18%
Horiba	CH4												

		Phase 1			Phase 2			Phase 3			Total		
Test Date & #	Constituent	Actual (g)	Profile (g)	% Diff	Actual (g)	Profile (g)	% Diff	Actual (g)	Profile (g)	% Diff	Actual (g)	Profile (g)	% Diff
Nov 5, 2001 - 2	C3H8	1.19054	1.190054	0.04%	1.19344	1.192608	0.07%	1.191429	1.186893	0.38%	3.575409	3.569554	0.16%
Overall Test #	CO	2.411399	2.40922	0.09%	2.419101	2.41439	0.20%	2.405559	2.402819	0.11%	7.236059	7.226429	0.13%
14	NOx	0.768461	0.7689	-0.06%	0.770496	0.77055	-0.01%	0.766611	0.766857	-0.03%	2.305568	2.306307	-0.03%
Calibration	CO2	1165.444	1165	0.04%	1168.786	1164.349	0.38%	1163.679	1162.5	0.10%	3497.909	3491.849	0.17%
Horiba	CH4												

		Phase 1			Phase 2			Phase 3			Total		
Test Date & #	Constituent	Actual (g)	Profile (g)	% Diff	Actual (g)	Profile (g)	% Diff	Actual (g)	Profile (g)	% Diff	Actual (g)	Profile (g)	% Diff
Nov 5, 2001 - 3	C3H8	1.191371	1.190054	0.11%	1.193154	1.192608	0.05%	1.191215	1.186893	0.36%	3.57574	3.569554	0.17%
Overall Test #	CO	2.412731	2.40922	0.15%	2.419561	2.41439	0.21%	2.405974	2.402819	0.13%	7.238266	7.226429	0.16%
15	NOx	0.768857	0.7689	-0.01%	0.770413	0.77055	-0.02%	0.766451	0.766857	-0.05%	2.305721	2.306307	-0.03%
Calibration	CO2	1165.487	1165	0.04%	1168.821	1164.349	0.38%	1163.679	1162.5	0.10%	3497.988	3491.849	0.18%
Horiba	CH4												

		Phase 1			Phase 2			Phase 3			Total		
Test Date & #	Constituent	Actual (g)	Profile (g)	% Diff	Actual (g)	Profile (g)	% Diff	Actual (g)	Profile (g)	% Diff	Actual (g)	Profile (g)	% Diff
Nov 5, 2001 - 4	C3H8	1.191874	1.190054	0.15%	1.193539	1.192608	0.08%	1.191559	1.186893	0.39%	3.576972	3.569554	0.21%
Overall Test #	CO	2.412812	2.40922	0.15%	2.419583	2.41439	0.22%	2.406016	2.402819	0.13%	7.238411	7.226429	0.17%
16	NOx	0.768947	0.7689	0.01%	0.770441	0.77055	-0.01%	0.766473	0.766857	-0.05%	2.305861	2.306307	-0.02%
Calibration	CO2	1165.462	1165	0.04%	1168.85	1164.349	0.39%	1163.601	1162.5	0.09%	3497.912	3491.849	0.17%
Horiba	CH4												

		Phase 1			Phase 2			Phase 3			Total		
Test Date & #	Constituent	Actual (g)	Profile (g)	% Diff	Actual (g)	Profile (g)	% Diff	Actual (g)	Profile (g)	% Diff	Actual (g)	Profile (g)	% Diff
Nov 6, 2001 - 1	C3H8	1.19179	1.190054	0.15%	1.193154	1.192608	0.05%	1.191464	1.186893	0.39%	3.576407	3.569554	0.19%
Overall Test #	CO	2.411583	2.40922	0.10%	2.419272	2.41439	0.20%	2.405714	2.402819	0.12%	7.23657	7.226429	0.14%
17	NOx	0.768943	0.7689	0.01%	0.771016	0.77055	0.06%	0.76713	0.766857	0.04%	2.307088	2.306307	0.03%
Calibration	CO2	1165.406	1165	0.03%	1168.933	1164.349	0.39%	1163.621	1162.5	0.10%	3497.96	3491.849	0.18%
Horiba	CH4												

TABLE 3. VEES VS. CVS RECOVERY FOR SEVEN TESTS

FORD APTL SITE 30

		Phase 1			Phase 2			Phase 3			Total		
Test Date & #	Constituent	VEES (g)	CVS (g)	% Diff	VEES (g)	CVS (g)	% Diff	VEES (g)	CVS (g)	% Diff	VEES (g)	CVS (g)	% Diff
Nov 2, 2001 - 3													
Overall Test #	CO	2.4106	2.421693	0.46%	2.4137	2.399816	-0.58%	2.4131	2.392605	-0.85%	7.2374	7.214115	-0.32%
11	NOx	0.7683	0.755422	-1.68%	0.7688	0.749642	-2.49%	0.7684	0.742063	-3.43%	2.3055	2.247127	-2.53%
Calibration	CO2	1164.878	1172.93	0.69%	1166.352	1175.401	0.78%	1166.32	1171.26	0.42%	3497.55	3519.591	0.63%
Horiba	CH4												
	C3H8	1.1902	1.185976	-0.35%	1.1911	1.180017	-0.93%	1.1926	1.177692	-1.25%	3.5739	3.543685	-0.85%

		Phase 1			Phase 2			Phase 3			Total		
Test Date & #	Constituent	VEES (g)	CVS (g)	% Diff	VEES (g)	CVS (g)	% Diff	VEES (g)	CVS (g)	% Diff	VEES (g)	CVS (g)	% Diff
Nov 2, 2001 - 4													
Overall Test #	CO	2.4115	2.413096	0.07%	2.4141	2.411684	-0.10%	2.4136	2.399875	-0.57%	7.2392	7.224655	-0.20%
12	NOx	0.7685	0.748891	-2.55%	0.7689	0.746967	-2.85%	0.7685	0.746458	-2.87%	2.3059	2.242315	-2.76%
Calibration	CO2	1164.969	1164.856	-0.01%	1166.414	1168.136	0.15%	1166.365	1166.48	0.01%	3497.748	3499.471	0.05%
Horiba	CH4												
	C3H8	1.19	1.187112	-0.24%	1.1911	1.180012	-0.93%	1.1924	1.177403	-1.26%	3.5735	3.544526	-0.81%

		Phase 1			Phase 2			Phase 3			Total		
Test Date & #	Constituent	VEES (g)	CVS (g)	% Diff	VEES (g)	CVS (g)	% Diff	VEES (g)	CVS (g)	% Diff	VEES (g)	CVS (g)	% Diff
Nov 5, 2001 - 1													
Overall Test #	CO	2.4108	2.413858	0.13%	2.4138	2.403798	-0.41%	2.4132	2.381783	-1.30%	7.2378	7.19944	-0.53%
13	NOx	0.7682	0.747927	-2.64%	0.7687	0.739061	-3.86%	0.7682	0.735761	-4.22%	2.3051	2.22275	-3.57%
Calibration	CO2	1165.163	1170.768	0.48%	1166.509	1160.634	-0.50%	1166.422	1157.73	-0.75%	3498.094	3489.132	-0.26%
Horiba	CH4												
	C3H8	1.1905	1.190745	0.02%	1.1905	1.186666	-0.32%	1.191	1.17832	-1.06%	3.572	3.555731	-0.46%

		Phase 1			Phase 2			Phase 3			Total		
Test Date & #	Constituent	VEES (g)	CVS (g)	% Diff	VEES (g)	CVS (g)	% Diff	VEES (g)	CVS (g)	% Diff	VEES (g)	CVS (g)	% Diff
Nov 5, 2001 - 2													
Overall Test #	CO	2.4096	2.409087	-0.02%	2.4136	2.381273	-1.34%	2.4128	2.390626	-0.92%	7.236	7.180986	-0.76%
14	NOx	0.7682	0.751915	-2.12%	0.7689	0.742224	-3.47%	0.7685	0.737782	-4.00%	2.3056	2.23192	-3.20%
Calibration	CO2	1165.057	1158.133	-0.59%	1166.391	1160.934	-0.47%	1166.452	1159.977	-0.56%	3497.899	3479.045	-0.54%
Horiba	CH4												
	C3H8	1.1897	1.189453	-0.02%	1.1908	1.183549	-0.61%	1.1914	1.179941	-0.96%	3.5719	3.552943	-0.53%

		Phase 1			Phase 2			Phase 3			Total		
Test Date & #	Constituent	VEES (g)	CVS (g)	% Diff	VEES (g)	CVS (g)	% Diff	VEES (g)	CVS (g)	% Diff	VEES (g)	CVS (g)	% Diff
Nov 5, 2001 - 3													
Overall Test #	CO	2.4109	2.409277	-0.07%	2.4141	2.395805	-0.76%	2.4133	2.389584	-0.98%	7.2383	7.194665	-0.60%
15	NOx	0.7685	0.750818	-2.30%	0.7688	0.742679	-3.40%	0.7684	0.742508	-3.37%	2.3057	2.236005	-3.02%
Calibration	CO2	1165.117	1156.682	-0.72%	1166.421	1160.409	-0.52%	1166.437	1160.51	-0.51%	3497.975	3477.602	-0.58%
Horiba	CH4												
	C3H8	1.1905	1.183823	-0.56%	1.1905	1.176669	-1.16%	1.1912	1.178898	-1.03%	3.5722	3.539391	-0.92%

		Phase 1			Phase 2			Phase 3			Total		
Test Date & #	Constituent	VEES (g)	CVS (g)	% Diff	VEES (g)	CVS (g)	% Diff	VEES (g)	CVS (g)	% Diff	VEES (g)	CVS (g)	% Diff
Nov 5, 2001 - 4													
Overall Test #	CO	2.411	2.42945	0.77%	2.4141	2.407784	-0.26%	2.4133	2.391135	-0.92%	7.2384	7.228369	-0.14%
16	NOx	0.7686	0.763586	-0.65%	0.7689	0.745594	-3.03%	0.7684	0.740813	-3.59%	2.3059	2.249992	-2.42%
Calibration	CO2	1165.083	1159.471	-0.48%	1166.446	1161.836	-0.40%	1166.369	1160.918	-0.47%	3497.898	3482.225	-0.45%
Horiba	CH4												
	C3H8	1.191	1.191119	0.01%	1.1908	1.182405	-0.71%	1.1916	1.175211	-1.38%	3.5734	3.548735	-0.69%

		Phase 1			Phase 2			Phase 3			Total		
Test Date & #	Constituent	VEES (g)	CVS (g)	% Diff	VEES (g)	CVS (g)	% Diff	VEES (g)	CVS (g)	% Diff	VEES (g)	CVS (g)	% Diff
Nov 6, 2001 - 1													
Overall Test #	CO	2.4098	2.422568	0.53%	2.4137	2.407314	-0.26%	2.413	2.395176	-0.74%	7.2365	7.225058	-0.16%
17	NOx	0.7682	0.754672	-1.76%	0.7689	0.748441	-2.66%	0.7685	0.74546	-3.00%	2.3056	2.248573	-2.47%
Calibration	CO2	1165.029	1165.095	0.01%	1166.489	1164.49	-0.17%	1166.43	1162.896	-0.30%	3497.948	3492.481	-0.16%
Horiba	CH4												
	C3H8	1.1909	1.189273	-0.14%	1.1905	1.181971	-0.72%	1.1915	1.17788	-1.14%	3.5729	3.549124	-0.67%

TABLE 4. COMPARISON OF RECOVERY RESULTS, SAO VS. LFE
FORD APTL SITE 30

CVS RECOVERY (g)

HORIBA - MFCs Calibrated by SAOs / FORD - MFCs Calibrated by LFE

CO	NOX	CO2	C3H8		CO	NOX	CO2	C3H8
7.214	2.247	3519.591	3.544		7.234	2.246	3419.787	3.539
7.225	2.242	3499.471	3.545		7.246	2.236	3426.406	3.526
7.199	2.223	3489.132	3.556		7.251	2.266	3471.288	3.545
7.181	2.232	3479.045	3.553		7.254	2.234	3427.502	3.550
7.195	2.236	3477.602	3.539		7.221	2.231	3424.511	3.531
7.228	2.250	3482.225	3.545		7.204	2.255	3432.060	3.534
7.225	2.249	3492.481	3.549		7.189	2.238	3410.914	3.512
7.210	2.240	3491.364	3.547	AVG	7.228	2.244	3430.353	3.534
0.018	0.010	14.682	0.006	Std. Dev.	0.025	0.013	19.261	0.013
0.252	0.448	0.421	0.163	%Dev.	0.344	0.569	0.561	0.358

%DIFF (CVS vs. VEES)

HORIBA - MFCs Calibrated by SAOs / FORD - MFCs Calibrated by LFE

CO	NOX	CO2	C3H8		CO	NOX	CO2	C3H8
-0.32	-2.53	0.63	-0.85		-0.08	-2.51	-2.13	-0.95
-0.20	-2.76	0.05	-0.81		0.08	-2.95	-1.94	-1.24
-0.53	-3.57	-0.26	-0.46		0.20	-1.65	-0.63	-0.74
-0.76	-3.20	-0.54	-0.53		0.18	-3.07	-1.91	-0.63
-0.60	-3.02	-0.58	-0.92		-0.24	-3.20	-2.00	-1.16
-0.14	-2.42	-0.45	-0.69		-0.47	-2.09	-1.77	-1.08
-0.16	-2.47	-0.16	-0.67		-0.67	-2.86	-2.35	-1.70
-0.39	-2.85	-0.19	-0.70	AVG	-0.14	-2.62	-1.82	-1.07
-0.76	-3.57	-0.58	-0.92	MIN	-0.67	-3.20	-2.35	-1.70
-0.14	-2.42	0.63	-0.46	MAX	0.20	-1.65	-0.63	-0.63
0.62	1.15	1.21	0.46	RANGE	0.87	1.55	1.72	1.07

TABLE 5. BMD AND CVS RECOVERY FOR SULEV LEVELS

FORD APTL SITE 31

VEES Nitrogen Dilution

	CO			%Diff	%Diff
Test	VEES(g)	CVS(g)	BMD(g)	CVS/VEES	BMD/VEES
1	0.0275	0.0427	0.0278	55.138	1.091
2	0.0276	0.0304	0.0286	10.150	3.623
3	0.0276	0.0383	0.0271	38.592	-1.812
4	0.0275	0.0345	0.0295	25.410	7.273
Ave.	0.0276	0.0365	0.0283	32.308	2.541
Std. Dev.	0.0001	0.0052	0.0010		
%Dev.	0.2096	14.3677	3.6616		

CO2			%Diff	%Diff
VEES(g)	CVS(g)	BMD(g)	CVS/VEES	BMD/VEES
593.795	585.811	580.825	-1.345	-2.184
593.901	585.373	572.222	-1.436	-3.650
593.933	586.524	563.28	-1.247	-5.161
593.889	585.844	562.112	-1.355	-5.351
593.8795	585.8880	569.6098	-1.346	-4.087
0.0593	0.4752	8.7348		
0.0100	0.0811	1.5335		

	NOX			%Diff	%Diff
Test	VEES(g)	CVS(g)	BMD(g)	CVS/VEES	BMD/VEES
1	0.0120	0.0128	0.0120	6.724	0.000
2	0.0120	0.0119	0.0118	-0.606	-1.667
3	0.0120	0.0116	0.0116	-3.059	-3.333
4	0.0120	0.0120	0.0115	-0.300	-4.167
Ave.	0.0120	0.0121	0.0117	0.690	-2.292
Std. Dev.	0.0000	0.0005	0.0002		
%Dev.	0.0000	4.1791	1.8911		

CH4			%Diff	%Diff
VEES(g)	CVS(g)	BMD(g)	CVS/VEES	BMD/VEES
0.0029	0.0034	0.0033	17.241	13.793
0.0029	0.0032	0.0033	10.345	13.793
0.0029	0.0031	0.0033	6.897	13.793
0.0029	0.0034	0.0033	17.241	13.793
0.0029	0.0033	0.0033	12.931	13.793
0.0000	0.0001	0.0000		
0.0000	4.5802	0.0000		

	NMHC			%Diff	%Diff
Test	VEES(g)	CVS(g)	BMD(g)	CVS/VEES	BMD/VEES
1	0.0117	0.0097	0.0129	-17.094	10.256
2	0.0117	0.0108	0.0118	-7.692	0.855
3	0.0117	0.0110	0.0116	-5.983	-0.855
4	0.0117	0.0109	0.0115	-6.838	-1.709
Ave.	0.0117	0.0106	0.0120	-9.402	2.137
Std. Dev.	0.0000	0.0006	0.0006		
%Dev.	0.0000	5.7125	5.4017		

VEES Air Dilution

	CO			%Diff	%Diff
Test	VEES(g)	CVS(g)	BMD(g)	CVS/VEES	BMD/VEES
1	0.0282	0.0395	0.0286	40.0709	1.4184
2	0.0281	0.0372	0.0275	32.3843	-2.1352
3	0.028	0.0382	0.027	36.4286	-3.5714
4	0.028	0.0382	0.0282	36.4286	0.7143
Ave.	0.0281	0.0383	0.0278	36.3281	-0.8935
Std. Dev.	0.0001	0.0009	0.0007		
%Dev.	0.3407	2.4620	2.5668		

CO2			%Diff	%Diff
VEES(g)	CVS(g)	BMD(g)	CVS/VEES	BMD/VEES
596.204	598.217	582.325	0.3376	-2.3279
596.192	590.295	576.285	-0.9891	-3.339
596.204	587.99	578.272	-1.3777	-3.0077
596.23	587.729	576.824	-1.4258	-3.2548
596.2075	591.0578	578.4265	-0.8637	-2.9823
0.0160	4.9101	2.7311		
0.0027	0.8307	0.4722		

	NOX			%Diff	%Diff
Test	VEES(g)	CVS(g)	BMD(g)	CVS/VEES	BMD/VEES
1	0.0183	0.0197	0.0183	7.6503	0.0000
2	0.0183	0.0188	0.0182	2.7322	-0.5464
3	0.0183	0.0188	0.0183	2.7322	0.0000
4	0.0183	0.0186	0.0183	1.6393	0.0000
Ave.	0.0183	0.019	0.0183	3.6885	-0.1366
Std. Dev.	0.0000	0.0005	0.0001		
%Dev.	0.0000	2.5918	0.2732		

CH4			%Diff	%Diff
VEES(g)	CVS(g)	BMD(g)	CVS/VEES	BMD/VEES
0.00653	0.00694	0.0069	6.279	5.666
0.00655	0.00696	0.0069	6.260	5.344
0.00653	0.00682	0.0069	4.441	5.666
0.00658	0.00709	0.0070	7.751	6.383
0.00655	0.00695	0.00693	6.186	5.766
0.00002	0.00011	0.00005		
0.36089	1.59032	0.72202		

	NMHC			%Diff	%Diff
Test	VEES(g)	CVS(g)	BMD(g)	CVS/VEES	BMD/VEES
1	0.0123	0.0119	0.0125	-3.252	1.626
2	0.0123	0.0115	0.0124	-6.504	0.813
3	0.0122	0.0117	0.0121	-4.098	-0.820
4	0.0122	0.0115	0.0119	-5.738	-2.459
Ave.	0.0123	0.0117	0.0122	-4.898	-0.204
Std. Dev.	0.0001	0.0002	0.0003		
%Dev.	0.4713	1.6437	2.2526		

OTHER INVESTIGATIONS

- Horiba has procured two STEC digital MFCs, in the flow range of 30 slpm and 3 slpm. We have installed them in a VEES system and are beginning to investigate whether the upgraded electronics package will result in improved performance in this application.
- More analysis of differences in results when using air vs. nitrogen VEES dilution is required.
- Investigation of BMD CO_2 recovery offsets.
- Investigation of CVS recoveries when using a Dilution Air Refinement System (DAR).
- Building modal profiles using actual vehicle data and running the simulation.

CONCLUSIONS

1. The VEES-100 Vehicle Exhaust Emission Simulator has evolved to the point that it is capable of checking all of the components of a vehicle test cell and also checking correlation between test cells
2. A sub-sonic venturi (SAO), calibrated by CEESI, can be used to calibrate an MFC to achieve an accuracy of ±1% of reading or better. Achieving this accuracy for the CO_2 MFC requires that the SAO be dual calibrated with Nitrogen and CO_2 gas flowing.
3. Using the dual-calibrated SAO for calibration of the CO_2 MFC produces better recovery data than using an LFE system with theoretical gas composition corrections.
4. Maintaining constant CO_2 gas temperature is very important for MFC stability.
5. The STEC and the Brooks MFCs do not exhibit any appreciable drift phenomena.
6. The interference factors for NDIR analyzers only have a small impact on VEES test results.
7. Numerous bag-only fixed and transient injection tests have demonstrated that the VEES MFCs maintain the requested flow-rates and CVS recoveries within ±1% can be attained with very good repeatability.
8. The VEES provides a means to evaluate tailpipe modal to bag results. These results can be influenced by the selection of the VEES diluent
9. BMD recovery results can also be influenced by the selection of the VEES diluent but it is very clear that the VEES can be used to evaluate any sampling techniques used to measure SULEV level emissions.

ACKNOWLEDGMENTS

The authors would like to thank the following individuals for their help and support with the work discussed in this paper:

Bill Boles, Karl Oestergaard, Mike Panzer, Chris Pearson, Rick Rooney, and Frank Slavik from Horiba Instruments, Inc.

Kim Isbrecht, Michael Landry, Larry Nobles, Emeka Oputa, and Gary Stevens from Ford Motor Co.

Bill Johansen and Walt Seidl from CEESI

REFERENCES

1. Landry M., Guenther M., Isbrecht K., Stevens G.: "Simulation of Low Level Vehicle Exhaust Emissions for Evaluation of Sampling and Analytical Systems", SAE Paper No. 2001-01-0211
2. Inoue K., Ishihara M., Akashi K., Adachi M., Ishida K.: "Numerical Analysis of Mass Emissions Measurement Systems for Low Emissions Vehicles". SAE Paper No. 1999-01-0150

CONTACT

Author Information:

1. Daniel Whelan, Systems Engineer, Engine Measurement Division, Horiba Instruments, Inc. 5900 Hines Dr., Ann Arbor, MI, 48108, phone: (734) 213-6555x598, fax: (734) 213-6525, e-mail: dan.whelan@horiba.com
2. Gary Lewis, Senior Product Engineer, Engine Measurement Division, Horiba Instruments, Inc., 17671 Armstrong Avenue, Irvine, CA, 92614, phone: (949) 250-4811x182, fax: (949) 250-0924, e-mail: gary.lewis@horiba.com
3. Scott Kiyabu, Project Engineer, Engine Measurement Division, Horiba Instruments, Inc., 5900 Hines Dr., Ann Arbor, MI, 48108, phone: (734) 213-6555x705, fax: (734) 213-6525, e-mail: scott.kiyabu@horiba.com
4. Travis Henney, Technical Standards Engineer, Correlation Engineering, Vehicle Environmental Engineering, Ford Motor Co., Allen Park Test Laboratory, Suite 3W-100, 1500 Enterprise Dr., Allen Park, MI, 48101, phone: (313) 845-1579, fax: (313) 594-2044, e-mail: thenney@ford.com
5. Darius Harrison, Ford Motor Co., Fairlane Business Park IV, Suite 145, 17225 Federal Drive, Allen Park, MI, 48101, phone: (313) 594-2785, e-mail: dharri70@ford.com

6. Jennifer Farrugia, Ford Motor Co., Fairlane Business Park IV, Suite 145, 17225 Federal Drive, Allen Park, MI, 48101, phone: (313) 323-6716, e-mail: jfarrug3@ford.com

7. Michael Gabbert, Technical Standards Engineer, Correlation Engineering, Vehicle Environmental Engineering, Ford Motor Co., Allen Park Test Laboratory, Suite 3W-100, 1500 Enterprise Dr., Allen Park, MI, 48101, phone: (313) 390-3282, fax: (313) 594-2044,e-mail:mgabbert@ford.com

2002-01-0051

Evaluation of New Bag Sampling Materials for Low Level Emissions Measurements

Jim Hill, Jeffrey F. Loo and Stephen J. Swarin
General Motors Corporation

ABSTRACT

Copolymer materials have been used for the collection of vehicle exhaust gas samples since the inception of regulatory standards. Some of these copolymers contain N,N-dimethylacetamide (DMA), which is added to improve the physical properties of the copolymer and eliminate manufacturing problems. DMA is highly soluble in water, and in effect is rinsed from the emission bag surface by humid exhaust gas samples. This study shows that DMA can thus incorrectly add to test vehicle overall hydrocarbon emissions. The DMA contribution can be significant for lower level emission vehicles. This study introduces a new bag material, KYNAR®, which significantly reduces this interference.

INTRODUCTION

Low-level vehicle emission testing now places much higher demands on sampling system integrity and cleanliness. This paper identifies and quantitates the effects of contamination that originates in the Tedlar® sample bags that are used throughout the emission testing industry. The contaminant is N,N-dimethylacetamide (DMA). Since the flame ionization detector (FID) responds to this compound, the total response of the FID is increased by the DMA contamination, and the hydrocarbon (HC) mass will be over-reported.

This study shows that sampling bags containing DMA should not be used for low level (i.e., ULEV, SULEV and Tier III) vehicle emission testing. As a result of our study, we have identified an alternate material, KYNAR®, which has superior sampling bag properties. The new material selection and testing process, including HC off-gassing, CO_2 permeation, and vehicle exhaust gas stabilization effects, will be discussed in this paper. It should be noted that we believe that KYNAR® is a superior sampling bag material. However, we are continuing to study this issue, and may change materials if a better material is identified and becomes available.

EXPERIMENTAL

Tedlar®, which is a registered trademark of DuPont, is polyvinyl fluoride (PVF). The Tedlar® bags that were tested have a film thickness of 2-mil and an approximate volume of 100 liters. The bags were manufactured by Plastic Film Enterprises (Royal Oak, MI). These bags had a KYNAR® central fitting and two ¼ inch Teflon® tubes. The distribution tubes aided in the evacuation of the bags.

Teflon® is a registered trademark of DuPont. The Teflon® bags that were tested were purchased from Alltech Associates (Deerfield, IL). The Teflon® type was FEP (fluorinated ethylene propylene). The film thickness was 5-mil. The bags had an approximate volume of 100 liters. The bags that were tested had a central stainless steel fitting, but with no additional distribution tubes.

KYNAR®, which is a registered trademark of Atofina Chemical, Inc., is polyvinylidene fluoride (PVDF). KYNAR® bags were made from 4-mil thick films by Plastic Film Enterprises. The bags had an approximate volume of 100 liters and contained a KYNAR® central fitting. They also contained two ¼ inch Teflon® tubes in the bottom half of the bags which helped to evacuate them.

The Tedlar®, Teflon® and KYNAR® films were each produced by the melt extrusion process.

A material safety data sheet for DMA is included as Attachment A. Key properties of DMA include its high boiling point, low vapor pressure, and 100% water solubility.

ANALYTICAL SETUP – Figure 1 shows the experimental set-up that was used to determine sample bag HC off-gassing.

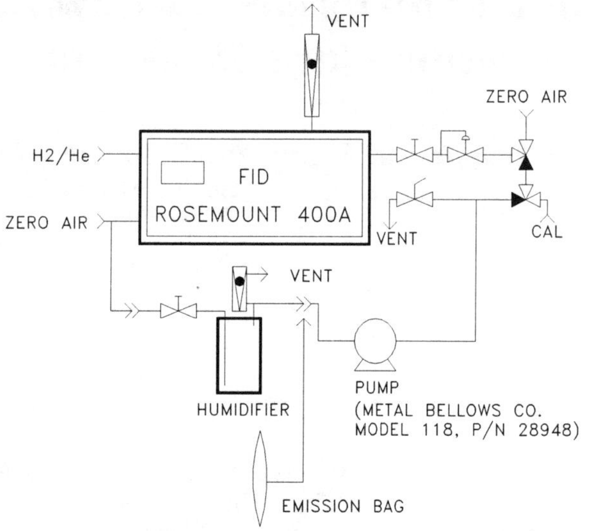

Figure 1
HC Sampling System

This system was designed to minimize the wetted surface area and the use of non-stainless steel components. Also, the system was designed so that switching from humidified zero air to dry zero air to reading the gas from a sampling bag was easily accomplished.

RESULTS AND DISCUSSION

TEDLAR® SAMPLING BAG RESULTS – The following is a brief overview of the procedure that is used in determining the HC off-gassing levels from an emission sample bag. It is important that the sample system be free of contamination prior to sampling. This is done by purging and checking the system with no bag using humidified zero air.

The sample bag is then attached and it is purged (with humidified zero air) and evacuated three times. Finally, the bag is filled with humidified zero air to approximately 80 percent of its maximum volume. This volume represents approximately the same volume used in a typical exhaust emission test. Readings are then taken from the bag at intervals of 15 minutes and at 34 minutes. These reading intervals were selected because they represent typical sample residence times during an FTP test for phases 1 and 2.

Typical values obtained for these time intervals were 60 and 66 ppbC, respectively. These initial values do not indicate a contaminated sample bag. This is because, in the clean (contamination-free) sample system, the internal surfaces have a large surface area that traps the DMA. It is not until the surfaces become saturated with DMA that the FID analyzer response starts to increase.

Figure 2 shows the measured HC off-gas level from Tedlar® bag that was sampled for an extended length of time (until it was emptied). The sample gas was humidified zero air.

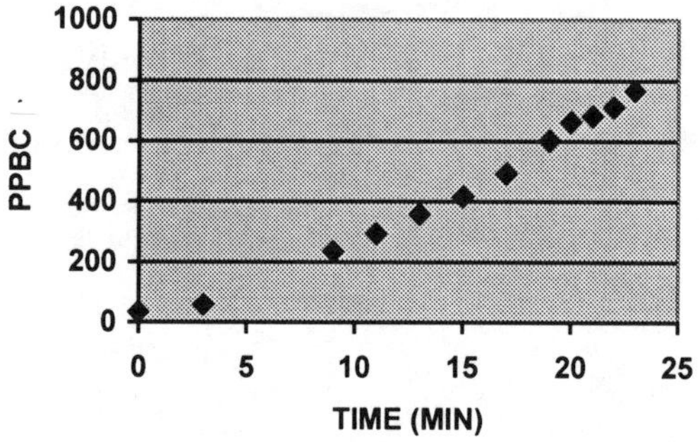

Figure 2
Extended Sampling Response

Figure 2 shows that HC (DMA) off-gassing continues to increase as more and more sample gas is taken from the Tedlar bag. This shows that the surfaces of the sampling system are becoming saturated with DMA, and the contaminant is increasing in the gas phase and migrating to the FID.

In another experiment that tested the persistence of DMA, a Tedlar® bag containing DMA was filled with humidified zero air and allowed to soak for an extended period of time (~5 days). The contents of the bag were then exhausted through the sample system. Next, a Teflon® bag which was known to be contaminant-free was filled with humidified zero air and sampled through the sample system. This FID response is shown in Figure 3. The initial FID response is extremely high indicating that the sampling system has been contaminated with DMA from the Tedlar® bag, and that this contamination has carried over to the Teflon bag sampling. The actual strip chart is shown in Attachment B.

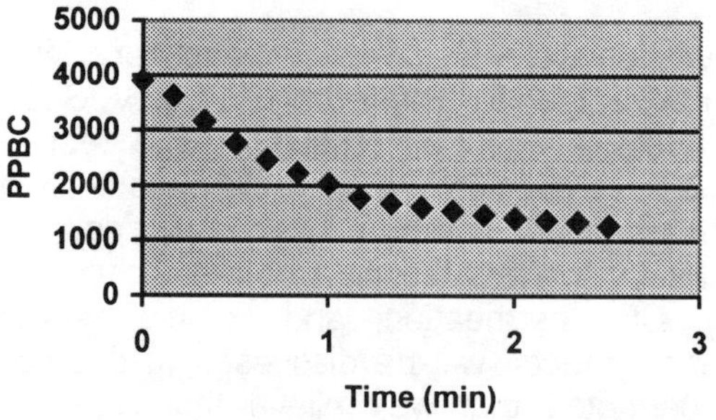

Figure 3
Sample System Contamination Response

It should also be noted that prior to and after sampling the Teflon® bag in the experiment described above, dry zero air was used to purge the sample system. However, the FID response to these purges was essentially zero. This means that purging with a dry zero gas did not remove the DMA from the sampling system. Since vehicle exhaust contains large amounts of water vapor, it will purge the DMA from the bag and the sample system to the FID, resulting in increased HC exhaust emission results.

Finally, the apparatus shown in Figure 1 has a wetted surface area that is much less (approximately 5 feet of ¼ inch stainless steel tubing) than a typical vehicle exhaust emission test site, where the tubing and associated plumbing may reach lengths in excess of 50 or 100 feet. As a result, DMA contamination and migration through the system can have a greater impact on the vehicle emission results with even longer time constants than were noted from the apparatus in Figure 1.

To summarize, DMA is added to Tedlar® film material during the melting prior to extrusion, as well as to the extrusion dies, to improve film thickness uniformity. This results in the DMA residing throughout the cross-section of the film as well as on the surface. DMA has a high boiling point so it is not easily removed by pre-heating. DMA is very soluble in water, so it is easily extracted from the Tedlar® bags by humidified air or automobile exhaust. A study performed by EPA in 1981 also reported sample bag contamination by DMA (1). They said, "efforts to clean or condition these bags by heat treatment and solvent washing procedures were unsuccessful." In our experience, even a vehicle producing zero emissions could fail the SULEV standard because of DMA contamination in the sampling bags.

ALTERNATIVE BAG MATERIALS - As a result of the previous findings, we decided to evaluate other potential bag materials. The goal was to find a material that would exhibit low HC off-gassing and minimal potential for CO_2 permeation. Alternative bag materials that were studied were Teflon® and 4-mil KYNAR®. Teflon® bags showed no detectable amounts of HC off-gassing, but did show significant CO_2 permeation and were subsequently dropped from consideration. In our preliminary testing with KYNAR® bags, we found that there was a HC contaminant in the bags we received. We identified the contaminant by GC analysis as iso-octane (2,2,4-trimethylpentane), which is a common gasoline component. Although we are working with the vendor to ensure that this contamination does not reoccur, we have also found that we can remove the iso-octane with a simple cleaning process. The cleaning process consists of placing the bags in a heated enclosure (88° C) and performing a series of purge (dry air) and evacuation cycles. Because iso-octane has limited solubility in water, humidified zero air was not used in the cleaning process or in the final reported HC results.

Five different KYNAR® bags, each cleaned as described above, were tested. These tests consisted of HC off-gassing analysis, vehicle exhaust gas stabilization and CO_2 permeation. Table 1 shows the results of the HC off-gas testing. The bags were filled with dry zero air and read at 15 and 35-minute intervals.

Table 1

KYNAR® Bag HC Off-gassing Results (Concentration units ppbC)

Bag Number	15 min. Reading	35 min. Reading	Notes
B8	18	27	Initial reading
	24	54	After exhaust test
B9	18	36	Initial reading
	18	51	After exhaust test
B11	21	45	Initial reading
	21	27	Recheck after 20 days
B12	18	36	Initial reading
	18	24	Recheck after 20 days
B13	18	26	Initial reading
	21	30	Recheck after 20 days
Overall Mean	20	36	
Std Dev	2	11	

KYNAR® bags that have not been cleaned can have background levels, consisting primarily of iso-octane, of about 750 ppbC. However, as shown in Table 1, these contaminants are easily removed to a low level, and once the contamination is removed, it does not reappear at a later time (see recheck analyses of bags 11, 12 and 13). Exposure to vehicle exhaust (bags 8 and 9) failed to recontaminate the bags, which means that the KYNAR® bags should show very low vehicle test-to- vehicle test memory effects.

The results of the exhaust gas stabilization test are shown in Attachment C. In this experiment, KYNAR® bags were filled with vehicle exhaust gas on a typical exhaust emission test site. The exhaust constituents were measured during the normal analysis sequence of the test. A second analysis was performed after the gas had been in the bag for approximately 35 minutes. The bar chart shown in Attachment C shows the percent difference between the first and second readings. The average constituent concentrations for THC, CO, NO_x, CO_2 and CH_4 were approximately, 15 ppmC, 53 ppm, 14 ppm, 1 percent and 1 ppm, respectively. A sufficiently high emitting vehicle was selected for these tests in order to minimize any contribution from DMA contamination. The reason for the increased variability in the data generated on test site 11 as compared to test site 9 results (see Attachment C) is unknown at this time. The differences between the two data sets (0 and 35 minutes) are not statistically significant for any of the measured components.

Additional CO_2 permeation tests were performed to determine the stability of dry CO_2 (~1.5%) over a predetermined length of time. A summary of the CO_2 retention data is shown in Table 2. Note that comparison data are also shown for Tedlar® and Teflon®. The readings were taken just after the bags were filled and at 15 and 35 minute intervals.

Table 2

KYNAR® Bag CO_2 Permeation Results

Time	Tedlar ®	Teflon ® (B2)	Teflon ® (B2)	Teflon ® (B3)	KYNAR ® (4mil)	KYNAR ® (4mil)*
0 min	14106	14084	14060	14089	14470	62.5
15 min	14077	14004	13981	13964	14462	62.5
35 min	14084	13900	13943	13938	14417	62.6
Chg 0-15 min	-0.2%	-0.6%	-0.6%	-0.9%	-0.1%	0.0%
Chg 0-35 min	-0.2%	-1.3%	-0.8%	-1.1%	-0.4%	0.1%

* Note that for this set of data, the values are in % full-scale (2.3%). All other values are ppm CO_2

CONCLUSIONS

- The results of this study show that unconditioned Tedlar® bags, which contain large amounts of DMA, should not be used for low level vehicle emission testing.
- KYNAR® bags were found to contain iso-octane as a contaminant, but it was easily removed. Efforts are being made to ensure that this contamination does not reoccur.
- The maximum HC level that was found in cleaned KYNAR® bags was 45 ppbC.
- The permeation rate of CO_2 in KYNAR® bags was not significantly higher than that of Tedlar®.
- There were no significant changes in other exhaust gas component levels when collected in KYNAR® bags.
- KYNAR® bags are thus a superior bag material for low level emission testing.

REFERENCES

(1) Lonneman, W. A.; Bufalini J. J.; Kuntz, R. L.; Meeks, S. A.; “Contamination From Fluorocarbon Films”, Environmental Science and Technology, 1981, Vol. 1, pp. 99-103.

CONTACT

Jim Hill
GM Powertrain
Mail Code 483-331-000
3300 General Motors Road
Milford, MI 48380-3726
Email: Jimmy.Hill@gm.com

Jeffrey F. Loo
GM Powertrain
Mail Code 483-331-000
3300 General Motors Road
Milford, MI 48380-3726
Email: Jeffrey.Loo@gm.com

Stephen J. Swarin
GM R&D and Planning
Mail Code 480-106-320
30500 Mound Road
Warren, MI 48090-9055
Email: Steve.Swarin@gm.com

Attachment A
MSDS for N,N-Dimethylacetamide

J.T.BAKER INC. 222 RED SCHOOL LANE, PHILLIPSBURG, NJ 08865
M A T E R I A L   S A F E T Y   D A T A   S H E E T
24-HOUR EMERGENCY TELEPHONE -- (908) 859-2151
CHEMTREC # (800) 424-9300 -- NATIONAL RESPONSE CENTER # (800) 424-8802

D5784 -06 N,N-DIMETHYLACETAMIDE PAGE: 2
EFFECTIVE: 08/31/95 ISSUED: 10/10/95

PRECAUTIONARY LABELING (CONTINUED)

INTERNATIONAL LABELING

HARMFUL BY INHALATION AND IN CONTACT WITH SKIN. IRRITATING TO EYES. IN CASE OF CONTACT WITH EYES, RINSE IMMEDIATELY WITH PLENTY OF WATER AND SEEK MEDICAL ADVICE. AFTER CONTACT WITH SKIN, WASH IMMEDIATELY WITH PLENTY OF SOAP AND WATER.

SAF-T-DATA* STORAGE COLOR CODE: RED (FLAMMABLE)

SECTION II - COMPONENTS

COMPONENT	CAS NO.	WEIGHT %	OSHA/PEL	ACGIH/TLV
N,N-DIMETHYLACETAMIDE	127-19-5	90-100	10 PPM	10 PPM

SECTION III - PHYSICAL DATA

BOILING POINT: 166 C (330 F) (AT 760 MM HG)

VAPOR PRESSURE (MMHG): 2.0 (20 C)

MELTING POINT: -20 C (-4 F) (AT 760 MM HG)

VAPOR DENSITY (AIR=1): 3.0

SPECIFIC GRAVITY: 0.94 (H2O=1)

EVAPORATION RATE: 0.17 (BUTYL ACETATE = 1)

SOLUBILITY(H2O): COMPLETE (100%)

% VOLATILES BY VOLUME: 100 (21 C)

PH: 4.7 (20% SOLUTION)

ODOR THRESHOLD (P.P.M.): N/A

PHYSICAL STATE: LIQUID

COEFFICIENT WATER/OIL DISTRIBUTION: N/A

APPEARANCE & ODOR: CLEAR, COLORLESS LIQUID. FAINT AMINE ODOR.

CONTINUED ON PAGE: 3

ATTACHMENT B

FID Strip-Chart Response from a Contaminated Sample System with Humidified Zero Air

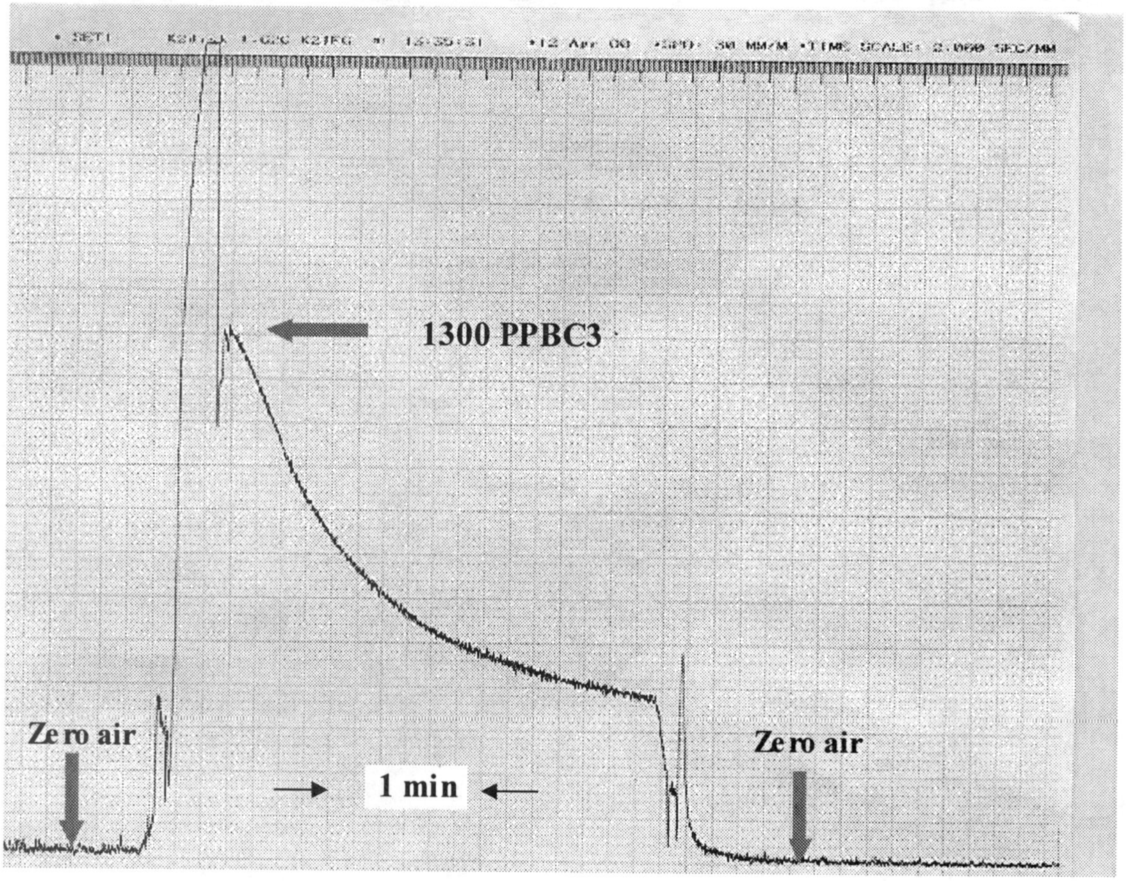

ATTACHMENT C

Exhaust Gas Stabilization

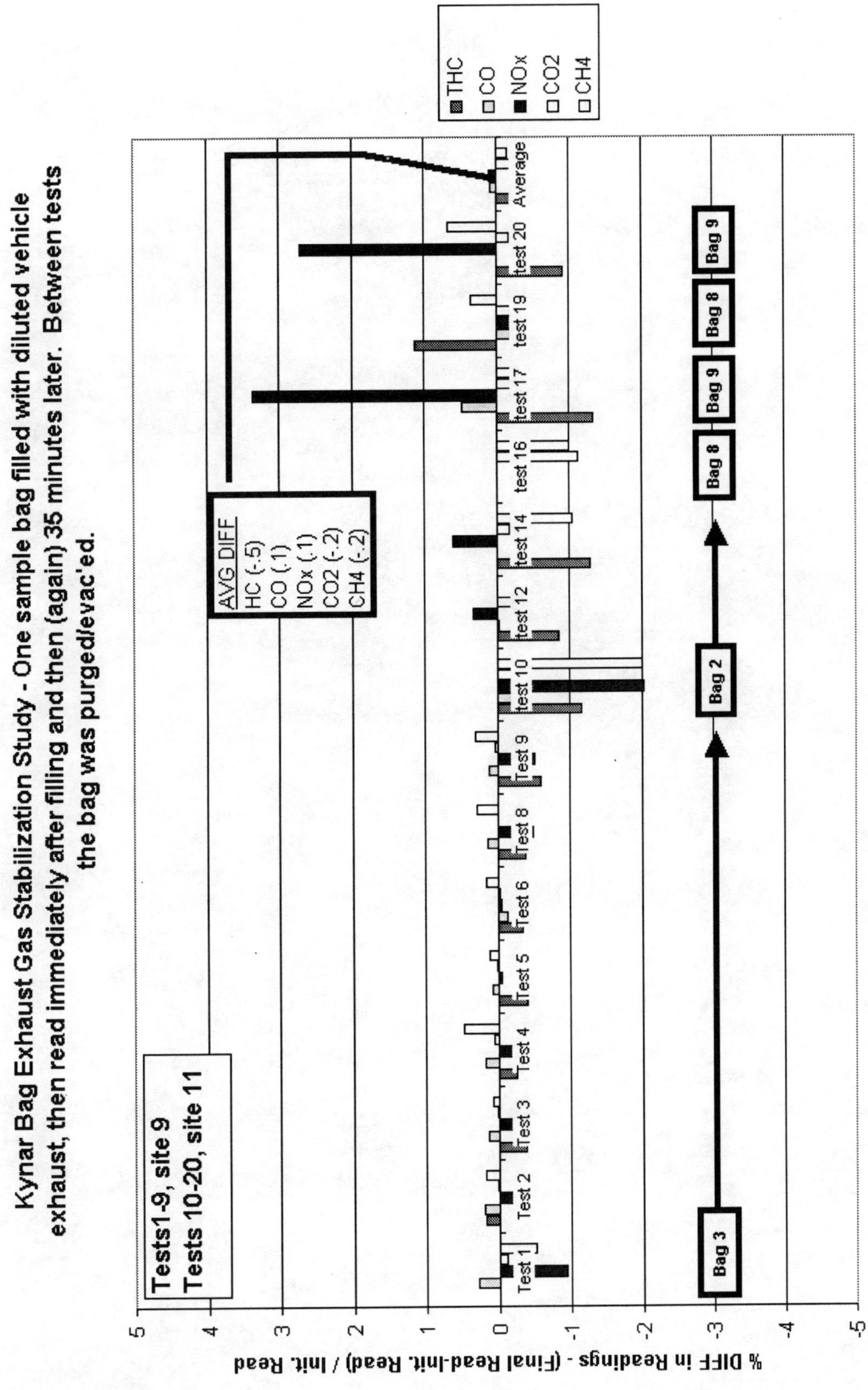

2002-01-0053

Gas Supply and Treatment Options for use in Emissions Measurement and Test

Larry Rabellino
SAES Pure Gas, Inc

ABSTRACT

The analysis of gaseous emissions from ULEV and SULEV automobiles requires very high levels of purity in the gases used for sample dilution, system purging, analyzer operation and analyzer calibration. These gases include as a minimum air, nitrogen, oxygen, hydrogen and helium.

The quality of these gases will vary depending on the "delivery system" - the method of generation, the quality of the distribution system and the type of storage. Methods of delivery range from "on site" gas generation to pressurized gas cylinders or cryogenic liquid storage. Distribution systems include tubing of carbon steel, copper, polymers or stainless steel.

Regardless of the method of delivery, the gas quality can vary. Increasingly stringent emission standards demand reduced variation in gas quality to improve test repeatability and accuracy. Variation can be reduced by either maintaining the quality at a fixed level of undesirable constituents, or by complete removal of all undesirable constituents via appropriate gas treatment.

Selected gas treatment methods were tested to demonstrate "part per billion" level concentrations of the undesirable constituents in gases relevant to emissions analysis.

INTRODUCTION

Exhaust gas analysis for ULEV automobiles includes the measurement of oxides of nitrogen (NOX), hydrocarbons (HC), carbon monoxide (CO) and carbon dioxide (CO_2) in a background of nitrogen and oxygen. Methods for the measurement include the constant volume sampler (CVS) and the Bag Mini-diluter (BMD). The BMD and CVS systems need "clean" dilution air for accurate measurement of the exhaust gas. The analysis equipment for exhaust gas measurement requires "clean" gases as well for the analyzer "zero" or calibration, the flame ionization detector (FID) fuel and burner air. Contamination in the dilution, zero gas or analyzer support gases (house gases) add variation to the measurement results.

Several different materials, processes and operating characteristics were explored for the removal of contamination from the house gases. The application of the materials must then take into account the method in which the house gas is prepared, stored and delivered to the measurement system. Additional considerations have been explored in the application of these materials to gas supply system design.

MATERIALS AND PROCESSES

There are several materials that use multiple processes for the removal of contaminants. Each material has unique performance that is dependent on the house gas type (air, nitrogen, and oxygen...) the contaminant that must be removed and the type of process.

MATERIALS - The types of materials tested to treat the house gases are differentiated by their composition and physical structure.

Getters

The getters tested were a metal based alloy. These getters are in pellet form, approximately 3 mm cylinders. The pellets were highly porous, with tens of square meters of surface area per gram of pellet. Typical metals used in getters include zirconium, titanium and vanadium.

Supported metals

The supported metals tested are differentiated from the getter family in that the metal is disbursed over a high surface area support. The materials tested were of pellet form, again approximately 3 mm cylinders. These pellets were also very porous with surface areas as high as hundreds of square meters per gram. Several different metals such as copper, nickel, palladium or platinum can be used. Several different preparation technologies and processes are available: metal concentration can range from <1% to more than 60%. Also, metal dispersion and sample surface area can be tuned.

Zeolites

The zeolites tested are a homogeneous material, in spherical form approximately 3 mm diameter. The zeolites are porous, with several hundreds of square meters of surface are per gram. Zeolite is composed of aluminum silicate. The Zeolites are also known as "molecular sieves".

PROCESSES - These materials are used in the purification processes of either catalytic oxidation, physical adsorption or chemical adsorption.

Catalytic oxidation process

In this application, he catalytic oxidation process is used to convert H_2, CO, CH_4 and NMHCs into CO_2 and H_2O. This reaction can occur in any inert gas (nitrogen, argon and helium), oxidizing gas (oxygen) and even a reducing gas (hydrogen) providing the correct conditions are present. The conditions needed to support the catalytic oxidation are the catalyst, oxygen and heat (primarily as a function of the catalyst material - some catalytic reactions can occur at room temperature).

The catalyst tested was a palladium metal of less than 5% palladium by weight on an alumina support. Operating temperatures were in the range of 250 - 350°C. Other catalysts include platinum with operating temperatures above 400°C In the case of use in an inert gas, it is necessary to add stoichiometric oxygen to be certain that complete oxidation will occur. For example, in the case of 1 part per million (ppm) each of hydrogen, methane and carbon monoxide, 3 ppm of oxygen will be needed as follows:

Relationships:

$$1\ H_2 + 1/2\ O_2 = 1\ H_2O$$

$$1\ CO + 1/2\ O_2 = 1\ CO_2$$

$$1\ CH_4 + 2\ O_2 = 1\ CO_2 + 2\ H_2O$$

Physical adsorption

Zeolites or molecular sieves use the process of physical adsorption to remove contaminants. The sieves have many benefits as a purification media. They can effectively remove CO_2 and H_2O from all the house gases. The sieves operate at room temperature, and they can be fully cleaned or "regenerated" very effectively with reverse flushing at elevated temperatures and low pressures using very clean gas. The weakness of the sieves is that they have a relatively low capacity.

Chemical adsorption - supported metals

In this case, some supported metals can be used as a remover (absorber). The type metal determines the end use. Supported copper or nickel, when processed correctly, can react with specific impurities forming a new compound. For example, using nickel,

$$2Ni + O_2 = 2NiO$$

$$Ni + 3CO = Ni(CO)_3$$

$$2Ni + H_2 = 2NiH$$

All house gases can be treated except oxygen because the catalyst will remove oxygen. These materials can also remove H_2O, CO_2 and many non-methane hydrocarbons. As with the zeolites, they can be regenerated. This regeneration requires the use of heat to release most impurities, and the addition of hydrogen to release the oxygen retained as follows:

$$NiO + H_2 + Heat = Ni + H_2O$$

These catalysts operate at room temperature and have a limited capacity. It is important to fully research the behavior of purification materials to apply them correctly. An example is supported nickel - it is a good oxygen remover, but can also support some catalytic processes at ambient temperature. CO and H_2 can combine with oxygen to form CO_2 and H_2O depending on the conditions. A detailed understanding of the materials is essential to make the appropriate application.

Chemical adsorption - Getters

Getters also use the chemical adsorption process to remove impurities. As an example, zirconium will combine with oxygen, nitrogen, carbon and hydrogen to form zirconium oxide, zirconium nitride, zirconium carbide and zirconium hydride. The significant difference is that the getter chemical reaction is irreversible - as an example, when oxygen combines with the zirconium, there is no possibility for the oxygen to be released - the chemical bond is permanent.

Getters can operate in all house gases except air and oxygen.

Getters also provide the unique ability to be the only single step process for the removal of methane and nitrogen (from non-nitrogen house gas) as impurities from rare gases.

Getters normally operate heated at temperatures of 300°C to 600°C depending on the house gas type and desired performance. The capacity of heated getters is as much as 40 times that of the supported metals. In fact, while the chemical reaction of a supported metal occurs only on the surface of the metal, the getter makes use of both the surface and the bulk structure. Impurities chemisorbed on the metallic surface of the getter diffuse in the bulk thus leaving the external getter surface always active to react and trap impurities.

OPERATING CHARACTERISTICS

The operation of these different materials is a function of many independent parameters. Ultimately, the goal in characterizing the materials is to provide clear design constraints for desired application. Measurable criteria needs to be established to evaluate effects of variables.

Purification material capability can be measured by the *"capacity"* and *"efficiency"* of the material under controlled conditions. The effects of variables can then be quantified with respect to the material capacity and efficiency.

CAPACITY AND EFFICIENCY

Capacity and efficiency are common terms in the description of purification materials and usage of gas purifiers. While simple in concept, the application of capacity and efficiency to real world situations becomes a complex equation involving interdependencies of inlet pressure, pressure drop, temperature, flowrate, type of impurities and concentration of impurities, media type and bed design. The testing considered the factors involved in understanding capacity and efficiency, as well as the relative sensitivity of each factor.

We should first define capacity and efficiency, and then identify the factors affecting performance such as operating temperature, operating pressure, flowrate, impurity spectrum and vessel design.

Capacity

The capacity can be considered as the measurement of the amount of impurities that can be "held" in a purification material. This can be measured as difference in "percent" of weight from manufactured state to saturated state. For example, a certain getter may include 0.1% nitrogen by weight as manufactured, and reach 5% nitrogen when fully saturated. The capacity for nitrogen (or all other impurities) can then be expressed in terms of "liters per liter" or liter of impurity at standard temperature and pressure (STP) per liter of volume of process material. It is the capacity of the materials used that dictates the ultimate process vessel volume for a given application.

Efficiency

The efficiency of a material can be considered as the "*rate of removal* (or *conversion* in the case of an oxidizing catalyst)" of impurities. The efficiency can be defined as the maximum impurity loading that a newly activated purifier can remove to the required outlet purity. The efficiency can be expressed as the impurity residence time (liter of impurity / liter of media*time). It is the efficiency of the material that dictates the vessel diameter for a given application (notwithstanding considerations for particulation or pressure drop).

Factors affecting performance

Capacity and efficiency are used to develop vessel design, and establish estimated "lifetime" as a result of the inlet conditions Changes in the conditions can drastically affect the performance. It is essential to fully understand the excursions in the conditions to be confident specifications are met. Ultimately, both capacity and efficiency must be fully evaluated before a material can be correctly applied.

The "Efficiency" testing must be planned to vary the key design elements and quantify any independent and interdependent responses. The methods include the use of multiple vessel configurations allowing space velocity characterization.

The test vessels are configured in 7 linear steps. The first vessel (1) is undersquare in length to diameter (L/D). The L/D ratio increases to the maximum for vessel 7. For example:

Several materials were tested including heated getter, heated palladium catalyst as well as supported metals and zeolite materials. Test parameters such as temperature, flow, impurity distribution and pressure were varied while continuous measurements were taken.

The test equipment needs to provide a source of purified test gas, methods for adding the selected impurities, control of test parameters and precision measurement of resulting gas quality. Figure 1 shows the experimental test setup.

Testing includes efficiency of heated getter, heated oxidizing catalyst and ambient temperature materials. In each case, the inlet gas contamination blend is chosen to give readily measurable outlet levels - the goal in these tests is to characterize trends.

Figure 1: Test Setup

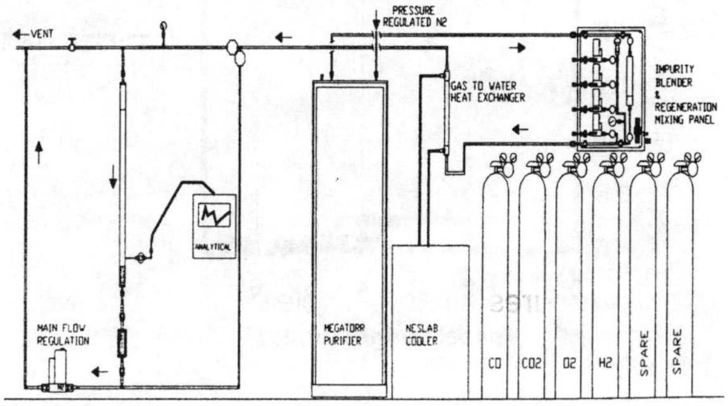

Heated getter test results

Figures 2,3 and 4 show heated getter sensitivity to changes in operating pressure for methane, nitrogen and carbon monoxide. It is clear that the removal improves with additional length. It is also clear that methane is the biggest challenge to remove, followed by nitrogen and then carbon monoxide. The effects of operating pressure are consistent with the physics of adsorption - as the pressure increases, the efficiency to remove impurities also increases.

This is due to the increased partial pressure of the impurities - the more impurities present for all other conditions fixed, the more likelihood that the impurities will contact active removal sites.

Figure 2: Getter Efficiency for CH4 vs. Pressure in Rare Gas

Inlet Impurities: 1ppm CO, 1ppm CO2, 4ppm H2, 3.5ppm O2, 1ppm CH4, & 5ppm N2.

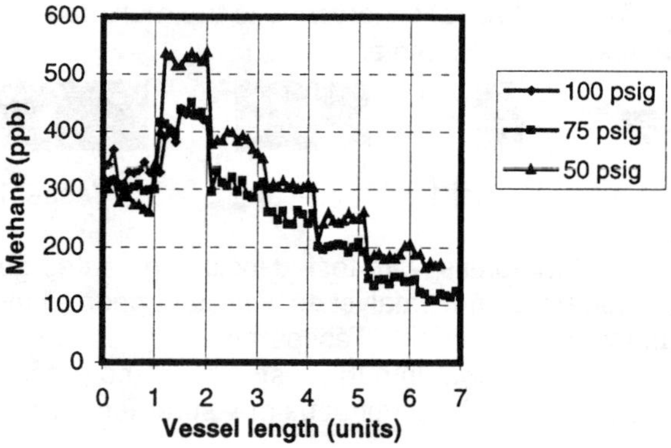

Figure 3: Getter Efficiency for N2 vs. Pressure in Rare Gas

Inlet Impurities: 1ppm CO, 1ppm CO2, 4ppm H2, 3.5ppm O2, 1ppm CH4, & 5ppm N2.

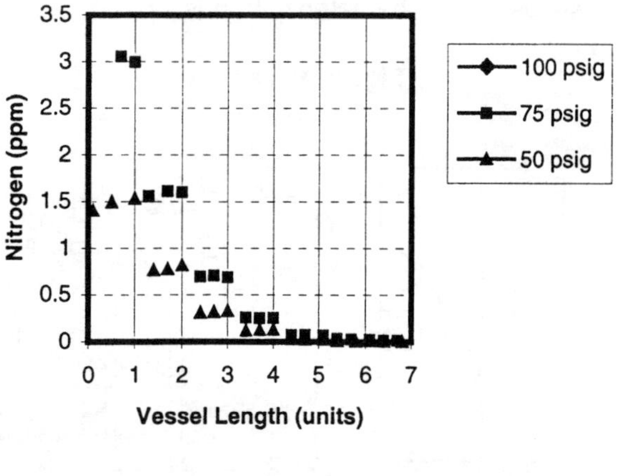

Figure 4: Getter Efficiency for CO vs. Pressure in Rare Gas

Inlet Impurities: 1ppm CO, 1ppm CO2, 4ppm H2, 3.5ppm O2, 1ppm CH4, & 5ppm N2.

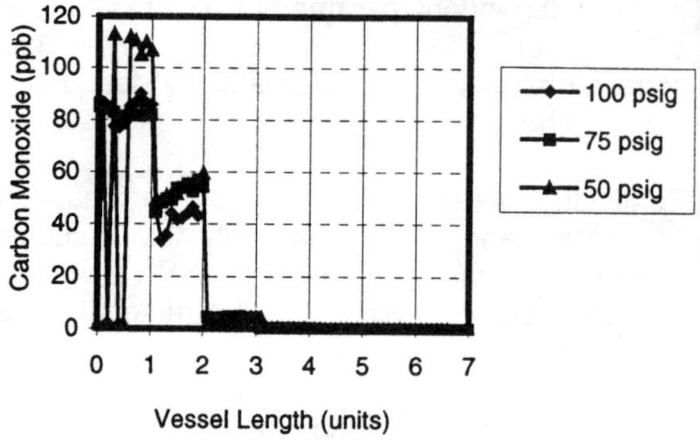

Figures 5,6,7 and 8 show heated getter sensitivity to changes in flowrate for methane, nitrogen, carbon monoxide and oxygen.

It is clear that again, removal efficiency increases with length - however, the rate of improvement as a function of length is different than the same rate for the previous testing at different operating pressures.

This data confirms that the removal efficiency of each impurity is uniquely affected by changes in operating conditions - and that these characteristics must all be considered in the design of purification vessels.

Figure 5: Getter Efficiency for CH4 vs. Flow Rate in Rare Gas

Inlet Impurities: 1ppm CO, 1ppm CO2, 4ppm H2, 3.5ppm O2, 1ppm CH4, & 5ppm N2.

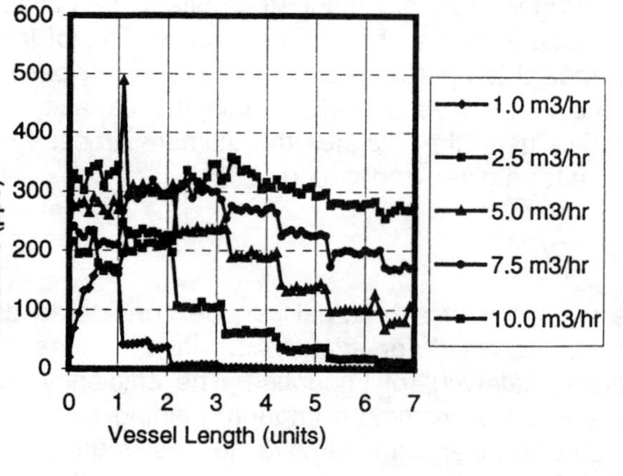

Figure 6: Getter Efficiency for CO2 vs. Flow Rate in Rare Gas

Inlet Impurities: 1ppm CO, 1ppm CO2, 4ppm H2, 3.5ppm O2, 1ppm CH4, & 5ppm N2.

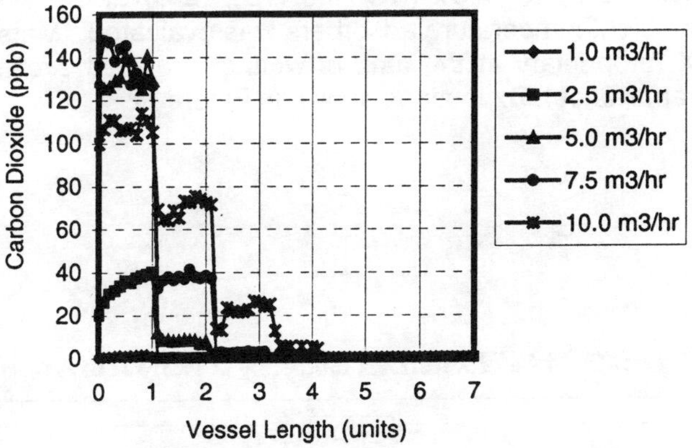

Figure 7: Getter Efficiency for N2 vs. Flow Rate

Inlet Impurities: 1ppm CO, 1ppm CO2, 4ppm H2, 3.5ppm O2, 1ppm CH4, & 5ppm N2.

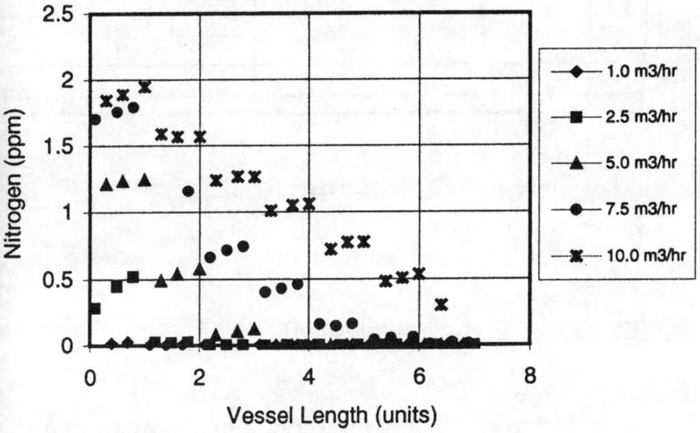

Figure 8: Getter Efficiency for O2 vs. Flow Rate in Rare Gas

Inlet Impurities: 1ppm CO, 1ppm CO2, 4ppm H2, 3.5ppm O2, 1ppm CH4, & 5ppm N2.

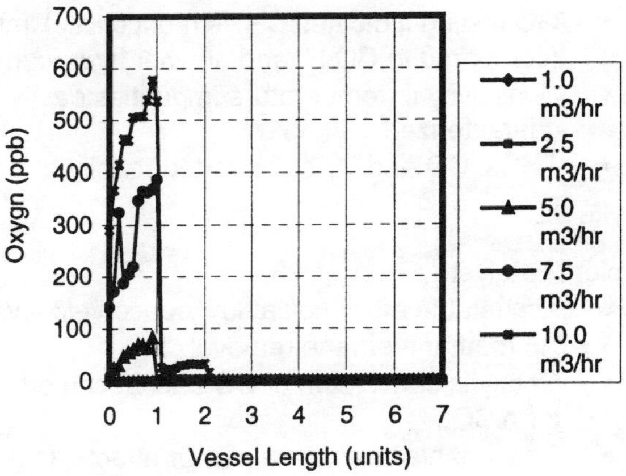

<u>Heated Oxidizing Catalyst</u>

The heated oxidizing catalyst has somewhat different criteria for evaluation. Instead of removal of impurities, the catalyst performance is measured by conversion efficiency. Figure 9 provides results from testing for sensitivity to operating temperature. It is clear from the data that the conversion efficiency at temperatures higher than 275°C begin to stabilize - indicating that under these operating conditions, the target minimum operating temperature would be 300°C.

Figure 9: Palladium Catalyst Efficiency for CH4 in Nitrogen Gas

10M3/HR, Inlet Impurities: CO = 2 ppm, CH4 = 1 ppm, O2 = 4 ppm

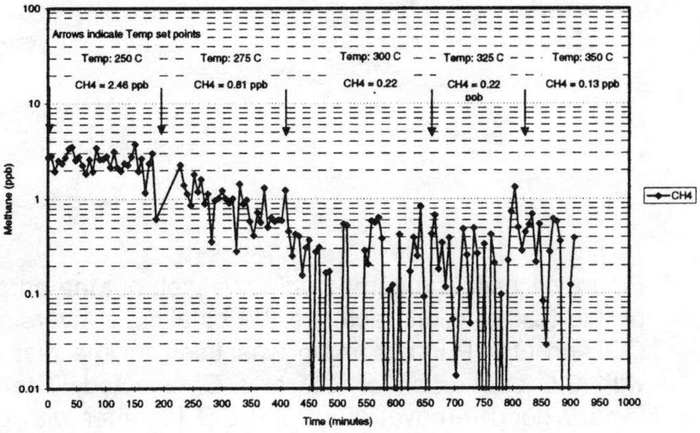

<u>Ambient temperature supported metal</u>

In this testing, tests are performed on low density, high surface area materials with nickel as the active metal. Testing conditions are using space velocities of approximately 1/3 that used in getter testing (lower flowrate per unit volume). Figure 10 shows a single vessel used to demonstrate typical performance. The inlet gas was liquid source argon gas, of a 5-9s rating - this gas was not purified, but instead represents typical gas supply quality.

Figure 10 test results show excellent removal of CO, CO_2, H_2, O_2 and H_2O. As expected, this media has no ability to remove CH_4.

Figure 10 Ambinet Temperature Nickel Catalyst efficiency using standard 5-9s grade argon

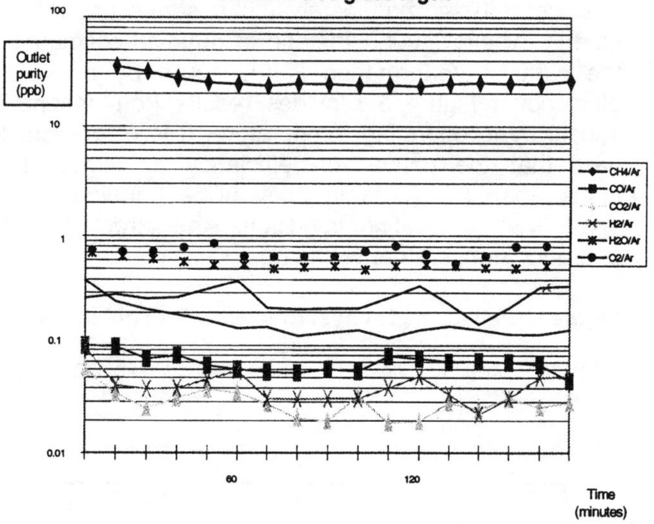

Figure 11 uses the same purifier vessel, but the purpose of this test is to characterize the catalytic conversion of CO and O_2. Purified argon was used as the test gas, with 3.5 ppm each of CO and O_2 injected. There is initially good removal of O_2 and CO together with good efficiency. O_2 combines with CO to make CO_2 (note that there is no CO_2 injected). Once the threshold is met for CO_2 capacity (at 6 days), CO_2 begins to break through the vessel.

Figure 11: Ambient Temperature Nickel Catalyst

Conversion of CO+O2=CO2

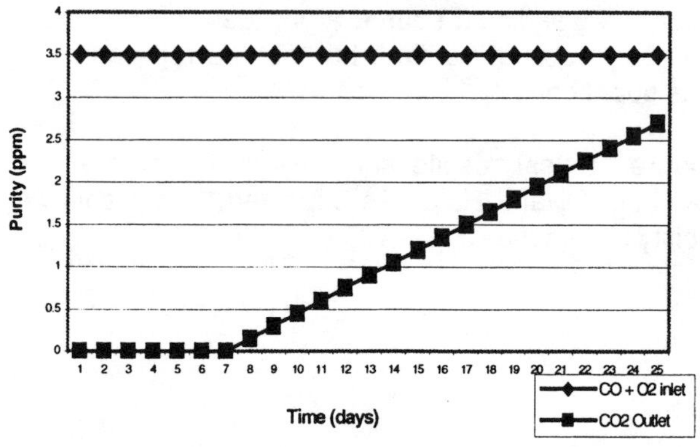

Ambient Temperature NOx removal

The ability for NOx (NO and NO_2) removal of sever ambient temperature adsorbers was evaluated. Materi X (proprietary at the time of writing) provided excelle removal of NOx in air as shown in Figure 12.

Figure 12: Material X removal efficiency of NO/NO2 in Air

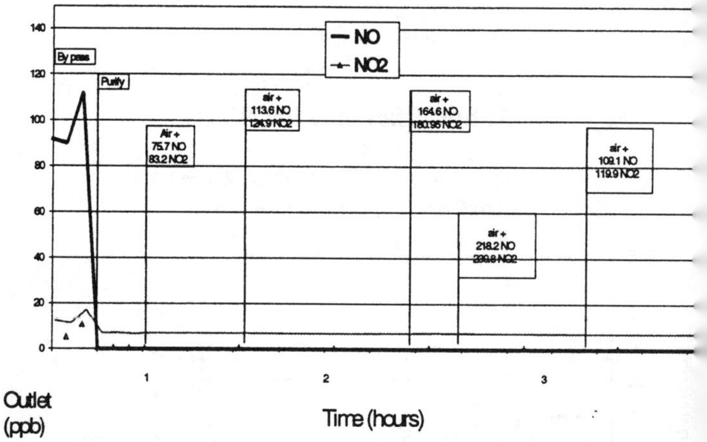

Summary of materials evaluation

Getter materials

- Methane (CH_4) is the most significant impurity
- Flowrate vs vessel design are most important parameters
- Operating temperature is significant
- Operating pressure is less significant, but mus be considered

Heated Catalysts

- GC testing indicates CH_4 is most critical impurity (compared to CO,H_2 and heavier hydrocarbons)
- Sensitivity to temperature/impurities easily characterized
- THCs, CO and H_2 is converted readily to below 1 ppb

Ambient catalyst

- Acceptable efficiency at low space velocity
- No methane/ethane removal
- Catalytic conversion of CO and O_2 can occur to form CO_2
- Reasonable to assume similar affects as getter on efficiency with changes in flow and pressure

- Reasonable to assume similar affects as getter on efficiency with changes in flow and pressure

Ambient NOx adsorber

- NOx is readily removed to less than 1 ppb

GAS SUPPLIES FOR EMISSION MONITORING

The evaluation of purification materials demonstrated an ability to remove undesirable contaminants from typical gases used in emission monitoring systems. Emission monitoring systems can use, as a minimum, the following gas supplies:

- Clean Air Supply
- Nitrogen Gas Supply
- Oxygen Gas Supply
- Hydrogen/Helium FID Fuel Supply

The practical application of the gas treatment technology was then evaluated with consideration for the gas quality requirements, types of gas generation, methods of storage and distribution.

Clean Air Supply

Clean Dry Air (CDA) can be generated by an on-site compression of the ambient air, often with some treatment (such as moisture removal) and storage as required.

Alternately, air can be purchased directly in high-pressure cylinders, or blended on site with house supplies of nitrogen and oxygen.

The most common method to obtain large flowrates of CDA is to use an industrial compressor followed by a "drier" based on a "pressure swing adsorption" (PSA) drying process. The PSA process has limitations for hydrocarbon, CO and NOX removal, and can provide variation in oxygen/nitrogen content.

Ultra High Purity (UHP) air from cylinders is often used when higher purity is required - in this case, purity can be extremely good, as this air is often "blended" from purified nitrogen and oxygen. The desired purity in this case is a directly proportional to cost.

Alternately, in place of the PSA type treatment system, ambient compressed air can be treated with materials tested in this study. The result would be air that has 1 ppb concentrations of H_2O, THCs, CO_2, CO and NOx and no change to the ambient O_2/N_2 ratio.

This air would be processed using the following treatments:

- Gross moisture removal using water trap
- Heated catalytic conversion of hydrocarbons, CO and H_2
- Ambient adsorption of NOx, CO_2 and H_2O
- Continuous process has no affect on nitrogen/oxygen ratio
- All regenerable materials - no consumables

Specifications for this air treatment process would be summarized as:

Concentration (ppm)	H_2O	THC	CO_2	CO	NOx	H_2	O_2
Inlet (ambient)	55% RH	5	400	20	3	1	amb
Outlet (treated)	.001	.001	.001	.001	.001	.001	amb

Amb = ambient

Nitrogen Gas Supply

Nitrogen is produced by separation from air. Optional methods for generating nitrogen include cryogenic distillation, and pressure swing adsorption. The cryogenic distillation can produce very good quality nitrogen - from 5-9s (10 ppm total impurities) to 7.5-9s (50 ppb total impurities). PSAs used for nitrogen generation can reach 4-9s (100 ppm total impurities).

Unless the site has a nitrogen generator, the nitrogen must be stored for usage. Options for storage are based on the demand and desired purity - nitrogen can be stored in high pressure cylinders for low flow and high purity usage. Higher flowrates can be accommodated with liquid nitrogen storage with some equipment for vaporizing the liquid to gas. The liquid can be stored in variety of containers. Portable vacuum jacketed vessels (~100 gallon), trailer mounted vacuum jacketed vessels (~2000 gallon) or fixed vacuum jacketed vessels (+2000 gallon) can be used, with several intermediate options.

The resulting gas quality will be a function of the quality of the gas prior to storage, and the quality of the storage and distribution. The distribution system can add significant contaminants due to outgassing of hydrocarbons, moisture and hydrogen as a minimum.

Nitrogen gas treatment technology tested in this study has demonstrated the ability to purify nitrogen to a very good quality - less than 1 ppb for the undesirable impurities.

The method for nitrogen treatment would depend on the need to remove methane and other hydrocarbons. In this case, the treatment would include:

- Catalytic conversion of HCs, CO and H_2 into CO_2 and H_2O

- Ambient adsorption of CO_2, H_2O and O_2

If methane and ethane do not need to be removed, it is possible then to use only the ambient adsorber as follows:

- Ambient adsorption of CO_2, H_2O, O_2, CO, H_2 and heavy hydrocarbons.

Specifications for this nitrogen treatment process would be summarized as:

Concentration (ppm)	H_2O	THC	CO_2	CO	H_2	O_2
Inlet 5-9s quality	3	0.5	0.5	2	2	3
Outlet (treated)	.001	.001	.001	.001	.001	.001

Oxygen Gas Supply

Oxygen is prepared in the same process as nitrogen, separated from air and cryogenically distilled. Storage and distribution is similar to nitrogen as well (with considerations that include that nitrogen is inert and oxygen is an oxidizer).

Oxygen gas treatment technology tested in this study has demonstrated the ability to purify oxygen to a very good quality - less than 1 ppb for the undesirable impurities.

The method for oxygen treatment generally includes the removal methane and other hydrocarbons. In this case, the treatment would include:

- Catalytic conversion of HCs, CO and H_2 into CO_2 and H_2O
- Ambient adsorption of CO_2 and H_2O

Specifications for this oxygen treatment process would be summarized as:

Concentration (ppm)	H_2O	THC	CO_2	CO	H_2
Inlet 4.5-9s quality	5	30	1	1	2
Outlet (treated)	.001	.001	.001	.001	.001

Hydrogen/Helium FID Fuel

Hydrogen can be produced from a variety of processes - electrolysis, steam reformation or recycled as a byproduct of a petrochemical process. Helium is mined gas - generated deep in the earth by radioactiv decay of uranium and thorium.

Both hydrogen and helium are stored as gases in hig pressure cylinders for most applications of low t medium flow demands, and stored as liquids for hig flow demands.

Both gases are available in a wide range of ga qualities. In all cases, several of the treatmer technologies evaluated are appropriate for both gases and provide more than one method for removal fc several of the undesirable impurities. The treatment would include:

- Heated Getter adsorption of N2, HCs, O_2, H_2O, CO_2, CO, H_2 (in helium)
- Catalytic conversion in helium of HCs, CC and H_2 into CO_2 and H_2O
- Ambient adsorption of CO_2, H_2O, O_2, CO, H (in helium) and heavy hydrocarbons.

Specifications for the hydrogen treatment process woulc be summarized as:

Concentration (ppm)	H_2O	THC	CO_2	CO	N_2	O_2
Inlet 5-9s quality	2	0.2	1	1	5	1
Outlet (treated)	.001	.001	.001	.001	.001	.001

Specifications for the helium treatment process can be summarized as:

Concentration (ppm)	H_2O	THC	CO_2	CO	N_2	O_2	H_2
Inlet 5-9s quality	1	0.2	1	1	5	2	1
Outlet (treated)	.001	.001	.001	.001	.001	.001	.001

CONSIDERATIONS IN GAS SUPPLY SYSTEM DESIGN

Typical of most design activities, there are several variables that must be characterized to optimize the solution. The most important task is to clearly identify the problems to be solved, and the relative importance.

In the case of gas supply systems, the following must be clearly defined, with priorities based on the agreed constraints.

- Gas type
- Source quality, pressure
- Final quality, pressure, flow (peak/average)

Next, identify multiple paths to achieve the requirements. If possible, characterize the risk or benefit of the variables to determine the sensitivity of change in the variable to the overall goal. These variables include:

- Pressurized gas supply
- Liquid supply
- Purity of supply before treatment (if any)
- Method of treatment (if any)
- Main supply or distributed supplies

Optimize the solution by manipulating the variables of greatest impact.

CONCLUSIONS

The analysis of gaseous emissions from ULEV and SULEV automobiles requires very high levels of purity in a variety of gases. Several methods are available to produce, store and deliver these gases. Treatment methods have been tested which confirm that many types of gases used can have the undesirable contamination reduced to less than 1 ppb. Specifically:

- Air can be treated to contain less than 1 ppb of all hydrocarbons, carbon dioxide, carbon monoxide, oxides of nitrogen, moisture and hydrogen
- Nitrogen can be treated to contain less than 1 ppb of all hydrocarbons, carbon dioxide, carbon monoxide, oxygen, moisture and hydrogen
- Oxygen can be treated to contain less than 1 ppb of all hydrocarbons, carbon dioxide, carbon monoxide, moisture and hydrogen
- Hydrogen can be treated to contain less than 1 ppb of all hydrocarbons, carbon dioxide, carbon monoxide, oxygen, moisture and nitrogen
- Helium can be treated to contain less than 1 ppb of all hydrocarbons, carbon dioxide, carbon monoxide, oxygen, ,hydrogen moisture and nitrogen

Careful evaluation of all the significant variables should be performed in the consideration of gas supply systems.

CONTACT

Larry Rabellino
Purifier Product Manager
SAES Pure Gas, Inc
4175 Santa Fe Rd.
San Luis Obispo, Ca 93401
Larry_Rabellino@SAES-Group.com

ADDITIONAL SOURCES

Chuck Applegarth
Development Engineering Manager
SAES Pure Gas, Inc
San Luis Obispo, Ca USA

Dr. Giorgio Vergani
Chief Scientist, Central Research
SAES Getters S.p.A
Lainate, Italy

2002-01-0054

Diesel Particulate Measurement with Partial Flow Sampling Systems: A New Probe and Tunnel Design that Correlates with Full Flow Tunnels

William M. Silvis, Gerald Marek and Norbert Kreft
AVL North America, Inc.

Wolfgang Schindler
AVL List GmbH

ABSTRACT

Partial flow sampling methods in emissions testing are interesting and preferred because of their lower cost, smaller size and applicability to engines of all sizes. However the agreement of the results obtained with instruments based on this method to those obtained with the traditional, large tunnel full flow sampling systems needs to be achieved, and the factors of construction that influence this agreement must be understood. These issues have received more attention lately in connection with ISO and WHDC standardization efforts underway to achieve a world-wide harmony in the sampling methods for heavy duty diesel engines, and with the introduction of similar Bag-minidiluter techniques into light duty SULEV gaseous pollutant measurement.

This paper presents the theory and practice of a partial flow probe and tunnel design that addresses and minimizes the undesirable effects of the necessary differences between the two sampling methods. Data are presented to show the efficacy of the design changes and the attendant improvement in the correlation between full and partial flow systems. As a consequence, results from the partial flow system described can be compared with confidence to results from the more expensive full-flow equipment for development and homologation purposes.

INTRODUCTION

Partial flow systems have been used for many years to measure the particulate matter emitted from diesel engines. European test procedures and the ISO 8178 test procedures have prescribed how such systems may be constructed and used for measurement on steady state engine tests. The unique nature of particulate matter introduces issues into the design that do not need to be considered for gaseous pollutant measurements. The result of a measurement of particulate matter is much more affected by the particulate's volatile/non-volatile composition, the temperature, the rate of dilution and the nature of the sample handling path, among other factors. It is a more recent interest in applying this type of system for transient testing, using a realistic and very dynamic driving schedule, that has renewed interest in what is actually an old problem: What is it that we measure and call PM, and how can measurement techniques be made repeatable, test to test and lab to lab, for effective engineering design and consequent pollution abatement?

All sampling systems for engine exhaust measurement must manage to collect a sample of the exhaust, and condition the sample to deal with its high temperature and moisture content. Partial flow systems for particulate matter like the SPC-472 differ in principle from full flow systems like the CVS in the order these steps are done. A full flow or CVS system first gathers all the engine's exhaust and then dilutes it at a variable rate before taking a flow-proportional sample. A partial flow first takes a proportional sample, and then dilutes it. As a consequence, a full flow system for a large engine can be large and expensive since all the engine's exhaust is treated. A partial flow system is small and inexpensive, but needs a more sophisticated controls system to insure proportionate sampling.

Although the partial flow system has been used for steady state homologation with success and satisfaction, there have always been issues with correlation in PM testing, even when one just considers the kinds of different results on gets using only full flow CVS tunnels. More recently, the attention to transient testing has caused these correlation issues to be more closely examined.

It has turned out that the measurement methodology issues that needed to be addressed were not always the ones expected. Instead of issues about the ability of higher tech systems to control flows dynamically, there were also important issues applying generally to the sampling of a hot, wet aerosol. It is also somewhat surprising that it is harder to resolve all these issues unequivocally on steady state tests than it is for transient tests.

In this paper, we describe in detail the following five areas that have affected differences between partial and full flow system PM measurements and the measures we have taken to address them:

- Deposition losses from sample handling differences causing PM sample to be lost before it could be measured.
- Anisokinetic sampling errors arising from the need to sample directly in the exhaust stack.
- Gas composition effects resulting from the effects of exhaust gas on flow metering devices.
- Time alignment issues affecting the validity of the pneumatically integrated total PM result.
- Soluble fraction changes from the different dilution and temperature treatment of the sample.

In each case, a theoretical basis for understanding the mechanisms that are affecting the measurement is described, as well as the means we've applied to manage them. We present as well correlation results from transient tests that show close agreement to CVS and excellent repeatability. These indicate that with collection of additional similar data, the statistical t-test standard for the acceptance of alternative measurement systems as specified in the CFR for 2007 can be met.

THE CORRELATION PROBLEM: THEN AND NOW

CORRELATION ISSUES: A PERSPECTIVE FROM PREVIOUS WORK – In the late 1980's the worldwide move towards tightening the emissions limits for particulate from Diesel engines, especially heavy-duty (HD) engines, resulted in new legislation in Europe, Japan and other Asian countries. The European Directive 91/542 EEC [1] defines emission limits, a steady-state test cycle, and test equipment. An important novelty in this directive – as well as in the Japanese legislation– was the specification of Partial-Flow Dilution (PFD) systems for particulate sampling as an alternative to the widely accepted full-flow CVS systems.

Although not explicitly stated, the specifications practically required that a probe flow (G_P), which must be a constant fraction (r) of the exhaust mass flow (G_{EXH}), is sampled and mixed with a flow of clean ambient air (G_{DIL}), in a dilution tunnel with a constant total flow (G_{TOT}). Such a PFD system can be regarded as a CVS system with all flows downsized by the factor r. The factor r can be as low as 10^{-4}, or 0.01%.

<u>Correlation between different full-flow CVS systems and between full-flow CVS and PFD</u> - The European and Japanese legislation was issued after tests had shown that PFD systems could give results similar to full-flow CVS systems [2]. Correlations at that time were around 5%, with a tendency to low results.

The differences were not considered a matter of serious concern, especially in view of the fact that the correlation among different full-flow CVS systems was, and still is, in the range of ± 10% of the mean – see [3]. Despite the fact that it has been observed that two different labs may measure up to 20% different PM on the same engine, while gaseous measurements agree, no systematic investigations of the reasons for the variability of full-flow CVS systems have been carried out. From a theoretical point of view this variability is not very surprising, because a dilution tunnel is not just a measuring tool but rather a complex chemical reactor. Some of the particulates – the black carbon part - are fed with the exhaust gas into the tunnel, where the cooling by the dilution air provokes chemical reactions, as well as condensation and adsorption processes of volatile substances – mainly hydrocarbons and sulfates. The result of these reactions depends on the parameters of the dilution process. It was also often observed in round-robin programs that the partial flow systems measured lower than the full flow systems.

The number of variable parameters of the dilution process is even larger for PFD systems than for FFD systems, but they are generally easier to adjust, directly or indirectly. Their influences were investigated by several researchers [5,12,13,14], but this did not result in a consistent picture or well-defined procedure for the layout or adjustment of PFD systems. This is largely due to the complexity of the problem. Figure 1: illustrates and summarizes the complicated structure of this parameter interdependence.

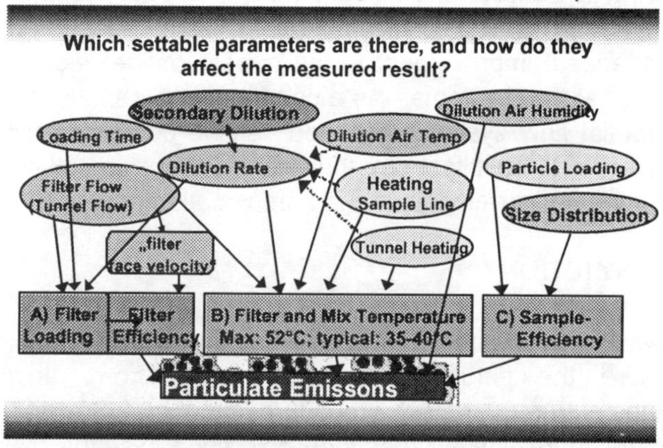

Figure 1: Parameters that influence the PM measurement

<u>The need for Transient Partial-Flow Dilution Systems</u> – In the late 1990's new legislation with even tighter emissions limits was being considered in the US, Europe and Japan. It was clear to the combustion engine community that this

would require more particulate emissions testing than before, including testing during transient cycles. The world-wide WHDC group was established by GRPE, which mandated the International Standardization Organization (ISO) to work out a standard applicable for future Worldwide Harmonized HD emissions measurement.

In the ISO TC 22/SC5/WG2 group the interest quickly focussed onto the possibility of using PFD for emissions testing in transient cycles. This was not considered problematical, since HD engine test-beds for emissions development and certification usually already provide a value of G_{EXH} – usually by adding air intake and fuel flow – which can serve to control the flow in PFD's for constant r, i.e. guarantee G_P to be proportional to G_{EXH}. A "Mini-CVS" is thereby obtained.

The ISO group was aware that the layout of a PFD system and its parameters may influence the particulate results and therefore had systematic parameter studies, as well as a correlation studies, conducted at the Swiss EMPA[4], at JARI in Japan and at the RWTUV Essen in Germany. It was surprising to find little influence of most parameters varied. As an example: the length of the sample line between the exhaust duct and the dilution tunnel was varied between 0.1 and 1.5 m. A decrease of ≈5% in measured particulate mass was found for the 1.5 m sample line. This is surprisingly small in light of the theoretical particle deposition calculations presented below.

The EMPA correlation study, employing equipment from 3 manufacturers, gave good agreement between full-flow CVS and PFD systems, see Figure 2. In the JARI study the PFD was low by 15±5% in both steady state and transient tests. The RWTUV study showed good correlation in the transient tests but deviations were similar to the JARI study for steady-state tests [5].

At this point, PFS/FFS correlation measurements were added to a much more comprehensive study of fuel, test cycle, and engine issues being conducted by EPA and ARB and performed at the Southwest Research Institute. In the course of this testing, less satisfying and more confusing results were observed. Without going into too much detail, the results with the first engine were similar to those observed in Europe, but soon tests on subsequent engines showed lack of agreement of as much as 40%. Of course, this triggered considerable concern from regulators, engine manufacturers and test equipment suppliers, as well as a renewed interest in more thorough investigations into possible reasons for the low measurements.

Our efforts to improve our product and understand the influences of the parameters affecting its set-up and use were augmented in no small part by the cooperative joint testing project directed by EMA/EPA/ARB, supported by AVL, Sierra and Horiba, and conducted by the laboratories at Southwest Research Institute [6]. The resulting system improvements brought the PFD to correlate with the full-flow CVS within 5%. The details of our investigations and improvements are the topic of the following paragraphs.

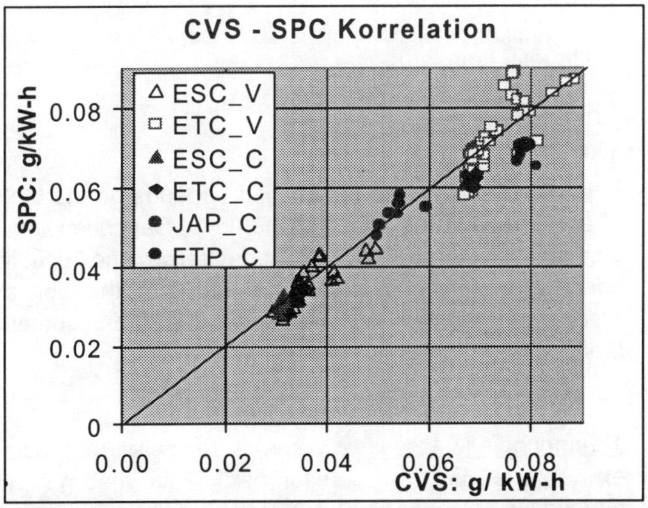

Figure 2: Correlation to the full-flow CVS to one of the PFD's tested (AVL SPC 472) in the EMPA study

ERROR SOURCES AND COUNTERMEASURES

DEPOSITION LOSSES – There are two deposition mechanisms that cause a fraction of the particulate matter to be lost inside the sampling system before it is collected on the sampling filters. These mechanisms are caused by the motion of the particulate in a concentration gradient and in a temperature gradient.

Diffusion losses - In 1827 Robert Brown first observed a phenomenon in pollen grains that is the mechanism behind what we call diffusion losses. Molecules of a gas, or the particles in an aerosol, move from a region of higher concentration to a region of lower concentration. Einstein later showed that this Brownian motion of the particles in an aerosol is equivalent to that of a giant molecules. However, unlike gas molecules, aerosol particles do not rebound when they collide with a surface.

Diffusion losses are often described by defining the Penetration, i.e. the fraction of molecules that are not deposited. The following approximation for the penetration, P, for flow in a pipe is found in reference [9]:

$$P = \frac{n_{out}}{n_{in}} = 0.819 \cdot e^{-11.5 \cdot \mu} + 0.0975 \cdot e^{-7.01 \cdot \mu} \qquad (1)$$

$$\mu \geq 0.009$$

Here μ is the deposition parameter for laminar flow conditions through a circular tube, given by:

$$\mu = \frac{D \cdot L}{Q} \qquad (2)$$

Where:

D = Diffusion coefficient for particulates, m^2/sec
L = Length of the tube, meters
Q = Volumetric flow rate, m^3/sec

Thus it's clear that the penetration depends on the length of the tube, and the speed of the flow. Dependencies on particle size come through effects on the diffusion coefficient. Curious to note is the fact that for this mechanism, the losses do not depend on the diameter of the tube.

Thermophoretic losses - When a temperature gradient exists in a gas, an aerosol particle in that gas will experience a force in the direction of the decreasing temperature. This occurs simply because the collisions it feels coming from the "hot" side are more energetic. Again, when the particle encounters the colder surface, it doesn't rebound and is lost. This effect is called thermophoretic loss.

As with diffusion losses, this effect can be modeled and calculated, albeit not as easily. Without presenting detail beyond the scope of this paper, the equations describing the thermophoretic forces in the continuum regime (particle size greater than mean free path) are summarized below:

$$F_{th} = \frac{-9\pi\mu^2 d_p H \nabla T}{2\rho_g T}$$

$$H \cong \left(\frac{1}{1+3Kn}\right)\left(\frac{k_a/k_p + 2.2Kn}{1+2k_a/k_p + 4.4Kn}\right) \qquad (3)$$

k_p = thermal conductivity of the particulate
k_a = thermal conductivity of air
∇T = Temperature gradient
Kn = the Knudson number

The magnitude of the force depends on the temperature gradient, which can be surprisingly large when hot gas is carried through a small tube with cooler walls. In order to calculate estimates of these losses, some information about the particle size distribution and some information about the temperature gradient in the sample lines are needed. For the former, we used the usual log normal size distribution with a mass median diameter of 170 nanometers and a geometric standard deviation of 2. For the latter, we used a typical heat loss model for gas transport along a pipe. The details of this work are found in reference [7]. We present some results here as illustration of the parameters of a design to minimize deposition losses.

Deposition losses for the previous tunnel design – A typical sampling arrangement had been a probe in the exhaust stack with a relatively long transfer tube conveying the sample to a dilution tunnel, and then to a collection filter. It had been the common practice to heat the sample line and tunnel, but not usually to the same temperature as the hot exhaust. An estimate of the temperature profile along such a transfer tube of one meter length is presented in Figure 3.

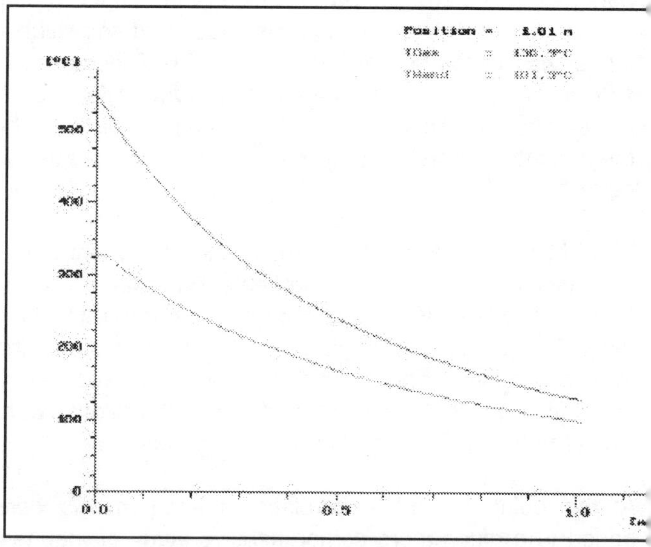

Figure 3: Gas and Wall temperature gradient along 1 meter transfer tube

Note that there is a significant cooling of the gas along the tube, and there is a strong temperature gradient between the sample gas and the tube wall. This leads to significant particulate losses to the walls, as shown in the following Figure 4. Here the particle size range is divided into bins and the losses estimated for each bin. The losses do depend on particle size, with greater loss of smaller ones, but not as strongly as one might expect.

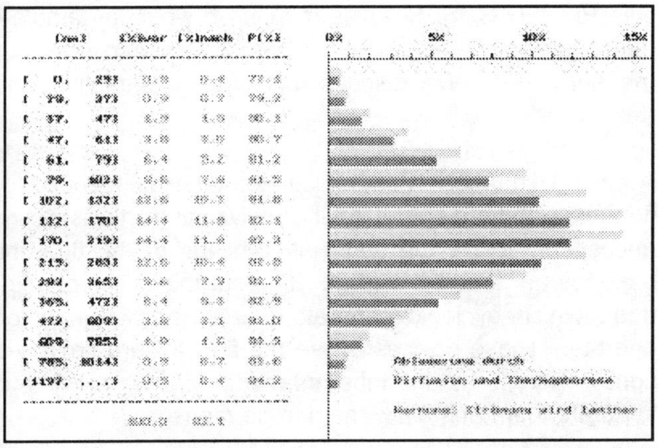

Figure 4: Partilcle losses by size along 1 meter transfer tube

Note that the overall deposition losses in this (admittedly extreme) case are approximately 18%.

<u>New tunnel and probe design</u> - To minimize these losses, the design of the sampling probe, transfer tube and tunnel were modified in the several ways. The transfer tube was made as short as possible by arranging to mount the tunnel as close as possible to the exhaust stack. The tube itself was made of very thin wall material, so that it could assume the temperature of the exhaust it carries as quickly as possible. For mechanical stability and to provide for insulation, the probe/transfer tube is encased in an outer tube, with a dead airspace.

Another improvement was made to the mixing zone. When pipe flow experiences an abrupt change in diameter, it creates re-circulation zones (Figure 5). Typical partial flow tunnels have such re-circulation zones where the mixing orifice connects to the Tunnel. These areas create a region where aerosol particulate is decelerated and forced against a surface due to the flow patterns. Indeed, examination of tunnels after use always shows deposits in these areas.

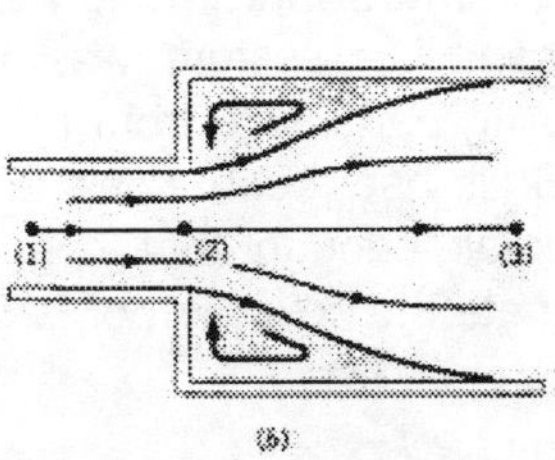

Figure 5: Recirculation zones from pipe diameter change

To avoid this, a diffuser cone with 15 degree angle was added to the outlet of the mixing area to prevent the formation of the recirculation zones. Figure 6 shows in cross section the tunnel with the modifications described above.

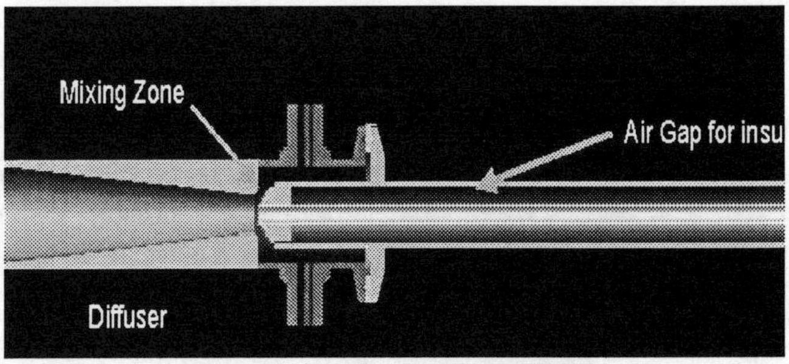

Figure 6: Modified Tunnel and Probe

The estimated temperature and deposition losses for this improved design are depicted in Figure 7 and Figure 8. The calculations show that temperature losses and gas to wall gradients are much reduced and deposition losses reduced to a small fraction of those from the previous design.

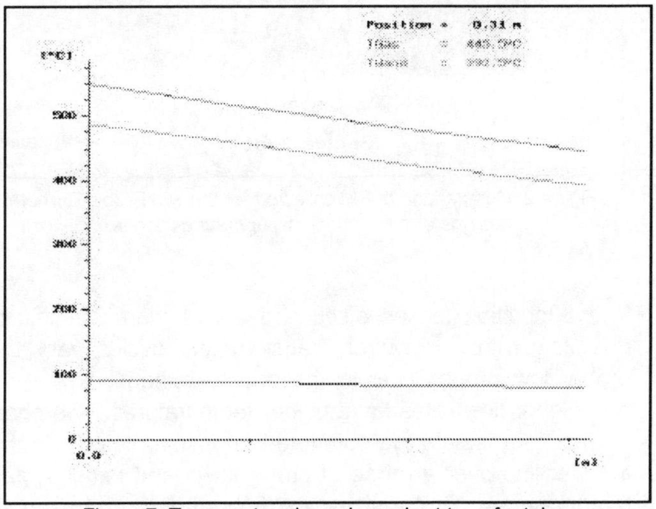

Figure 7: Temperature loss along short transfer tube.

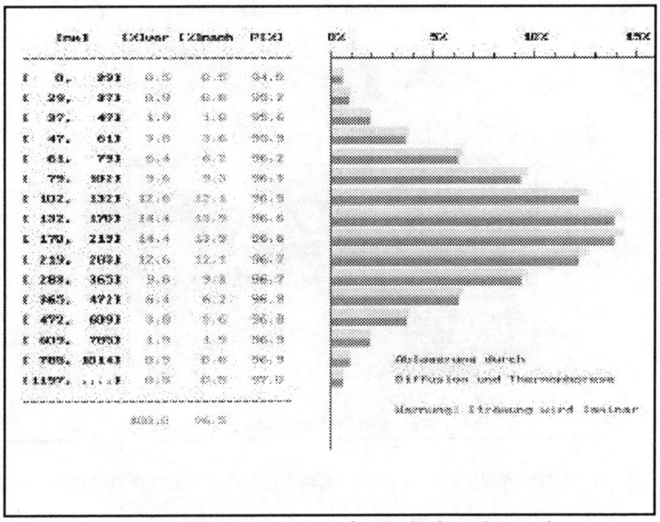

Figure 8: Deposition losses for re-designed tunnel.

We compared the PM measurements from the same engine from the former probe/tunnel and the redesigned one. The results are shown in Figure 9 below. Engine operating modes at 100% load, 25% load and no load were used. Higher loads on the engine produce hotter exhaust and correspondingly higher deposition losses with the older design.

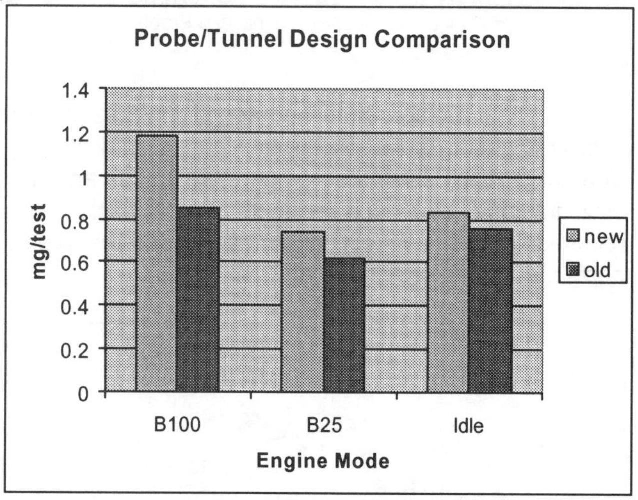

Figure 9: Comparison of PM collected for the two Probe/Tunnel designs at several engine-operating modes.

The deposition losses discussed above were specific to one flow rate. However, transient test cycles vary the probe flow proportional to the engine exhaust flow rate, and higher flow rates have higher temperatures. The chart in Figure 6 depicts the estimated deposition losses in the transfer line over a range of probe flows and exhaust gas temperatures. Probe sizes and flow rates are chosen so that the partial flow system operates in the shaded region.

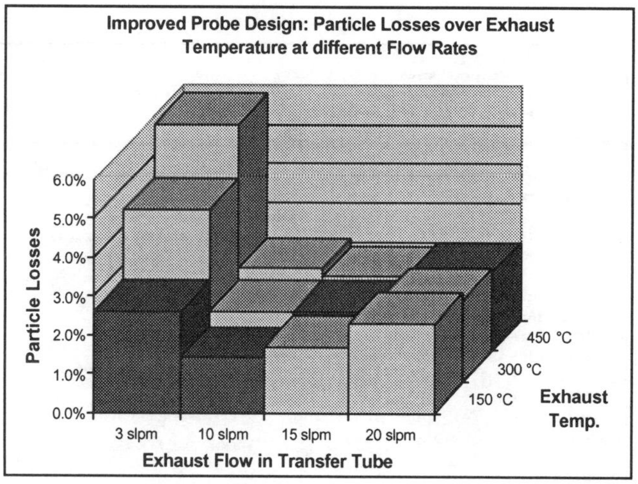

Figure 10: Deposition losses by flow and temperature

SAMPLING LOSSES – The manner in which a sample particulate is extracted from a larger flow can effect th measurement. We discuss this type of sampling err next.

Aerosol Dynamics - The particulate matter suspended in flow has inertia. Therefore the individual particles do n necessarily follow the same streamlines of the flow th gaseous parts of the fluid do. Indeed, this is the princip upon which cyclonic separators and particle impactor operate. These devices cause the flow to turn sharply tightly enough that the inherent inertia of any particles w separate them and allow them to be measured.

The parameter that characterizes the condition of a flo with respect to this phenomenon is called the Stoke number, a relative indicator of how differently th particulate may behave compared to the gas in which it embedded. For our purposes here, the Stokes numb can be thought of as the ratio of the time, τ, it takes for particle to adjust to a change in flow direction, to the tim d_c/U_0, available for adjustment around and obstacle of siz d_c.

$$Stk = \frac{\tau U_0}{d_c} \qquad \tau = \frac{\rho_0 d_a^2 C_c}{18\eta} \tag{4}$$

For a few particle sizes in the range of diesel smok relaxation times are listed in the table below.

Particle Diameter	Relaxation time
nanometers	nanoseconds
10	7
100	90
1000	3500

For particulate sampling it had been generally felt tha diesel particulate matter was so small, and correspondin relaxation times so short, that the flow field of the usua plumbing of a sampling system was not problemati Also, when sampling particulate from the full flow dilutio tunnel of a CVS, it had been considered that Stoke numbers were small enough that even hatted probes which could eliminate the collection of large re-entertaine particles, could be used [8].

However, when particulate is sampled directly from th exhaust stack there are larger differences in flow velocit between the bulk exhaust flow in the stack and th sampled exhaust flow in the sample probe. Thi difference can produce a severe enough flow field tha sampling errors do arise when the particulate matter i unable to follow the gas streamlines.

Sampling Geometry in the Exhaust Stack - Because of this, when a partial flow system extracts its sample from the exhaust stack of an engine, some concern must be given for the flow field of the sampling arrangement to insure that the target particles are able to be properly sampled. Flow fields for a sample stream facing into a flow and facing away from a flow are depicted in Figure 11 and Figure 12. The condition for which the velocity in a sample probe exactly matches the velocity in the sampled exhaust stream is called isokinetic sampling. Inasmuch as true isokinetic sampling is cumbersome to achieve and may be unnecessary, it is useful to estimate the effects of anisokinetic sampling and to design the sampling geometry to control its effect on the measured result.

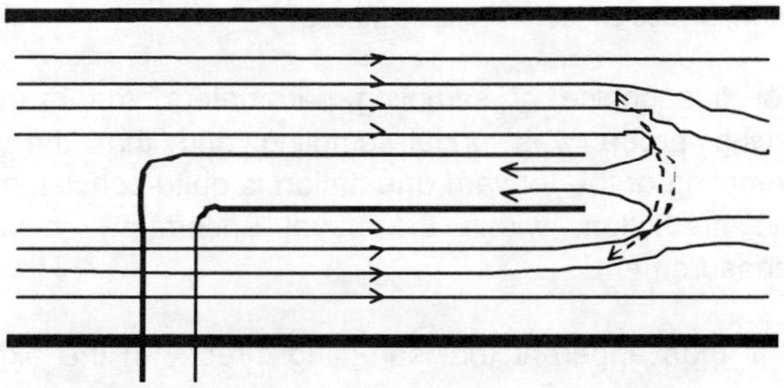

Figure 11: Anisokinetic sampling - rearward facing

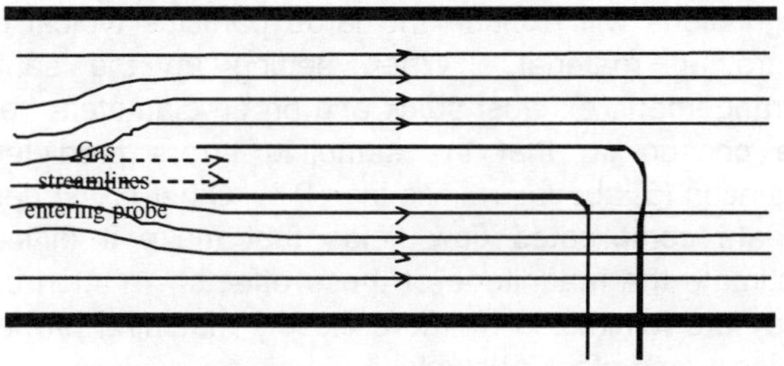

Figure 12: Anisokinetic sampling - forward facing

Sampling Error Estimations - For a partial flow sampling system, today's higher dilution ratios and relatively low probe flow rates lead to a sub-isokinetic sampling condition. That is, the velocity into the probe is lower than the velocity in the exhaust stack. In such a case, a forward probe orientation would overestimate the particulate, since some particles not intended to be sampled would not adjust to the flow field in time to be diverted around the probe. A backward facing probe would underestimate the particulate, since some of the particles would not be able to follow the severe flow field, reverse direction, and enter the probe.

Very useful Information is available in the literature for the forward facing probe orientation. The following empirical relationship for super and sub-isokinetic sampling is described in reference 9.

$$P := 1 + \left(\frac{U_0}{U} - 1\right)\cdot\left[1 - \frac{1}{1 + \left(2 + 0.62\cdot\frac{U}{U_0}\right)\cdot \mathrm{Stk}}\right] \qquad (5)$$

Here P is the penetration, U_0 is the stack velocity and U the probe velocity. For a typical exhaust stack sampling geometry (a 5" exhaust stack, 5mm probe, rated speed and load), the following chart, Figure 13, illustrates the magnitude of errors predicted by this equation as a function of particle size for bulk stream to probe velocity ratios ranging from .1 to 10. It can be seen that even when particles and Stokes numbers are small, significant sampling errors still arise when the velocity mismatch is great.

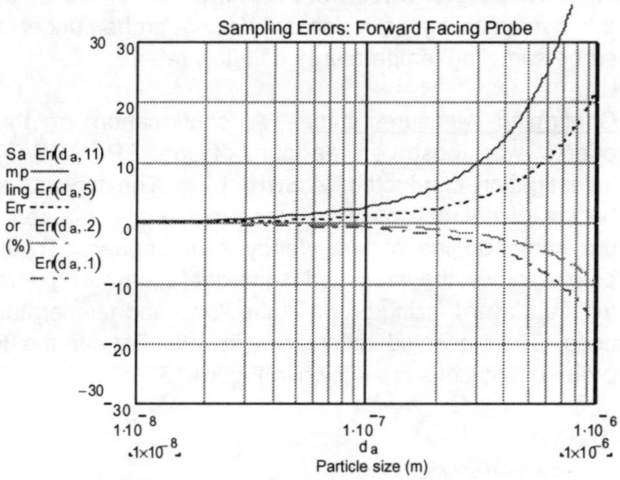

Figure 13: Sampling loss as a function of particle size for several speed ratios

In a practical sampling set-up on the test bed, it is not velocities and particle sizes that are adjusted, but rather the diameters of the exhaust stack and sample probe. In the Figure 14: below, we see the errors predicted for a 170 nanometer particle at rated speed and load as stack diameter and probe diameter are changed.

It is aparent from the figure that small stack sizes and larger probe diameters can lead to sampling errors well above 5%.

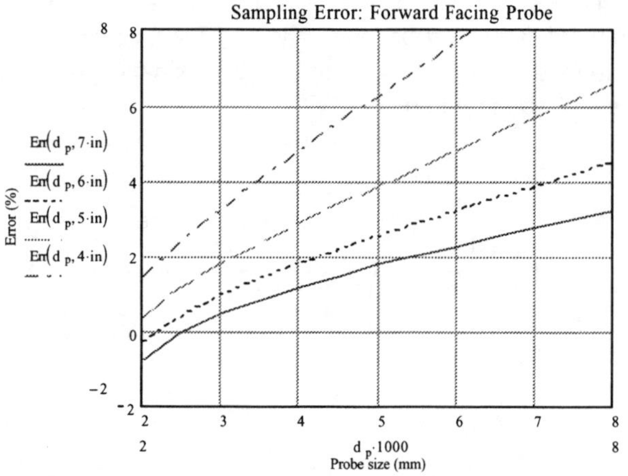

Figure 14: Sampling error as a function of probe diameter for several stack sizes

For the rearward facing probe, the flow field is difficult to calculate and an empirical formula is not available. However, in this case it can be seen from the nature of the field as depcited in Figure 11 that significant particulate losses could be expected as the particualte tries to follow the more contorted streamlines, reverse direction, and enter the probe. It is reasonable to assume that the undersampling of a rearward facing probe is greater than the oversampling of a forward facing probe, under the same sampling temperatures and flow rates.

Confirming Measurements - A confirmation of these effects was observed as part of the EPA/ARB/EMA investigation conducted at SwRI [5]. The forward and backward orientations were measured and compared for the same engine at two steady state modes. Table 1 below shows the predicted anisokinetic sampling errors for the forward facing probe at the flows and temperatures used. The observed differences in total PM for the two probe orientations are shown in Figure 15.

Expected Anisokinetic Errors
probe 5.3 mm
stack 6"

mode	rpm	load (%)	Tex ('C)	Gex g/s	Gp g/s	U_0/U	Stk	Err (%)
A	2000	100	480	398	0.08	6	0.001	1.7
D	1400	100	554	279	0.05	5.9	0.001	1.3

Table 1: Anisokinetic sampling errors, forward facing probe, full load modes

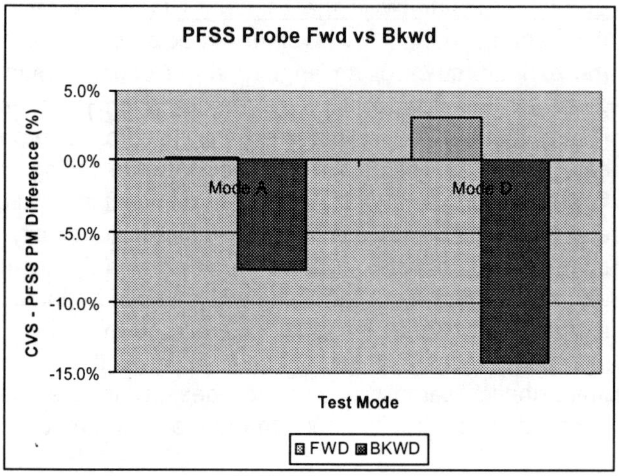

Figure 15: Observed differences between forward and backward facing probes on full load engine modes

For this choice of sampling parameters, the rearward-facing probe was under-sampling and that the over-sampling of the forward orientation is quite consistent with the prediction, within the usual uncertainty in a PM measurement.

It is thus apparent that sampling directly in the exhaust stack is more affected by an anisokinetic geometry than sampling in a dilution tunnel, and that a forward facing probe gave less error. With a probe of this type, the use of in-line particle classifiers as required in EPA's 2007 regulations will exclude the large particles typical of re-entrained material. When setting up the sampling arrangement, exhaust stack and probe diameters need to be chosen so that the sampling errors predicted by equation (5) above are less than 2%. Other probe designs create complicated flow fields that make it difficult to estimate the magnitude of these effects. In such cases, they are minimized by more closely matching probe gas velocity and stack gas velocity

ERRORS DUE TO GAS COMPOSITION - Exhaust gas is composed of significant and varying amounts of CO_2 and water vapor, depending on the operation mode of the engine. Therefore, the flow measurement and control elements of a sampling system that use physical properties of these components to detect a gas flow will be affected. We next discuss the nature and magnitude of these effects and our countermeasures.

Theory of Thermal Mass Flow Meters- Many partial flow particulate samplers utilize thermal mass flow controllers to control the flow rates. The thermal-based mass flow sensor uses the heat capacity of a gas to measure its mass flow rate:

$$Sensor\ output = Q = \{m \cdot Cp \cdot \Delta T\} / N \qquad (6)$$

Where Q is the thermal flux, m is the mass flow rate, C_p is the specific heat at a constant pressure, ΔT is the net change in gas temperature, and N is the "molecular structure" correction factor. N is an empirical factor which compensates for differences in variation of the specific heat with temperature for different types of gas molecules (i.e. monatomic, diatomic, etc.)[10]. In a thermal mass flow sensor, gas is heated in a sensor tube and resistor temperature devices measure the temperature gradient along the tube (Figure 16).

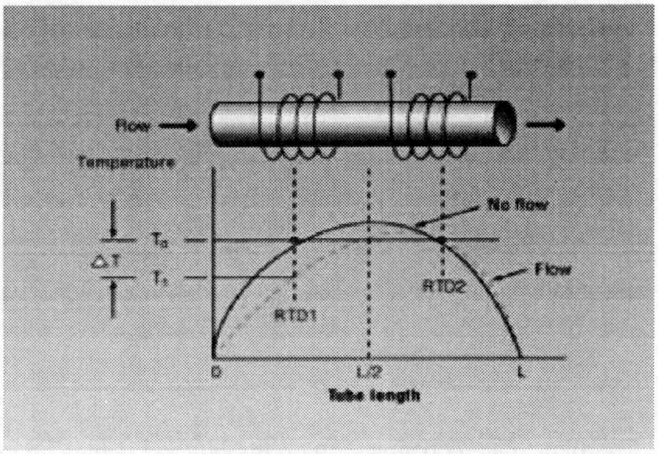

Figure 16: Thermal flowmeter principle

The mechanism creating the temperature difference is the mass transport of the gas in contact with the heated element by virtue of its heat capacity. For a given amount of heat applied to the gas, the temperature change is a function of mass flow and heat capacity [11].

Thermal mass flow sensors are usually calibrated with dry synthetic air or Nitrogen. The heat capacity of CO_2 and H_2O (vapor) is quite different from that of the typical calibration gas. This results in an inaccuracy of the reading when diluted or undiluted exhaust gas is measured. If the composition is known, a correction can be computed by means K-factors in the following manner:

$$Q\ gas\ x = Q\ read \cdot (K\ gas\ x / K\ calibration\ gas) \qquad (7)$$

Published K-factors for pure gases are empirically determined by the meter manufacturer or calculated using the physical properties of the gases. An example calculation method to determine K-factors is:

$$K_{gas\,x} = N_{gas\,x} / (C_{P\,gas\,x} \cdot \rho_{gas\,x}) \qquad (8)$$

To predict the K-factor of a mixture that represents typical engine exhaust the following formula can be used.

$$K_{exhaust} = \frac{1}{\left(\frac{c_{H2O}}{K_{H2O}} + \frac{c_{CO2}}{K_{CO2}} + \frac{c_{N2}}{K_{N2}} + \frac{(1 - c_{H2O} - c_{CO2} - c_{N2})}{K_{O2}}\right)} \qquad (9)$$

Thus, the effect can be quantified once the amount of CO_2 and H_2O in the exhaust is determined. The resulting mass flow error can be as high as negative 5% in some operating modes of a diesel engine.

Comparison of measured data to theoretical values - In order to verify the usefulness of the K-factor corrections, raw exhaust CO_2 and diluted CO_2 measurements were made in parallel with the collection of the particulate data for a number of engine operating conditions. The actual dilution ratios as measured by CO_2 tracing were compared to the requested dilution ratios, as corrected by K-factors (H_2O values were implied from the H/C ratio for the fuel). The data in Figure 17 show good agreement between predicted and measured sampling errors, well within the specified flow accuracy of the instrument.

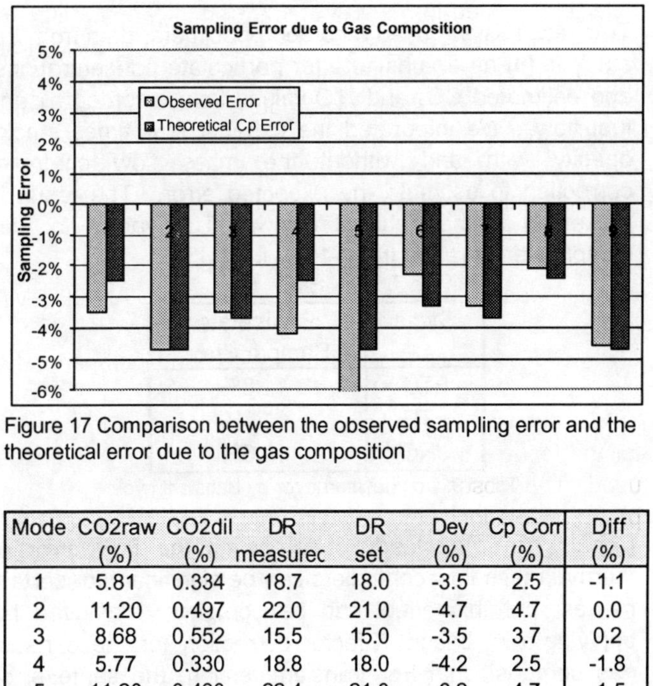

Figure 17 Comparison between the observed sampling error and the theoretical error due to the gas composition

Mode	CO2raw (%)	CO2dil (%)	DR measurec	DR set	Dev (%)	Cp Corr (%)	Diff (%)
1	5.81	0.334	18.7	18.0	-3.5	2.5	-1.1
2	11.20	0.497	22.0	21.0	-4.7	4.7	0.0
3	8.68	0.552	15.5	15.0	-3.5	3.7	0.2
4	5.77	0.330	18.8	18.0	-4.2	2.5	-1.8
5	11.20	0.490	22.4	21.0	-6.2	4.7	-1.5
6	7.67	0.388	20.5	20.0	-2.3	3.3	1.0
7	8.70	0.555	15.5	15.0	-3.3	3.7	0.3
8	5.76	0.336	18.4	18.0	-2.1	2.4	0.4
9	11.21	0.498	22.0	21.0	-4.6	4.7	0.1

Table 2: Gas composition and induced MFC error by engine mode

It can be seen from Table 2 that the deviation from requested dilution ratio depends on the engine operating mode (heavier loads have higher CO_2 content in the exhaust). The last column also illustrates that the correction factor is well able to account for the difference between the requested and actual dilution ratios.

When direct measurements of CO_2 concentrations are not available, CO_2 and H_2O are calculated based on measured air and fuel flow, and the chemical composition of the fuel. This agrees well with the CO_2 online measurement.

Considerations for transient modes - During transient operation average CO_2 and H_2O concentrations are relatively low and one might expect flow errors in the range of negative 2...2.5%. However, due to the highly dynamic characteristics of particulate concentration in a transient test (see Figure 18 Opacimeter and exhaust flow data, in the next section) the effect is actually smaller.

Most engines have by far more particulate emissions during the short acceleration phases than during the rest of the cycle. An acceleration usually follows either a deceleration or a low power phase, both characterized by a high air-fuel ratio and therefore low CO_2 and H_2O concentration. This means that while sampling the most significant particulate emissions of the cycle the flow error is relatively low.

To check this, we used recorded opacimeter data from the study in [6] as a substitute for particulate concentrations, and estimated CO_2 and H_2O values from recorded air and fuel flow. We integrated the exhaust flow times smoke opacity with and without the mass flow controller correction to estimate the expected error. The result is presented in the Table 3 below. The values for this sampling error range from -1.2% to -1.4%.

Cycle	Particulate Sample Error
ETC	-1.38%
FTP	-1.20%
JRC	-1.36%

Table 3: Composition error by transient cycle

Conclusion - In steady state testing the flow error of thermal mass flow controllers can be significant for certain modes, but this effect can be completely removed by applying a C_p based *K*-factor correction for the exhaust gas composition. For transient testing, the same error mechanisms apply, but are much less significant than the other sources of sample error.

TIME ALIGNMENT - Partial flow systems collect the correct sample by keeping their sample flow in direct proportion to the engine's exhaust flow. That is, they compute a pneumatic integral of the product of two highly dynamic signals: the exhaust flow and the particulate concentration. In order that this integral be accurate, these two signals must represent the same moment in time when they are multiplied. We turn attention next to errors that are caused when this is not the case.

Sensitivity of particulate sampling to time shift – On th initial transient tests done on US engines in the EPA/ARE programs, partial flow systems measured less than fu flow systems, even when the error sources mentione above had been addressed. For a closer analysis of th conditions during transient cycles, we once agai investigated this using the opacimeter data.

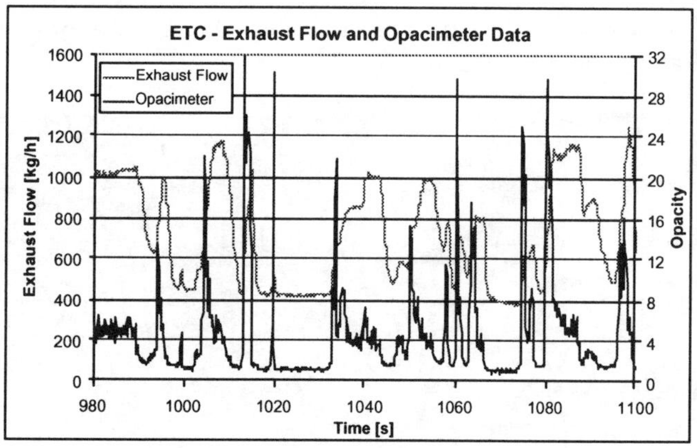

Figure 18 Opacimeter and exhaust flow data

The particulate concentration on the off-road engine testec varied in a highly dynamic manner, as seen in Figure 18. I under these circumstances the probe flow of a partial flow system is following the exhaust flow with a certain time delay, the particulate emissions will not be samplec proportionally and the collected particulate matter wil typically lead to underestimation of the emissions.

Possible causes for time delays are:

- Response time of the flow controlling circuit
- Response time of the dilution air MFC
- Pneumatic response time of the partial flow sampling system
- Exhaust (Air) flow meter response

With the opacity as an estimation for the particulate concentration, a calculation with different positive and negative delay times for the exhaust flow data was performed. In Figure 19 the results of the calculation as well as experimentally obtained data are displayed for 3 different transient cycles.

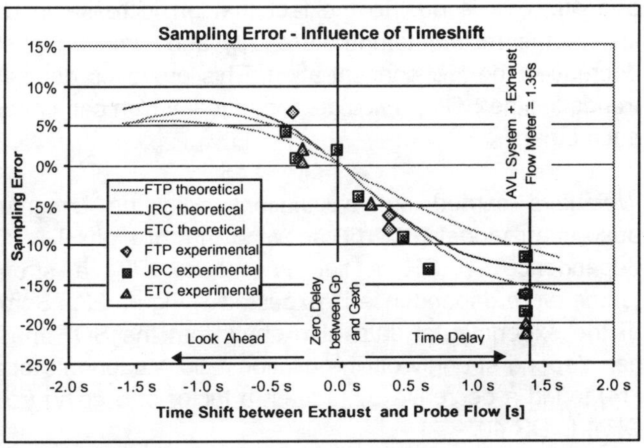

Figure 19 Calculated sampling error based on measured opacimeter and exhaust flow data

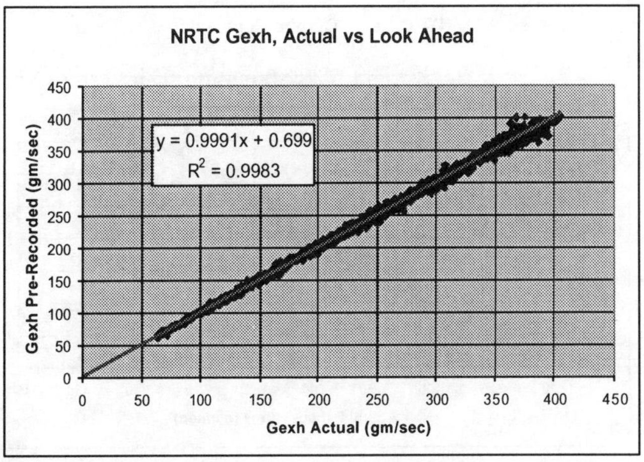

Figure 20: Regression of actual exhaust flow against look ahead exhaust flow.

"Look Ahead" controlling – The data presented in Figure 19 indicate that in order to maintain a particulate sampling accuracy of ±5%, the probe flow of the partial flow sampling system has to follow the exhaust flow within a delay time of 300ms or less. Given the time delays inherent in intake air flow meters, fuel flow meters, data acquisition systems and flow control devices, this quick a response is impractical to achieve. Furthermore, the pneumatic delays inherent in quickly changing flows of a compressible fluid in tubes and tunnels cannot be completely eliminated. Finally, even a substantial improvement could not reduce the error from this source to zero.

Comparisons of recorded transient exhaust flow data from test to test show an excellent repeatability for today's engine test beds. The regression analyses required to validate control performance on certification tests have R^2 values typically above 98%. This makes the use of a method to compensate for system time delays, the so-called "Look Ahead" controlling, quite feasible. "Look Ahead" uses a prerecorded exhaust flow signal to control the partial flow sampling system rather than the online signal. For the emission calculations the actual on-line flow is used. The prerecorded signal is time shifted in a way to compensate for all the cumulative time delays. An example of a typical regression of a pre-recorded look-ahead exhaust flow against an actual exhaust flow is shown in Figure 20. It shows the same high regression coefficients observed between un-normalized FTP drive cycle speed and torque against actual test speed and torque.

To obtain the correct time delay value for look-ahead controlling, the response of the partial flow sampling system to a step change of the controlling signal (exhaust flow signal) is determined by means of an measurement of the flow at the probe inlet. The t_{60} or transformation time was found to characterize this time delay most accurately (Figure 21).

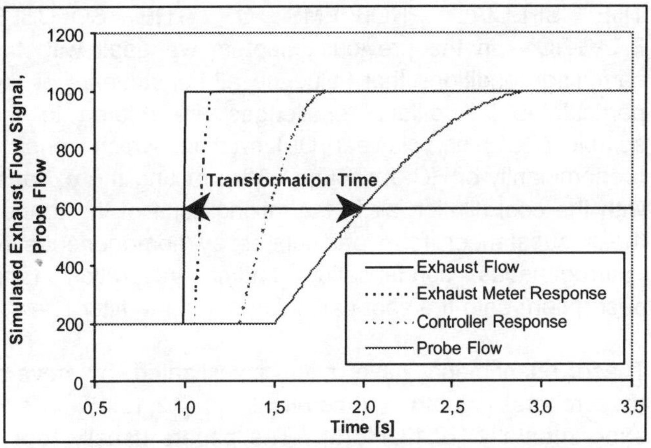

Figure 21: Determination of the probe flow transformation time

When the probe flow transformation time plus the t_{60} time of the exhaust flow meter are used as "Look Ahead" time, the sampling error can be completely removed, as the experimental results in Figure 19 illustrate. That the method maintains the actual probe flow correctly in proportion to the actual on-line exhaust flow is illustrated in Figure 22, where the regression of these two signals has an R^2 of .985.

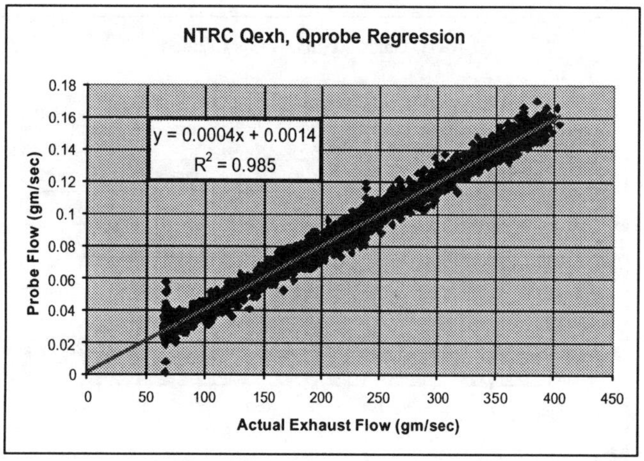

Figure 22: Regression of actal exhaust flow and actual probe flow with look-ahead controlling.

Conclusion – With "Look Ahead" controlling it is possible to compensate for inherent time delays and measure particulate emissions during transient cycles accurately. To insure the data quality, the actual exhaust flow data recorded with the test are compared, using appropriate time delays, to the actual probe flow. A high R^2 value for the regression of these signals insures that the probe flow was proportional to the exhaust flow with the desired *r* factor, creating the "downsized CVS".

THE SPECIAL PROBLEMS OF THE SOLUBLE FRACTION- In the previous chapters we dealt with the sampling conditions that influence all constituents of the particulates. Additional challenges are related to the soluble (SOL) or volatile (VOL) fraction, which consists predominantly of HC and SO_4^{--}. After mixing the exhaust with the cool dilution air in the mixing zone of the tunnel, these substances form particulates by homogeneous or heterogeneous condensation and/or adsorption (and desorption) onto the soot particulates and the filter fibers.

These phenomena have been investigated by several researchers, both theoretically [12,13,15] and experimentally [12,13,14,15]. The papers usually report good correlation between theory and experiment. But the theoretical predictions differ substantially between the papers. This indicates that the SOL-forming emissions from different engines cannot be considered chemically equal. As a consequence the adjustable parameters used to make the theoretical model agree to the experiments vary from one experiment to the other, making quantitative predictions very difficult.

However, some qualitative model results seem to enable a procedure for optimizing the tunnel parameter adjustment. Plee et al. [13] and Reichel et al. [15] claim that the fraction F_{ads} of SOF relative to the total HC in the exhaust is practically constant for dilution ratios (DR) between 6 and 40. This is due to the fact that an increase in DR lowers the mixing temperature T_{mix} and simultaneously decreases the HC concentration. This provokes opposite trends for the SOL particulate formation, which can cancel each other.

We have carried out calculations with the Langmuir adsorption model [13, 15], showing that at high T_{mix} the dependence on DR is flat, but at low T_{mix} a strong temperature dependence is expected – Figure 23. Some of the experimental data also show that the SOF mass can depend strongly on the dilution ratio – Lapuento et.al [14] found a decrease by at least a factor of 5 going from DR 4 to DR 40.

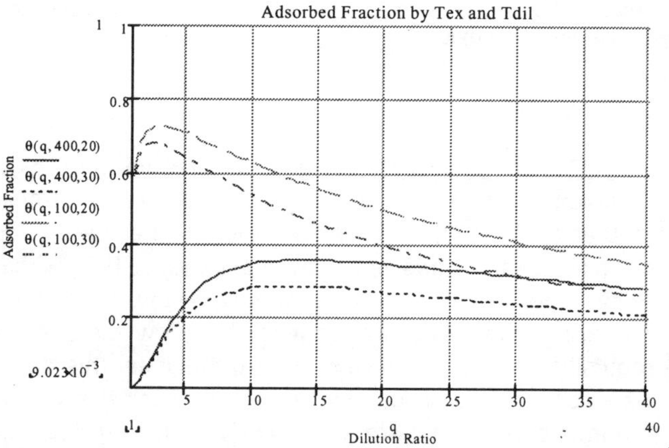

Figure 23: Calculation of the dependence of F_{ads} on DR for different mixing temperatures Tmix

In the correlation tests from [6], we found a marked dependence on the dilution conditions for low-load conditions, where SOF content is very high, especially for Mode C (rated speed 10% load) with a SOF content of ≥90%. Results of 3 configurations are given in Table 4. Each result shown is the mean of 3 tests

Run #		DR	Tmix	Tfil	PM (g/hr)	Diff %
104	CVS	29.0	31	34.5	15.67	
	SPC	6.5	48	45.0	10.21	- 36
152	CVS	29.0	31	32.8	17.14	
	SPC	16.0	30	31.0	14.38	- 16
155	CVS	7.0	48	45.8	12.81	
	SPC	6.5	46	38.0	13.10	+2.3

Table 4: Result of DR variation of CVS and SPC for mode C (2000 rpm, 10% load)

In the test runs at number 155, the low CVS dilution ratio could only be obtained by turning off secondary dilution. From the table it seems clear that the partial flow and full flow measurements are the same when the dilution ratio and mixing temperatures are the same.

Another tests series, which incorporated the comparison of a short and a long residence time tunnel, revealed that matching the residence time brought both systems to the closest agreement.

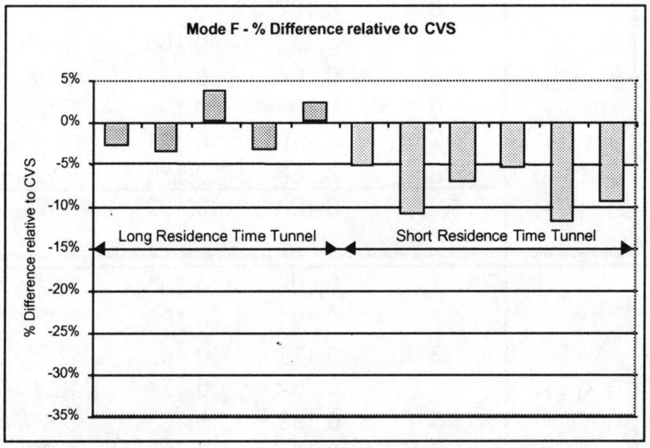

Figure 24: Effect of longer residence time (. 0.5 sec) partial flow tunnel

It appears that only equivalent dilution and temperature conditions can guarantee equivalency of the SOF results. At the levels of accuracy and repeatability needed today, such measures as introduced in the CFR for 2007, requiring narrower filter temperature limits, are clearly justified. In addition, since the DR independence of the PM result seen in the past does not appear to hold for cool high SOF modes, some of the unexplained variability among different labs is more understandable. Although this may not be important for many modern engines, where SOF content is low, it must be considered for future engines equipped with PM traps. These engines will not only have low PM emissions, but most of it is expected to be SOF.

TEST RESULTS

APPLICATIONS – The measures described above were implemented on the system we provided for the partial to full flow PM correlation study sponsored by EPA, ARB EMA and reported in [6]:

- Optimized design for transfer tube and tunnel
- Forward facing probe
- CO_2/H_2O Correction
- Look-Ahead Control
- Parameter matching
  - Dilution ratios
  - Filter Temperature
  - Face Velocity

STEADY STATE RESULTS – The test program included a subset of the off-road HD ISO 8-Mode test. The modes were:

A	Rated engine speed, 100% load
B	Rated engine speed, 50% load
C	Rated engine speed, 10% load
D	Maximum torque engine speed, 100% load
E	Maximum torque engine speed, 50% load
F	Idle

Modes A and D are high temperature modes with a low soluble fraction. For modes C and F Filter analysis showed nearly 100% SOF content.

The steady state results in Figure 25 and Figure 26 show good agreement for modes A and D, while the deviation of the other modes is due to the differences of the soluble fraction. On modes C and F the result was improved significantly by matching the dilution ratio and mix temperature.

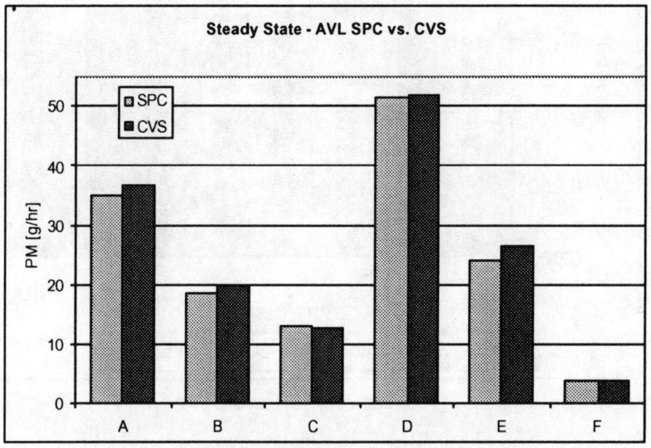

Figure 25 PFD results – consecutive tests (*C and F: dilution ratio of CVS and PFD matched)

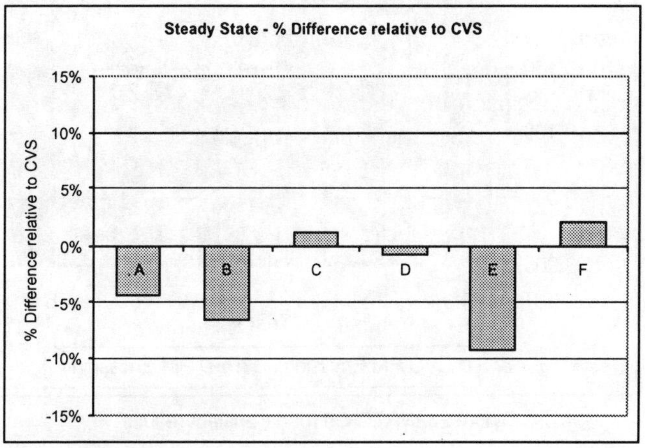

Figure 26 SPC-PFD results – consecutive tests (*C and F: dilution ratio of CVS and PFD matched)

TRANSIENT RESULTS – A series of tests were run using three different transient driving cycles, the US FTP, the European transient cycle ETC, and a proposed off-road cycle, the NRTC. "Look Ahead" controlling was applied, using the same transformation (t_{50}) time, determined by step function response time as described above.

These results are displayed below in Table 5 and Figure 27, Figure 28, and Figure 29. The overall agreement is 5% or better, which is very much in line with the expected small remainig offsets due to residual deposition losses (about 2%) and the as yet uncorrected gas composition error (about 1.5%). The variability of the partial flow system is about the same or slightly better than the full flow system.

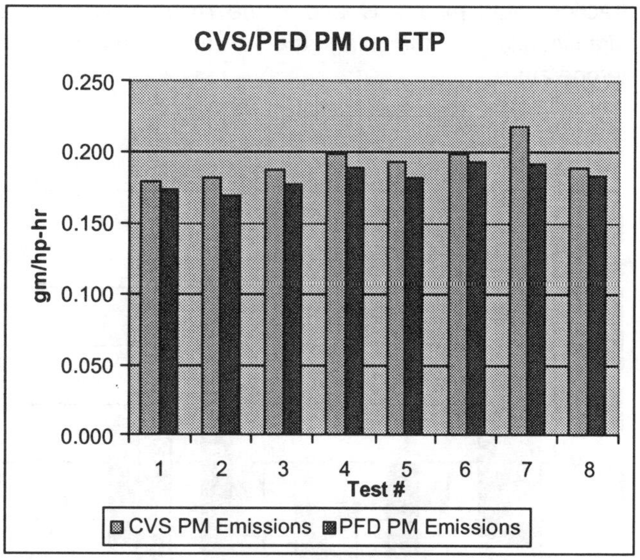

Figure 27: FTP Results

Cycle	Test Number	CVS PM gm/hp-hr	SPC gm/hp-hr	%
FTP	1	0.179	0.174	3.0%
	2	0.182	0.170	6.7%
	3	0.187	0.177	5.6%
	4	0.198	0.188	4.7%
	5	0.194	0.182	6.0%
	6	0.199	0.194	2.9%
	7	0.218	0.191	12.1%
	8	0.188	0.183	2.8%
	Ave	0.193	0.182	5.6%
	COV	6%	5%	
NRTC	1	0.199	0.201	-0.9%
	2	0.191	0.183	3.9%
	3	0.178	0.182	-2.0%
	4	0.202	0.177	12.1%
	5	0.189	0.183	3.6%
	6	0.194	0.190	2.0%
	Ave	0.192	0.186	3.2%
	COV	4%	4%	
ETC	1	0.250	0.255	-2.0%
	2	0.247	0.235	4.8%
	3	0.251	0.252	-0.3%
	4	0.280	0.267	4.8%
	5	0.280	0.262	6.3%
	6	0.287	0.274	4.5%
	7	0.251	0.249	1.1%
	Ave	0.264	0.256	2.9%
	Std	0.018	0.013	
	COV	7%	5%	

Table 5 Tabular results of transient tests

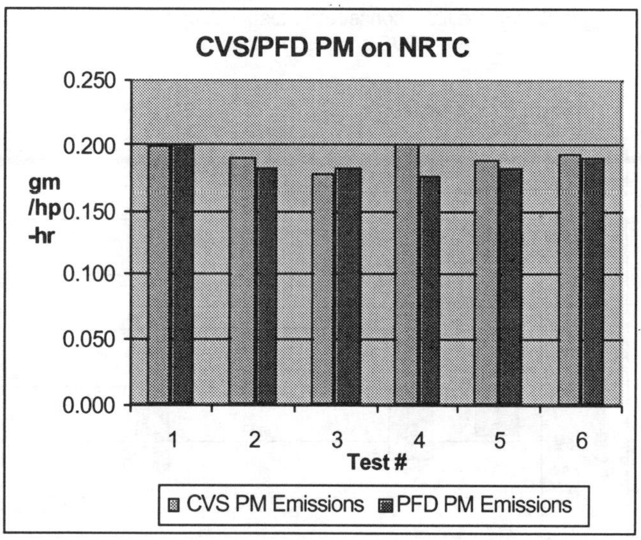

Figure 28: NRTC Off road transient results

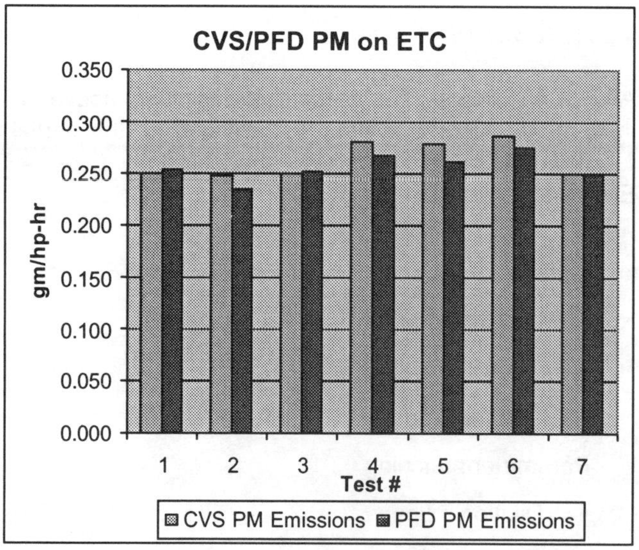

Figure 29: ETC transient results

CONCLUSION

In summary, we've applied the following measures to achieve a close agreement of values obtained from full and partial flow sampling systems:

- Deposition losses from diffusion and thermophoresis are greater in a partial flow system because of the larger surface to volume ratio, low flow rates and long sample tubes. These losses are reduced to low levels by using a short, thin walled transfer tube and insulating it right up to the mixing zone. Eliminating re-circulation zones in the mixing tunnel and the rest of the sample flow path prevents deposition losses at those locations.

- Sampling errors arising from the need to sample directly in the exhaust stack are minimized by using a forward facing probe, matching stack diameter and probe size to keep the ratio of stack to probe velocity lower, and using particle classifiers to avoid re-entrainment artifacts.

- The influence of the changing composition of exhaust gas on the accuracy of flow metering is correctable, but not significant for transient tests. For steady state tests, the effects are eliminated by applying a straightforward physics based on the heat capacity of H_2O, CO_2 and N_2, and calculations of actual exhaust composition from available measurement of fuel and air flow.

- For transient testing the timing of the proportioning of sample flow to the exhaust flow is critical for the proper function of the pneumatic integration. Because control elements can't be placed in the flow path between the sample point and the filters, an indirect flow proportioning is necessary. This method is intolerant of the delays inherent in the flow measurement and control systems of any sampling device. The problem is resolved by means of a look ahead control algorithm, using pre-recorded flow control signals, and statistical regression checks on the proportionality of the probe and exhaust flows.

- While the above measures assure correlation between the two systems for the insoluble fraction collected, the modes with high soluble fraction are the cooler ones for which the collected volatile particulate matter varies more strongly than expected with dilution ratio, temperature, and residence time. For these engine operating modes, when full and partial flow systems are set and controlled to the same conditions, they get the same result.

Finally, the results from the transient tests are exactly in line with the expectations from the theory and practice outlined above. It is expected that tests in the near future using engines and CVS units qualified for 2007 emissions levels will provide the needed data for the statistical paired t-tests required to demonstrate the equivalence of an alternative method by the CFR.

As a consequence, the partial flow system can now provide to industry and government a more repeatable and cost effective alternative for the measurement of PM in the exhaust from diesel engines.

ACKNOWLEDGMENTS

The authors are grateful for the opportunity to have participated with the ISO TC 22/SC 5/WG 2 committee members on the definition international test procedure ISO 16183, which is the driving force behind these advances in partial flow sampling. We are especially appreciative to our colleagues at the EPA, the ARB and the EMA, at Southwest Research Institute, and as well to those at our competitors Sierra Instruments and Horiba Instruments whose professionalism found the right mix of competitiveness and constructive cooperation to advance this technology to the benefit of our industry and our air.

CONTACTS

From AVL North America:

William M. Silvis
(734)414-9583
silvis@avlna.com

Gerald Marek
(734)414-9645
gerald.marek@avlna.com

Dr. Norbert Kreft
(734)354-3721
norbert.kreft@avlna.com

From AVL Graz Austria

Dr. Wolfgang Schindler
0043 316 787 535
wolfgang.schindler@avl.com

REFERENCES

---

[1] Counsil Directive 91/542/EEC, Oct. 1991

[2] Mollenhauer,K et al. 'Entwicklung eines Verfahrens zur Messung der Partikelemission von NFZ Motoren im Rahmen der ECE R49", Abschlußbericht FVV Vorhaben Nr. 379, FVV Fortschrittsberichte Verbrennungskraftmaschinen Heft 466, 1990

[3] Stein,H.J. Ekermo,A.L and Treiber,P.H.J „Emissions Correlations of Heavy-Duty Transient Test facilities", SAE 892492

[4] Schweizer,Th. and Stein,H.J., „A New Approach to particulate Measurement on Transient Test Cycles: Partial Flow Dilution as Alternative to CVS Full Flow Systems" SAE 2000-01-1134

[5] Stein,J, Informal Document No 3 for the 41st GRPE session, "Worldwide Harmonized Heavy Duty Emissions Certification Procedure- Exhaust Emissions Measurement. Iso 2nd Interim Report, Jan 2001

[6] I. Khalek, "Performance of Partial Flow Sampling Systems versus Full Flow CVS on PM Emissions Under Steady State and Transient Engine Operation", Report on results of EPA/ARB/EMA Study, December 6, 2001

[7] C. Berger, H: Horvath and W.Schindler "The Deposition of Soot Particles From Hot Gas Streams through Pipes", J.Aerosol Sci. 26 (2) p 211 (1995)

[8] Brooker, D. R. "Validation of the apparatus used for vehicle particulate emission measurement", Aerosol Science Centre, AEA Technology, Didcot, UK. Report C524/171/97

[9] Hinds, W. C., Aerosol Technology, Chapter 10, 2nd Edition, John Wiley & Sons, 1999

[10] J.E. Hardy, J.O. Hylton, T.E. McKnight: Empirical Correlations for Thermal Flowmeters Covering a Wide Range of Thermal-Physical Properties.

[11] N. Urdaneta, J. Krell, B. Brown: Thermal-based mass flow control for SDS gas delivery systems; Solid State Technology April, 1999

[12] Kittelson, D.B and Johnson,J.H., „Variabilty in Particulate Emission Measurement in the Heavy-Duty Transient Test" SAE 910738,

[13] Plee,S.L. and McDonald,J.S, „Some Mechanisms affecting the Mass of Diesel Exhaust Particulate Collected Folling a Dilution Process" SAE 800186

[14] Lapuerta, M., Armas, O., Ballesteros, R. and Duran, A. „Influence of Mini-Tunnel Operating Parameters and Ambient Conditions on Diesel Particulate Measurement Analysis" SAE 1999-01-3531

[15] Reichel, St., Pischinger, F. F., Lepperhof, G., "Influence on Particles in Diluted Diesel Engine Exhaust Gas", SAE 831333

2002-01-0055

Electrical Filter Stage for the ELPI

Marko Marjamäki, Leonidas Ntziachristos, Annele Virtanen, Jyrki Ristimäki and Jorma Keskinen
Tampere University of Technology. Institute of Physics

Mikko Moisio and Marko Palonen
Dekati Ltd.

Maija Lappi
VTT Energy, Engine Technology

ABSTRACT

Electrical low pressure impactor ELPI was modified to measure particles below 30 nanometers in aerodynamic diameter. This was accomplished by adding a filter stage to collect and measure nanoparticles. The charging unit of the instrument was modified to increase the charging efficiency of the smallest, nanometer sized, particles. The modified charging unit was calibrated and the new construction of the ELPI was tested in laboratory and in vehicle dynamometer test cell. Measurements performed in the engine test cell showed that modifications improve the size range and measurement capability of the ELPI for engine emissions.

INTRODUCTION

Electrical Low Pressure Impactor, ELPI (manufactured by Dekati Ltd., Tampere, Finland), measures particle size distribution in real-time combining a cascade low pressure impactor with a corona charger and electrical detection [4, 8]. ELPI is used in many engine laboratories to measure emitted particle size distributions and number concentrations [e.g. 1,7,9]. The nominal ELPI measurement size range is from 30 nm to 10 μm aerodynamic diameter with 12 stages. The lower limit of the measurement range is governed by the impactor physical characteristics and practical low pressure limitations. Modern engines emit a significant number of particles below this size [e.g. 5]. Therefore the extension of the instrument range below this size limit would be advantageous. In order to lower the detected particle size range without having to redesign the impactor, other solutions have been sought. In this paper a filter stage enclosed in an isolated Faraday's cage with electrical current measurement is added to the ELPI system downstream of the impactor to improve the ELPI measurement range. Main advantage of the construction is that the existing electrometer can be used for the detection of the particles and a minimal unit conversion or recalibration work is required.

IMPACTOR MODIFICATION

The easiest way to increase the size limit of the impactor is to add additional stages in the impactor. Instead of a new impactor stage, a filter stage with electrical detection was developed and built. In this construction there is no need for a new more powerful pump. The filter stage was constructed into the stage housing similar to the impactor stages used in ELPI. Required pressure lines to monitor the pressure downstream of the last impactor stage and a filter grid support were constructed to hold a 47mm standard filter. The pressure measurement system inside ELPI unit was modified to enable reversion to original configuration. Since the addition of the filter stage increases the overall length of the construction, stage 12 was removed to fit the impactor and filter stage into the instrument (Figure 1). Removal of the stage 12 increases the size range of stage 11. However adding a filter stage downstream of the impactor does not change the operation of the impactor in any way and therefore no additional calibration for the impactor is needed. Ideally the filter stage only collects particles that are smaller than the cut size (30nm) of the last impactor stage.

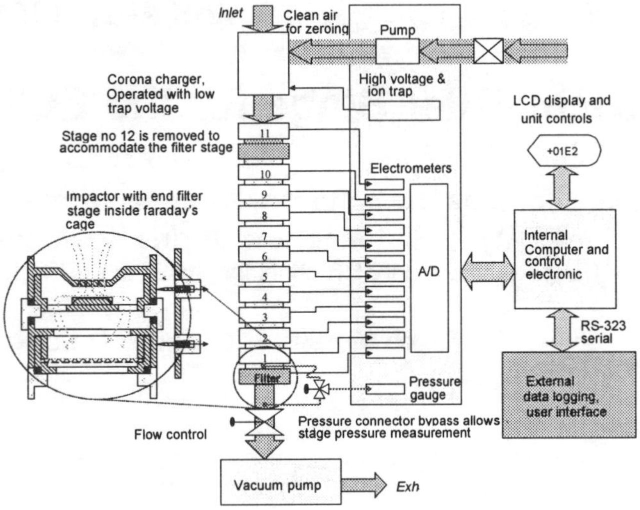

Figure 1. Schematics of ELPI unit with additional filter stage.

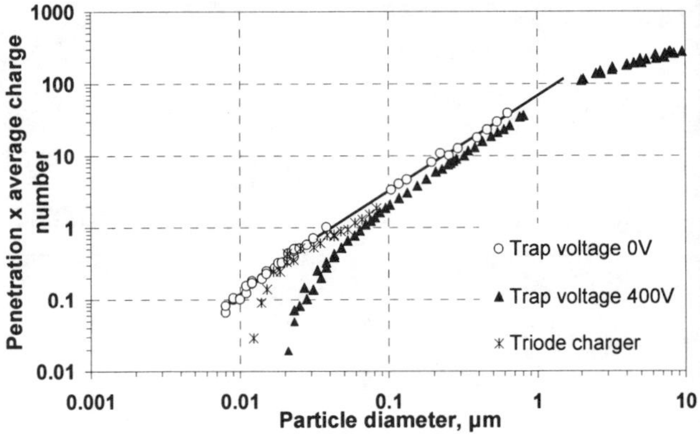

Figure 2. Charging efficiency with different configurations.

CHARGER MODIFICATION

In order to be able to measure particles below 30 nm ELPI charger has to be modified. The ELPI charger is positive corona type diode charger incorporating radial electrical field downstream of the charging zone. The trap field was originally designed to remove the ions and the smallest charged particles from the sample flow, so that they could not interfere with the measurement. This trap field acts as a zero order electrical precipitator with a nominal cut size of 20 nanometers. In order to operate in the sub-30 nm range – and hence achieve an improved charging efficiency - the trap field is switched off. Using a 10-lpm flow rate no ion escape was detected and even for higher flow rates, i.e. 30 lpm version, a weak electrical field in the order of 5 V/cm can be used to ensure sufficient ion removal without increasing significantly ultrafine particle losses. Additionally, a prototype triode charger similar to EAA [6] without the sheath air flow was constructed and tested. This construction offers better defined charging conditions, field charging is negligible due to absence of high electric field. In triode charger the ion concentration in the charging zone is lower than in the diode charger but the losses inside the charger are also lower. The charger efficiency at different configurations was measured using the method described by Marjamäki et al. [8]. Comparison between the charging efficiency of the original and modified configuration can be seen in the figure 2.

Figure 2 shows that the charging efficiency with the triode charger is not much improved over the diode one while its construction is not as simple as diode charger's construction. Therefore the triode charger was rejected for this work. The triode charger can be made more efficient by using sheath air flow or using alternating voltage in the triode grid, but these constructions would need major modifications in the current ELPI unit and therefore they were not used in this work. Instead the diode charger without the trap voltage was chosen to be used. This construction offers high charging efficiency with minimal conversion work. Charging efficiency increase for the small particles is substantial. For 20 nm particles increase is tenfold. The lower size limit for the charger was estimated to be 7 nm where the charging efficiency is only a few percent because of the low charging probability and the expected losses of charged particles in the high voltage charging zone.

When converting the measured electric currents to number concentrations, the cut diameter concept is used [3,4]. In this case each channel is assumed to collect particles between the cut size of the stage corresponding to the channel and the cut size of the stage before it, for example stage 1 is assumed to collect particles between 30 nm and 60 nm. The geometric mean size for the channel is calculated from the upper and lower cut sizes of the stage. The measured electrical currents are converted to number distribution by dividing the measured electric currents by the charging efficiency of the charger for the mean sized particle. As described in Marjamäki et al. [8], charging efficiency can be characterized by the product of penetration and the average charge, *Pn*. Following fitted functions of *Pn* can be used to convert the measured current signals to particle number concentrations.

$$Pn = 222.49 D_p^{1.637}, D_p \le 0.023 \mu m$$

$$Pn = 68.12 D_p^{1.32}, D_p > 0.023 \mu m$$

LABORATORY TESTS

The modified construction was evaluated in the laboratory using well defined aerosol. Polydisperse DOS (dioctyl sebacate) aerosol distributions with varying shape and mean diameters were generated with condensation/evaporation aerosol generator. Generated aerosol was further diluted and neutralized. Resulting aerosol distributions were measured simultaneously with the modified ELPI and SMPS (model 3934,TSI Inc., St. Paul, MN) system. Two cases are shown in the Figure 3, cases with and without ultrafine particle mode.

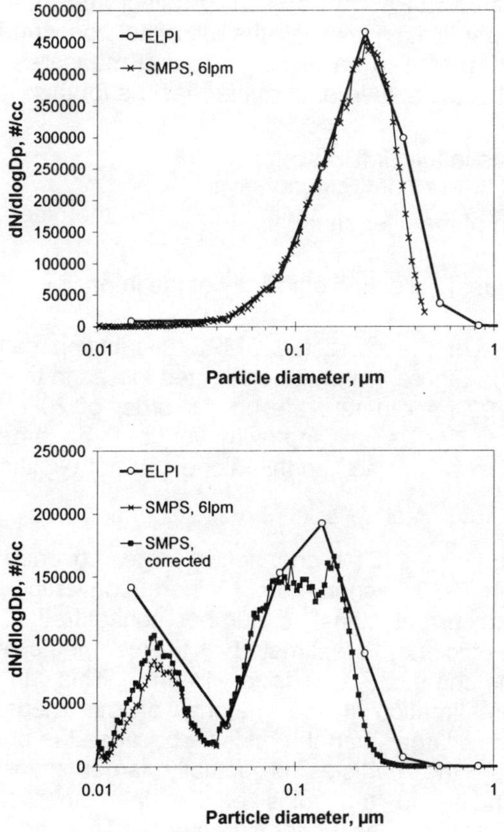

Figure 3. Comparisons between ELPI and SMPS. Dioctyl-sebacate aerosol.

ELPI measures aerodynamic diameter and SMPS measures mobility equivalent diameter. For comparison purposes measured ELPI diameters were converted to mobility equivalent diameters using the bulk density of the DOS (0.918 g/cm^3) and assuming that the particles are spherical. A corrected SMPS distribution is also shown in figure. This is because SMPS-software does not automatically compensate the losses. For the used DMA type, TSI long DMA (TSI 3071A), the losses were estimated according to paper by Birmili et al. [2]. The flows used in that study were not exactly the same as the flows in the paper, but they were very close and therefore a correction estimate can be attempted. Results show that generally the instruments agree. When there was no distinct ultrafine mode present, distributions agree very well. However when an ultrafine mode is present, the distributions generally agree but ELPI results indicate slightly higher ultrafine particle concentration than the SMPS.

ENGINE- AND VEHICLE DYNAMOMETER TESTS

In addition to laboratory tests, the modified ELPI was also tested in the VTT engine laboratory. Figure 4. shows a case where a large amount of ultrafine particles is present. These ultrafine particles are believed to result from nucleation of gaseous components in the exhaust. ELPI results are presented in aerodynamic diameter and SMPS results are presented in mobility equivalent diameter.

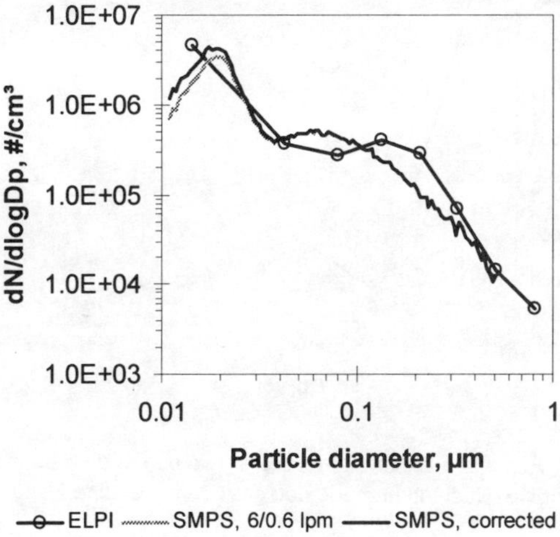

Figure 4. Heavy-duty diesel engine in engine test bench. Comparison between ELPI and SMPS. Measurements taken from the mini dilution tunnel, dilution ratio of 15.5.

Figure 5. shows a different case when there is practically no ultrafine particles present.

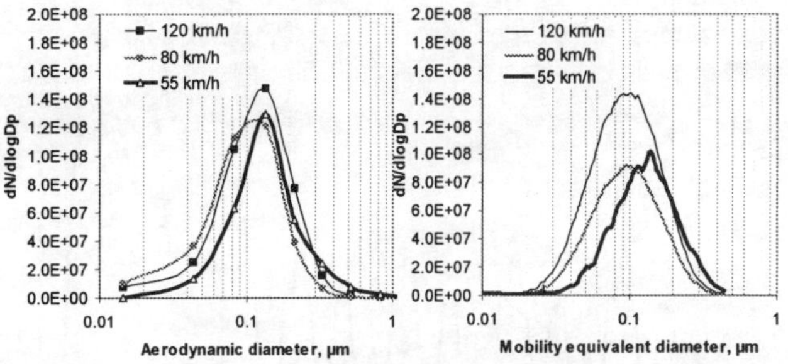

Figure 5. Comparisons between ELPI (left) and SMPS (right). Light duty diesel vehicle on vehicle dynamometer. Measurements taken from the dilution tunnel.

The same tendency as in laboratory measurements can be seen in these results. Agreement between the ELPI and SMPS distribution is good, although concentration estimated by the filter stage seems to be higher than in the SMPS. It has to be remembered that the size indicated by SMPS is mobility equivalent particle size and for ELPI the particle size is aerodynamic diameter.

The signal from the filter stage can be used to detect nucleation. Figure 6. presents measurements taken from the exhaust of small scale diesel powered electric generator. When idling engine produces less soot particles but small particles are formed by nucleation during the exhaust dilution. When in high power, nucleation does not occur.

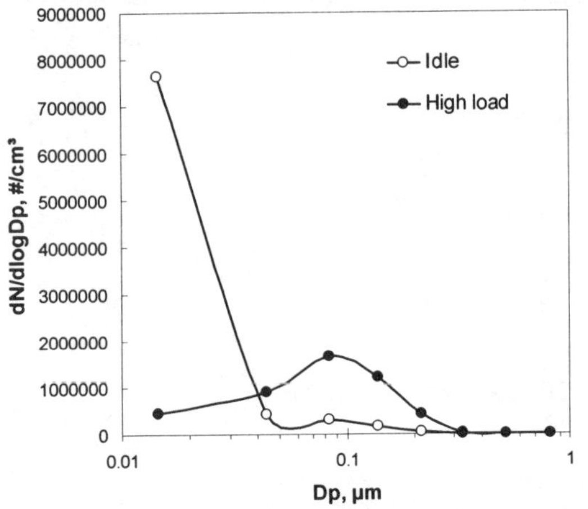

Figure 6. Exhaust number distribution in two operation modes of small diesel engine. Nucleation mode detected during engine idling.

Construction was tested also on a transient driving cycle (fig. 7).

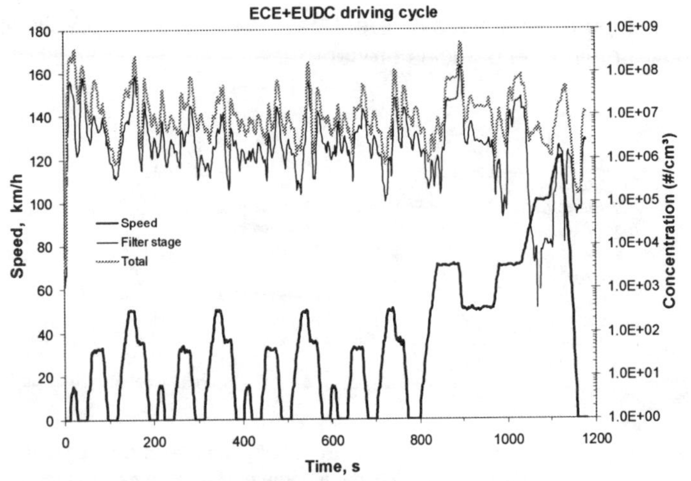

Figure 7. European transient driving cycle with GDI vehicle. The tested vehicle was Mitsubishi Carisma GDI. Measurements taken from the dilution tunnel.

Measurements taken during the transient New European Driving Cycle (NEDC) show that the modified ELPI is capable of real-time measurement and the filter stage brings additional information about the ultrafine particle sizes. In the case of the tested GDI vehicle ultrafine particle concentration generally follows the total particle concentration except in the high speed phase where concentration of the ultrafines decreases.

DISCUSSION

The modified ELPI can be used to detect and measure ultrafine particles. However the ultrafine concentration measured seems to be higher than that measured with SMPS. There are several possible reasons for this:

- Losses in the SMPS system
- Width of the filter channel
- Non-ideal collection efficiency of the impactor

The effect of the corrected SMPS distribution can be seen in the Figure 3 and the estimated losses in the size range of 20 nanometers are on the order of 20%. The used correction is an approximation but the order is expected to be correct, so this alone can not explain the difference.

The width of the filter channel is large, covering the range of 7-30 nanometers. When converting the measured current signal to number concentration, the charging efficiency is estimated at the geometric mean size of the channel (i.e. midsize ~14.5nm). This can lead into overestimation if the majority of the measured particles are larger than the midsize because the charge carried by the particles is actually larger than that corresponding to the midsize. On the other hand, concentration can be underestimated when majority of the particles is smaller than the midsize.

When calculating ELPI results the collection efficiency of the impactor is assumed to be ideal, i.e. all particles smaller than the cut size penetrate the stage and all particles larger than the cut size are collected. In practice some of the particles larger than the cut size are not collected and they will carry their charge to a lower stage or on the filter. These larger particles have a higher charge – compared to what normally would be expected for particles collected on the filter stage - and can distort the distribution especially in the case of a wide filter stage.

The reason for the concentration discrepancy is believed to be the combination of the wide size range of the filter stage, i.e. limited resolution, and non-ideal collection efficiency of the impactor. There is a similar problem when using diffusion batteries, where the cut-diameter concept cannot be used when interpreting the data, owing to shallow collection efficiency curves. Further investigations are needed to show whether this discrepancy can be corrected easily by modifying the midsize of the channel taking into account the kernel functions of the impactor stages, or if an improved method or inversion scheme is needed for determining the concentration in the filter stage.

CONCLUSION

The modified ELPI was constructed and tested in laboratory, engine test cell and vehicle dynamometer both in steady state and transient tests. The design criteria of easy conversion and minimal recalibration work to the ELPI units were enforced. A recalibration of ELPI charger operating without the ion-trap field was conducted. In steady state tests, where scanning mobility particle sizer (SMPS) could be operated, results were compared. Evaluation of the charger showed that the charging efficiency is sufficient for the measurement of small particles down to 7 nanometers. This is considered to be the lower size limit for the filter stage. Measurements performed in the engine test cell showed that the filter stage can be used for real-time detection of nucleation mode which has not been possible without the modification. Filter stage improves the size range and measurement capability of the ELPI for the engine emissions. Some deviations in the size distributions measured with filter stage ELPI and SMPS probably originate from the inappropriate use of the cut-diameter concept for the filter stage measurements. This requires additional investigation of the data reduction algorithm for the filter stage.

ACKNOWLEDGMENTS

We would like to acknowledge Mr. Leo Holma for the innovative construction work.

REFERENCES

1. Ahlvik, P., Ntziachristos, L., Keskinen, J., Virtanen, A. (1998) Real time measurements of diesel particle size distribution with an electrical low pressure impactor. SAE Technical papers 980410.
2. Birmili, W., Stratmann, F., Wiedensohler, A., Covert, D., Russell, L.M., Berg, O. (1997) Determination of differential mobility analyzer transfer functions using identical instruments in series. Aerosol Sci. Technol. 27., 215-223.
3. Cooper, D.W., Guttrich, G.L. (1981) A study of the cut diameter concept for interpreting particle sizing data, Atm. Env., 15, 1699-1707.
4. Keskinen, J., Pietarinen, K. and Lehtimäki, M. (1992) Electrical Low Pressure Impactor, J. Aerosol Sci. 23, 353-360.
5. Kittelson, D. (1998) Engines and nanoparticles: A review. J. Aerosol Sci. 29, 575-588.
6. Liu, B.Y.H. and Pui, D.Y.H. (1975) On the performance of the electrical aerosol analyzer, J. Aerosol Sci. 6., 249-264.
7. Maricq., M.M, Podsiadlik, D.H., Chase, R.E. (1999) Examination of the size-resolved and transient nature of motor vehicle emissions. Environ. Sci. Technol. 33., 1618-1626.
8. Marjamäki, M., Keskinen, J., Chen, D-R. and Pui, D. Y. H. (2000) Performance Evaluation of the Electrical Low-Pressure Impactor (ELPI), J. Aerosol Sci. 31. 249-261.
9. Pattas, K., Kyriakis, N., Samaras, Z., Pistikopoulos, P., Ntziachristos, L. (1998) Effect of DPF on particulate size distribution using an electrical low pressure impactor. SAE Technical papers 980544.

DEFINITIONS, ACRONYMS, ABBREVIATIONS

ELPI: Electrical Low Pressure Impactor

SMPS: Scanning Mobility Particle Sizer

DOS: Dioctyl sebacate

2002-01-0056

Effective Density of Diesel Exhaust Particles as a Function of Size

Annele Virtanen, Jyrki Ristimäki, Marko Marjamäki, Kati Vaaraslahti and Jorma Keskinen
Tampere Univeristy of Technology, Institute of Physics

Maija Lappi
VTT Energy

ABSTRACT

New method to define the particle effective density as a function of particle size has been applied to diesel vehicle exhaust particles. The results show that, the effective density of agglomerated diesel particles decreases as a function of particle size. The density of primary particles varies from 1.1 to 1.2 g/cm^3. Also the effect of used dilution method and fuel type on particle density was studied. The dilution effect seems to have stronger effect on particle effective density and structure than the fuel type.

INTRODUCTION

Diesel exhaust soot particles are formed through collisions of primary soot particles typically sized between 20 and 50 nm. The collision rate, i.e. the degree of agglomeration, depends on particle concentration and surrounding temperatures. The primary particle concentration is affected by completeness of combustion and fuel properties [1], [2]. As a result of this process, agglomerate particles are formed with a varying level of sintering of the primary particles.

Agglomerate particles, in general, have effective densities much lower than could be predicted from their chemical composition. This is because of large amount of void space in their agglomerated structure. Many particle characteristics, such as settling velocities and aerodynamic properties of particles, depend on their effective density. Thus the effective density of particles is an important property when health effects of the particles or their residence time in the atmosphere are evaluated.

Electron microscopy is a powerful technique to study the structure of individual agglomerate particles. However, the technique is off-line, and a large number of particles need to be analyzed. Furthermore, only a two-dimensional image of a three-dimensional object is obtained. Measurement of effective particle density as a function of particle size would offer an indirect way to gain information about particle morphology, and even indications of particle formation mechanism.

The aerodynamic and mobility equivalent diameters of an aerosol particle are connected by effective density [3]. Thus, effective density can be defined by measuring these two diameters. A new method, based on simultaneous distribution measurements by Scanning Mobility Particle Sizer (SMPS) and Electrical Low Pressure Impactor (ELPI) was introduced by Virtanen et al [4], [5]. A numerical fitting method is used to obtain effective density as a function of mobility diameter. In the case of fractal-like agglomerates, also the fractal dimension can be estimated. In this paper this new method is applied to the study of particles in diesel exhaust.

EFFECTIVE DENSITY AND MORPHOLOGY OF AGGLOMERATES

The aerodynamic and mobility equivalent diameters of an aerosol particle are connected by effective density according to equation (1):

$$D_a{}^2 C_c(D_a)\rho_0 = D_b{}^2 C_c(D_b)\frac{[C_c(D_m)]^3 \rho_p}{[C_c(D_b)]^3 \chi^3} \quad (1)$$

$$= D_b{}^2 C_c(D_b)\rho_e$$

Here D_a is the aerodynamic diameter, ρ_0 is the unit density, D_b is the mobility equivalent diameter, and D_m is the mass equivalent diameter of the same particle. The slip correction factor C_c is evaluated separately for each of the equivalent diameters. Effective density ρ_e is affected by the dynamic shape factor χ, bulk density ρ_p, and, through the slip correction factors, also by particle size. The dynamic shape factor χ takes into account both particle shape and porocity. Once the aerodynamic and mobility equivalent sizes of the particles are known, the effective density can readily be calculated using equation (1).

One way to define the structure of agglomerates is to treat them as fractal structures [6], [7]. As the effective density is affected by the structure of particle, it gives also important information about the morphology of particle. If the particles are agglomerates, it is possible to evaluate the fractal dimension of the agglomerate when the scaling of the effective density or the mass and the particle size is known.

There are a number of fractal dimension concepts [8]. Here the fractal dimension d_{fm} is defined through:

$$\rho_e \propto D_b^{\,d_{fm}-3} \qquad (2)$$

The fractal dimension, d_{fm}, defined according to equation 2, may have values from 1 to 3. Fractal dimension is 3 for spherical particles, resulting in a constant, size independent value of effective density. The fractal dimension of agglomerates is less than 3, and therefore effective density decreases as particle size increases. The looser the agglomerate structure, the smaller the fractal dimension is. For simple chain agglomerates the fractal dimension is 1.

EFFECTIVE DENSITY ESTIMATION METHOD

The method to estimate the effective density is based on measuring the aerodynamic and mobility size distributions of the sample aerosol and a consecutive numerical procedure to obtain agreement of the two distributions. In the procedure, size dependent effective density is varied to obtain best fit between the distributions [5].

INSTRUMENTATION – On-line instruments are used for the measurement of the size distributions. Aerodynamic size distribution is measured with Electrical Low Pressure Impactor (ELPI) and mobility distribution by Scanning Mobility Particle Sizer (SMPS).

In ELPI impactor particles are classified into 12 size channels from 30 nm - 10 µm, according to their aerodynamic diameter. Particles are first charged, and then the current carried by the charged particles is measured from each impactor stage by a multi-channel electrometer. [9]

SMPS includes a Differential Mobility Analyzer (DMA) and a Condensation Particle Counter (CPC). DMA classifies particles according to their electrical mobility. The classified particles are then counted with CPC. [10]

FITTING – As described by Virtanen et al. [4], the method is based on fitting of simultaneously measured ELPI and SMPS distributions. The primary output of ELPI measurement is the current distribution as a function of aerodynamic particle size. The SMPS number distribution can be transferred to current distribution corresponding ELPI measurement. First the SMPS number distribution is convoluted with the charger efficiency curve of ELPI. As the charger efficiency is a function of mobility equivalent size, at this point there is no role for the particle effective density (i.e. there is no need to make the size conversion according to equation 5). The result of the convolution is the current response $I'(D_b)$. Next the resulted response function $I'(D_b)$ is convoluted with ELPI kernel functions to imitate the particle collection in impactor. In other words, the SMPS current distribution is separated to size intervals corresponding ELPI stages with the help of the kernel functions. The impactor kernels are functions of aerodynamic diameter. Because the convolution is made now for kernel functions and response function $I'(D_b)$, it is evident that the conversion from aerodynamic particle size to mobility equivalent size is needed. The result of the convolution is the discrete current response $\Delta I'(D_a)$. At this point, the particle effective density comes involved to calculations and the fitting of the distributions begins. By varying the value of ρ_e, the difference of $\Delta I'(D_a)$ and measured ELPI current response is minimized.

Assuming that the particles are fractal like agglomerates, the particle density should depend on particle size according to equation 2. Thus a further restriction can be applied to the size dependence of effective density, which makes the fitting procedure more stable. The parameters of effective density used in fitting are the size limit of particles (point (a) in figure 2), which have constant density (i.e. are not agglomerates but primary particles), the density of primary particles (point (b)) and the slope of the curve in figure 2. The difference between ELPI distribution and calculated SMPS current distribution is minimized by minimizing difference of each size intervals corresponding ELPI impactor stages. This is done by varying parameters (shown in figure 2), which define the density profile. The density profile that gives the smallest difference between the two current distributions is the result of the fitting. [5]

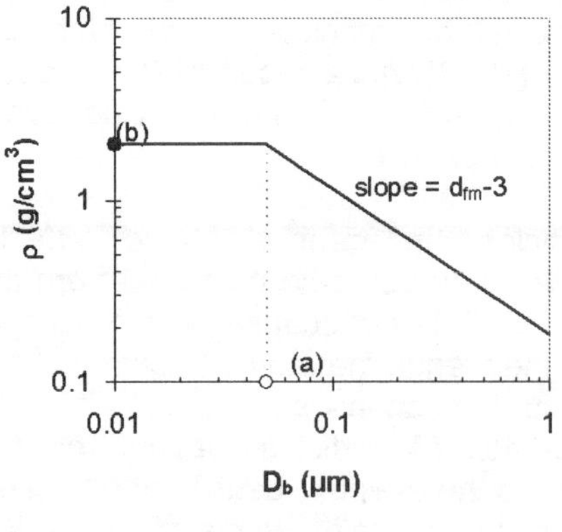

Figure 1. Parameters used in fitting are the maximun size of the primary particles (point (a)), density of primary particles (point (b)) and the fractal dimension d_{fm} (i.e. the slope of the curve) [5].

EXPERIMENTAL

VEHICLE AND FUELS – Steady state measurements with three different speeds (55 km/h, 80 km/h and 120 km/h) and normal road loads were done. The test vehicle used in measurements was Audi A4 1.9 TDI. Detailed information about the test car is given in table 1.

Model year	1995
Odometer reading, km	27 500
Transmissions	A4 automatic
Cylinders	4
Displacement, dm^3	1.89
Max power	66 kW at 4000 rpm
Max torque	202 Nm at 1900 rpm
Compression ratio	19.5:1
Combustion system	Turbo-charged, direct injection, Inter cooler
Exhaust control	EGR + oxidation catalyst

Table 1. Test vehicle Audi A4 1.9 TDI

Two types of diesel fuels were used: a low sulphur content EURO IV fuel, DIKC, and a higher sulphur content EURO II fuel EN590(Table 2).

FUEL ANALYSIS	DIKC City diesel	EN590 Euro
Density at 15 °C ASTM D4052; EN-ISO12185, kg/m^3	837,1	858,1
Viscosity 40 °C, ASTM D445; ISO 3104, mm^2/s	3,26	3,51
Sulphur content ASTM D3120/ASTM D4294 ;mg/kg	23	430
Cloud point ASTM D 2500; ISO 3015, °C	-9	-2
CFPP DIN EN 116 ; °C	-22	-6
Distillation T start ASTM D 86; ISO 3405 ; °C	194	192
5 vol-% evaporated; °C	214	223
10 vol-% evaporated; °C	224	235
20 vol-% evaporated; °C	239	248
30 vol-% evaporated; °C	254	262
40 vol-% evaporated; °C	267	275
50 vol-% evaporated; °C	282	288
60 vol-% evaporated; °C	296	301
70 vol-% evaporated; °C	310	314
80 vol-% evaporated; °C	326	327
90 vol-% evaporated; °C	344	342
95 vol-% evaporated; °C	358	354
Distallation T final ; °C	365	362
Water content (K-F) ASTM D1744 ; mg/kg	19	-
Cetane index D4737; ISO4264	55,4	48,7
Cetane number D613; ISO5165	54,0	49,4
Aromatic content Mod IP 391/90;vol-%	20,6	35,8

Table 2. Test fuels

DILUTION – Both dilution tunnel (constant volume sampler, CVS) and tailpipe size distributions were measured. Dilution tunnel parameters are shown in table 3. Dilution air was taken from the control room and it represented ambient indoor air without drying. There was an all-steel transfer tube with a metal connector between the tail pipe and the CVS.

Diameter ID, cm	27
Insulation of the tunnel	No
Heating of the tunnel	No
Maximum tunnel flow, Nm^3/min	11
Residence time, s	1.5 - 4
Dilution air	Indoor air
Dilution air RH	Ambient air RH
Heating of the dilution air	No
Transfer tube	non-insulated, 2 m
Interface to the tunnel	Insulated, < 1 m

Table 3. Dilution tunnel parameters

The residence time of the raw exhaust in the insulated, but non-heated transfer line varied between 0.5 – 2.1 s and in the dilution tunnel 0.9 – 1.8 s.

The exhaust gas temperatures and primary CVS dilution ratios are in table 4.

	120km/h road load	80 km/h road load	50 km/h road load
Exhaust gas temp. before Catalyst, °C	401	310	260
Exhaust gas temp. at the tailpipe	320	213	170
EGR-rate, %	45-48	70-72	81-83
Primary dilution ratio	4.1	7.2	8.0
Dilution factor	7.2	12	13

Table 4. Exhaust gas temperatures and CVS dilution ratios for different driving conditions.

In the alternative scheme, hot exhaust gas was diluted with two ejector diluters [11] immediately after the tail pipe (dashed line in figure 2). Both the first dilution unit and the dilution gas were heated up to 280 °C. The second diluter and the dilution air were at room temperature. The dilution ratio of both diluters was approximately 8. The delay time of the hot raw exhaust to the ejector diluter was short, 0.06 – 0.07 s, and the residence time in the primary ejector diluter less than 0.13 s. The setup minimizes changes in size distribution caused by agglomeration and gas-to-particle conversation (i.e. nucleation and condensation of gaseous components onto soot particles). Two stage ejector dilution systems have been applied to different combustion sources in several studies ([12],[13])

MEASUREMENT SET-UP- In both cases the diluted aerosol was measured by ELPI and SMPS. The measurement scheme is shown in figure 2. The SMPS sheath air and sample flow rates were 6 and 0.6 lpm, respectively.

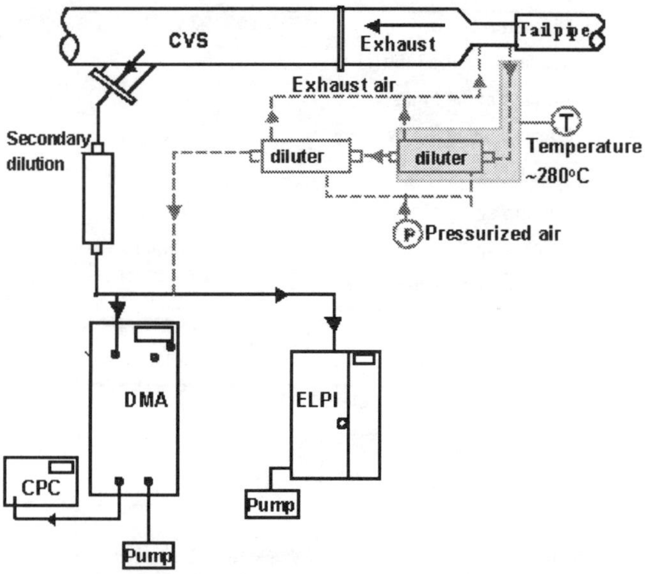

Figure 2. Measurement scheme for density measurement. CVS and ejector diluters were used alternatively.

RESULTS

SIZE DISTRIBUTIONS – The size distribution measurements were made using both CVS and ejector diluters to see the effect of dilution on measured particle distribution and density. The effect of the dilution system on the size distribution is evident at 55 km/h speed: the CVS size distribution is shifted towards larger particle sizes compared to the ejector measurement. Maricq and co-workers reported the same trend in tunnel measurement vs. ejector diluter measurements [13]. Possible reasons for the shift are condensation of volatile hydrocarbons onto the particles or the coagulation of soot agglomerates in the transfer line and tunnel. From density analysis it is possible to get some information which of the two phenomena is more probable.

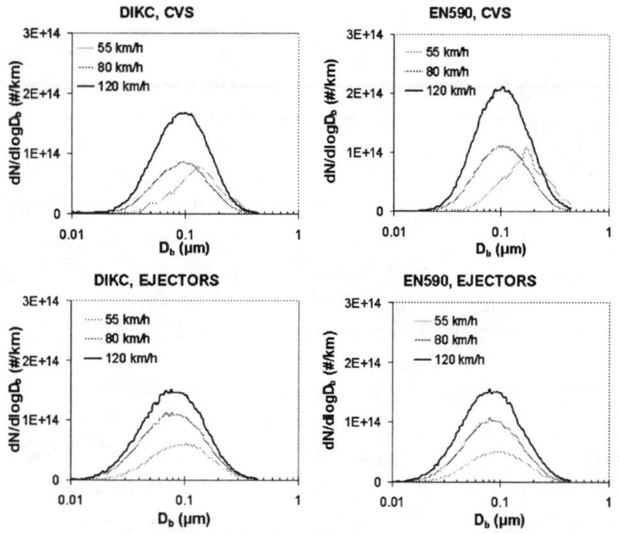

Figure 3. Measured SMPS distributions for EN590 Euro and DIKC fuels. The results in upper row are measured from CVS tunnel and the results in lower row are measured from tail pipe after two ejector diluters.

EFFECTIVE DENSITY – The measured distributions described above were submitted to the effective density fitting procedure. An example of the first phase current response I'(D_b) and the density fitted discrete current response $\Delta I(D_a)$ compared to measured ELPI current is shown in figure 4.

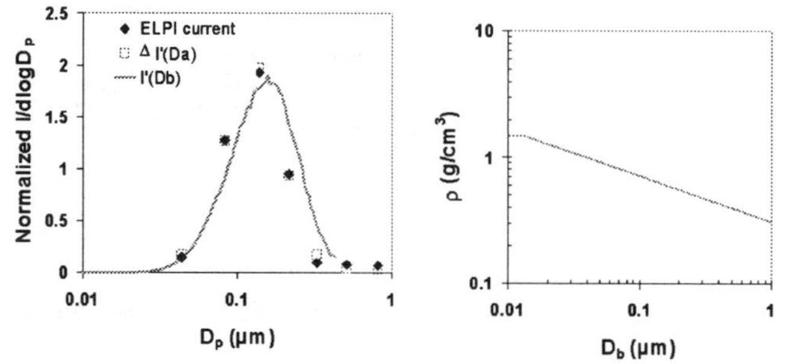

Figure 4. Example of density fitting. Left: Measured ELPI current (black diamonds) compared to first phase current response I('D_b) (solid line) and fitted current response $\Delta I(D_b)$ (open rectangles). Right: Effective density curve producing the fit shown left.

The fitting routine was applied to distributions measured at every speed point, with both of the dilution alternatives. In figure 5, the results for EN590 Euro fuel are shown. The ejector measurements are marked with dashed lines and tunnel measurements are marked with solid lines. In ejector measurements, the density variation of exhaust particles shows a similar trend for all the speeds points. In every case the primary particle density is between 1.1 and 1.2 g/cm^3, and the primary particle size is approximately 50 nm. With increasing particle size, density decreases. The fractal dimension varies from 2.77 to 2.84.

In tunnel measurements, the particles at 55 km/h evidently differ from other cases. As mentioned before (shown in figure 3.), also the number distribution of this driving point in tunnel measurements differs from the others. The density analysis shows that the effective density is constant as a function of particle size. At the same time the density of primary particles is low: 0.64 g/cm^3. This indicates that a remarkable amount of hydrocarbons have condensed onto the soot agglomerates. The peak size of the particle distribution increases because of condensation (in figure 3.). Also the effective density of primary particles decreases because of condensation of hydrocarbons, which have smaller density than soot. Because of condensation/adsorption of hydrocarbons and water, the particle structure may collapse [14],[15] and the whole agglomerate may be covered with volatile condensed materials [16]. Thus in this case the larger particles are almost spherical (in spite of the collapsed agglomerate core) and the density remains the same even if the particle size gets larger. Results for particles produced by DIKC fuel show similar trend.

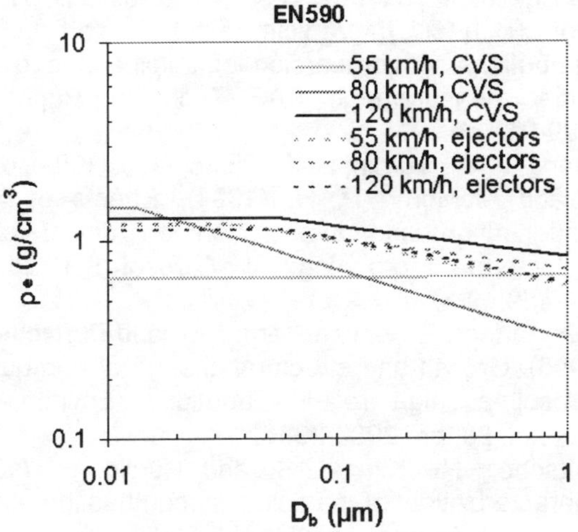

Figure 5. Effective density of diesel particles as a function of particle size: tunnel and ejector measurements.

In light of these results, the ejector dilution should be used when the structure of pure agglomerated soot particles is studied.

The effect of different fuel types on effective density of soot particles is shown in figure 6. The measurements were made with ejector diluters to prevent the "tunnel effect" i.e. the gas-to-particle conversion and its impact on particle structure. The results measured for DIKC fuel are marked with continuous lines and results for EN590 fuel with dashed lines. The results indicate that there is a small difference in the density profiles of particles produced by different fuels. The particles produced by higher sulphur content fuel (EN590) tend to have a higher fractal dimension, i.e. denser structure, than those produced by low sulphur content fuel (DIKC). This would be in accordance with the effect of sulphur content on particle hygroscopicity. Particles with larger amount of sulphur adsorb more water, resulting in a denser structure.

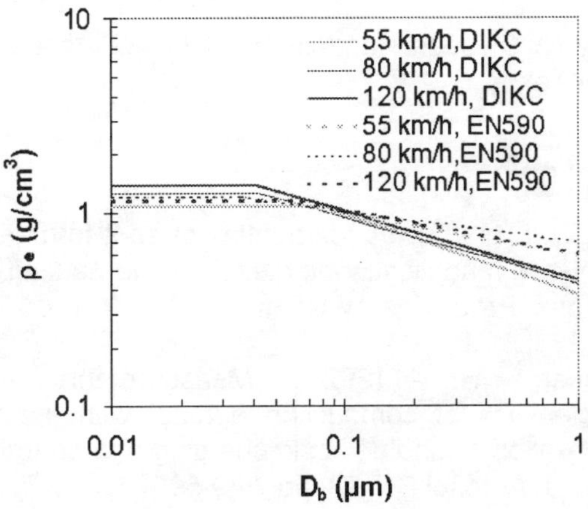

Figure 6. Effective density of soot measured using ejector diluters. Fuel sulphur contents are 23 mg/kg (DIKC) and 430 mg/kg (EN590).

CONCLUSION

A new method to define the particle effective density has been used for diesel particles. The method seems to be a powerful way to find information about the particle structure and its formation. Particle structure is strongly affected by gas-to-particle conversation during sampling. For this reason the dilution method should be carefully chosen to make sure that information got from measurements corresponds the original purpose. If the structure and concentration of only soot particles is under interest, the dilution scheme should prevent nucleation and condensation. One way to achieve this is to use a heated diluter with heated dilution air.

Effective density of diesel agglomerates decreases as a function of particle size. Small difference was detected between the structures of particles for the two fuel types. The fractal dimension of soot particles varies from 2.58 to 2.84, indicating that particles are more cluster-like than chain-like. Bérubé et al. [17] have found three different morphological types in diesel exhaust: small primary particles (~30 nm), small cluster or chain like particles and larger (1-2 µm) agglomerates. Skillas et al. [18] found fractal dimensions varying from ~2.3 to almost 3 for particles produced by diesel engine. Also Weingartner et al. [19] have found high values of fractal dimension for diesel exhaust particles. One should note that fractal dimension values estimated with different methods might disagree.

The density of primary soot particles varies from 1.1 to 1.4 g/cm^3, which is less than density of bulk soot. This might indicate that a layer of hydrocarbons has condensed on the soot particles in all cases. The size limit of which smaller particles are spherical primary particles or almost spherical small compact clusters varies from 20-80 nm.

ACKNOWLEDGMENTS

The study has been funded by Mobile2 research program of Tekes.

REFERENCES

1. Smith, O. I. (1981). Fundamentlas of soot formation in flames with application to diesel engine particulate emissions. Prog. Energy Combust. Sci., **7**, pp. 275-291.
2. Burtscher, H. (1992). Measurement and characteristics of combustion aerosol with special consideration of photoelectric charging and charging by ions. J. Aerosol Sci. **23**, pp. 549-595
3. Kelly, W. P. and McMurry, P. H. (1992). Measurement of Particle density by inertial classification of Differential Mobility Analyzer-generate monodisperse aerosols. Aerosol Sci. Tech. **7**, pp. 199-212.
4. Virtanen, A., Ristimäki, J., Rostedt, A. and Keskinen, J. On-line measurement of size distribution and effective density of submicron aerosol particles. Submitted to J. of Aerosol Sci.
5. Virtanen, A., Ristimäki, J. and Keskinen J. New method for measurement of effective density and fractal dimension of agglomerate particles. Submitted to J. Aerosol Sci.
6. Schmidt-Ott, A. (1988). New approach to in situ characterization of ultrafine agglomerates. J. Aerosol Sci. **19**, pp. 553-563.
7. Rogak, S. N. and Flagan, R. C. (1990). Stokes drag on self similar clusters of spheres. J. Colloid Interf. Sci. **134**, pp. 206-218
8. Mandelbrot, B.B. (1977). Fractals: Forms, Chance and Dimensions, W.H. Freeman, San Francisco
9. Keskinen, J., Pietarinen, K. and Lehtimäki, M. (1992). Electrical Low Pressure Impactor. J. Aerosol Sci. **23**, 353-360.
10. Wang, S. C. and Flagan, R. C. (1990). Scanning Electrical Mobility Spectrometer. Aerosol Sci. Tech. **13**, pp. 230-240.
11. Koch, W., Lödding, H., Mölter, W. and Munzinger, F. (1998). Verdünnungssystem für die Messung hochkonzentrierte Aerosol mit Optichen Partikelzählern. Staub-Reinhaltung der Luft **48**, pp. 341-344.
12. Kauppinen, E.I. and Pakkanen, T.P. (1990). Coal combustion aerosols: a field study. Environ sci. Tecnol. **24**, pp. 1811-1818.
13. Maricq, M. M., Chase, R. E., Podsialik, D. H. and Vogt R. (1999). Vehicle exhaust particle size distributions: A comparsion of tailpipe and dilution tunnel measurement. SAE Technical Paper No. 1999-01-1461
14. Huang, P.-F., Turpin, B. J., Pipho, M. J., Kittelson, D. B. and McMurry, P. H. (1994). Effects of water condensation and evaporation on diesel chain-agglomerate morphology. J. Aerosol Sci. **25**, pp. 447-459.
15. Weingartner, E., Baltensberger, U. and Burtscher, H. (1995). Growth and structural change of combustion aerosol at high relative humidity. Environ. Sci. Technol. **29**, pp. 2982-2986.
16. Burtscher, H., Künzel, S. and Hüglin, H. (1998). Characterization of particles in combustion engine exhaust. J. Aerosol Sci. **29**, 389-396.
17. Bérubé, K. A., Jones, T. P., Williamson, B. J., Winters, C., Morgan, A. J. and Richards, R. J. (1999). Physicochemical characterisation of diesel exhaust particles: Factors for assessing biological activity. Atmospheric Environment **33**, pp. 2599-1614.
18. Skillas, G., Künzel, S., Burtscher, H., Baltensberger, U. and Siegmann, K. (1998). High fracta-like dimensions of diesel soot agglomerates. J. Aerosol Sci. **29**, 411-419.
19. Weingartner, E., Burtscher, H. and Baltensberger, U. (1997). Hygroscopic properties of carbon and diesel soot particles. Atmospheric Environment **31**, pp. 2311-2327.

2002-01-0612

Development of a Wet-based NDIR and Its Application to On-board Emission Measurement System

Hiroshi Nakamura, Nobutaka Kihara, Masayuki Adachi and Kozo Ishida
Horiba, Ltd.

ABSTRACT

Due to a need for a robust measurement system for on-board real-world vehicle emission measurement, a heated ND-IR(h-NDIR) technique has been developed and evaluated for its potential. The h-NDIR is capable of measuring CO and CO_2 under wet-based condition by correcting interference from co-existing gas with an algorithm specially developed for the present study. The resulting H_2O interference to the CO_2 measurement is less than 0.01vol% for zero point and less than ±1% for span points and that of CO measurement is less than 0.001vol% for zero point and less than ±2% for span point against 0 to12vol% H_2O.

An on-board emission measurement system using the h-NDIR in combination with an Annubar® flow meter and an air to fuel ratio sensor has been evaluated. The result reveal correlation between the present system and a chassis test system to be within 7% for fuel consumption, within 5% for CO mass emission, and within 6% for CO_2 mass emission.

INTRODUCTION

On-board emission measurement of vehicles is getting much interest due to a fact that the laboratory testing is facing its limit for predicting what is taking place in real world. On the other hand, recent reported studies have been showing some difficulties of bringing the conventional laboratory instrument into a vehicle[(1),(2),(3)]. Major technical challenges required for such an application include vibration proof, low power consumption, small dimensions, and easy operation. Such requirements lead demands for wet-based NDIR with solid-state sensors.

Generally, CO and CO_2 in vehicle emission are measured by dry NDIR method due to a significant interference from water vapor. Such extraction of water from the sample gas is usually done by a cooler that could be a high power consumer if installed on-board. Also, in this measurement, measured value of CO and CO_2 concentration are higher than real concentration because of the decrease in sample gas volume. In order to obtain the real concentration, one has to measure the water concentration separately.

In the present study, a heated NDIR(h-NDIR) has been developed and evaluated. The h-NDIR consists of a heated cell, and solid-state detectors for CO, CO_2 and H_2O. Sample gas is introduced into the heated cell through a heated line without H_2O extraction. For such a wet-based measurement, care must be taken to correct two types of interference from water vapor, i.e. spectral overlap and molecular interaction. The spectral overlap is generally known as zero point interference that can be observed as an erroneous reading when nitrogen based water vapor is introduced to CO or CO_2 analyzers. The molecular interaction can be described by a model of energy level transition with a light absorption influenced by coexisting gases that can only be seen when CO or CO_2 is exposed to a change in bulk gas composition[(4)].

The h-NDIR has also been evaluated for on-board measurement application with its advantages of not needing a cooler and robustness for vibration effects. An on-board measurement system has been developed using the h-NDIR, an air to fuel ratio(AFR) analyzer and an Annubar® flow meter for measurement of exhaust gas flow rate. The following sections are showing the discussion of fundamental evaluation of the h-NDIR as a gas analyzer, evaluation of the Annubar® flow meter for the exhaust emission measurement, and actual performance of such a system using a chassis test cell.

FUNDAMENTAL EVALUATION OF THE HEATED NDIR

CONFIGURATION OF THE HEATED NDIR

The h-NDIR has been developed with a light source, a heated sample cell, an optical chopper, optical band pass filters and solid-state detectors for CO_2, CO, H_2O and reference wavelength. Figure 1 shows infrared light is passing through the heated sample cell, modulated by the light chopper, and then reaches to the detectors. The optical filters are selecting the wavelengths that are characteristic to the target compounds. When a sample gas that is carrying CO, CO_2 and H_2O is introduced to heated sample cell, light absorption takes place at each of the characteristic absorption wavelength and light intensity being brought to each detector is decreased. The decrease of light intensity are in direct relation to concentration of CO, CO_2 and H_2O in sample gas. Therefore CO, CO_2 and H_2O concentrations are calculated using output of each detector. The sample cell is heated up to 120 degreeC and sample gas introduced to sample cell through a heated line without cooler. Therefore, a sample gas such as vehicle emission can be measured in wet condition.

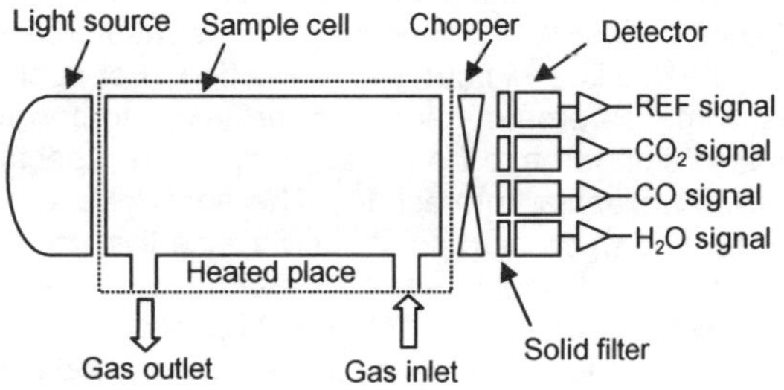

Figure 1. Configuration of the h-NDIR

CORRECTION OF H2O INTERFERENCES

<u>H2O interference & coexisting effect and correction of these effect</u> – As stated previously, there are two kinds of interference originated from a presence of water in sample gas. One is the spectral overlap and the other is a molecular interaction. Again the molecular interaction obviously changes the amount of absorption when both interfering and interfered compounds are present.

A use of water detector in the same optics of the h-NDIR enabled us to be successfully correcting these H_2O interference and molecular interaction.[5] Figure 2 shows a corrected H_2O zero point interference of CO and CO_2 detector when a nitrogen based H_2O is introduced to the h-NDIR. H_2O zero point interference (spectral overlap) of each analyzer has been observed to be within ±0.001vol% at 0~12% of H_2O.

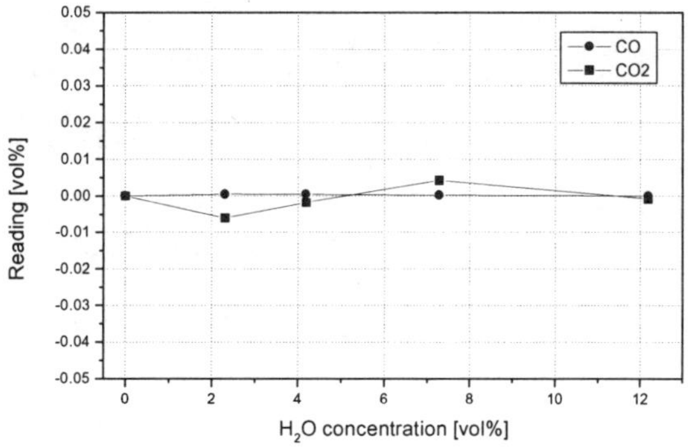

Figure 2. H_2O zero point interference of h-NDIR CO and CO_2 detector. (after corrected)

Figure 3 shows corrected H_2O span point influence (molecular interaction) of CO and CO_2 detector when span gases of CO and CO_2 with known amount of H_2O are introduced to the h-NDIR. The interference is evaluated as an error of point percent with the following equations.

$$(Error\%) = \frac{[Conc_{corr}] - [Conc_{real}]}{[Conc_{real}]} \times 100 \qquad \text{---(1)}$$

$$[Conc_{real}] = [Conc_{span}] \times \frac{100 - [H_2O(\%)]}{100} \qquad \text{---(2)}$$

where,

Error%: Error percentage relative to the point,

$Conc_{corr}$: CO and CO_2 readings after interference correction,

$Conc_{real}$: CO or CO_2 concentration in the span gas (after adding H_2O)

$Conc_{span}$: CO and CO_2 concentration of cylinder gas in use.

H_2O: H_2O concentration in the span gas.

Span points error of CO and CO_2 detectors inevitably includes zero point interference and molecular interaction. However, with a use of water detector, these error of CO detector are observed to be within ±2%, and these of CO_2 detector was also observed to be within ±1% at 0 to 12% H_2O.

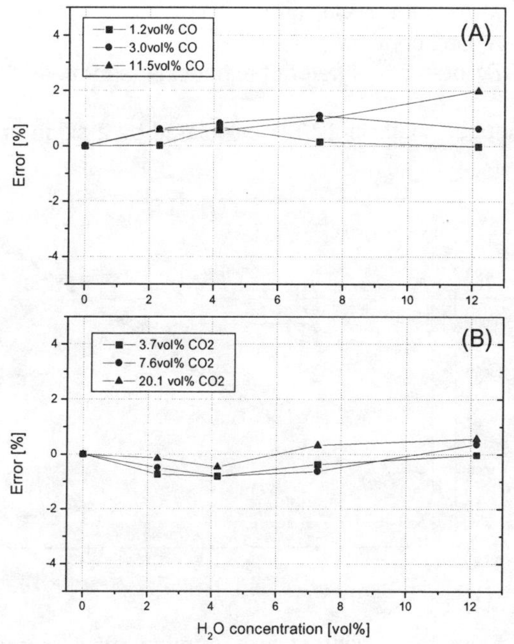

Figure 3. H_2O span point interference of h-NDIR CO and CO_2 detector. (after corrected)

Interference to the H2O detector – Interference from the coexisting compounds to the water detector has also been evaluated using typical gas compounds in exhaust emission. Figure 4 shows the measurement result of such interference. As a result, H_2O detector has less than 6% interference against these interference gases. This performance of water detector is enough to be used for H_2O interference correction of CO and CO_2 detector.

One can also raise a discussion of a difference of response time between the water and other compounds detection because, in general, water analysis is significantly slower than the CO/CO_2 measurement. However, the major factor of slowness of the water analysis is a gas transportation and is not the detection itself. So, the water coexistence can be precisely corrected with the optics used in the present study that is using only one gas sampling cell having multiple detectors.

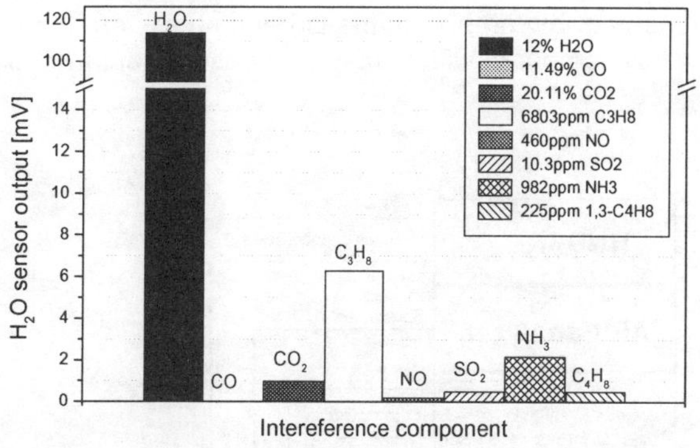

Figure 4. Interference of H_2O detector in h-NDIR.

THE SPECIFICATION OF HEATED ND-IR - The measured performance of the h-NDIR can be summarized in Table 1.

Table 1. The typical performance of the h-NDIR.

Item	CO Analyzer Performance	CO_2 Analyzer Performance
Range	0~12vol%	0~20vol%
Response time (T_{90})	0.9 second	0.9 second
Noise	within 1.5%PT	within 1.5%PT
Drift	±1.0%PT/8h	±1.0%PT/8h
H_2O interference (0~12vol%H_2O)	±0.001vol% or ±2%PT	±0.01vol% or ±1%PT

EVALUATION OF THE HEATED NDIR FOR ON-BOARD EMISSION MEASUREMENT SYSTEM

The h-NDIR has also been evaluated for the application of on-board emission measurement system. Tests are done to investigate for the performance of the Annubar® flow meter and the h-NDIR combined with it. Quantitative discussion has been carried out using chassis test equipment, i.e. CVS bag measurement.

CONFIGULATION OF ON-BOARD MEASUREMENT SYSTEM – The on-board measurement system consists of the h-NDIR, Annubar® flow meter and the AFR sensor. Annubar® flow meter is being used for measurement of exhaust gas flow rate, and it was mounted on an exhaust pipe as shown in Figure 5. The exhaust pressure and temperature have also been measured to correct an exhaust gas flow rate to standard condition (at 293.15 K and 101.3 kPa). To measure AFR, the ZrO_2 type sensor was installed to exhaust pipe directly. CO and CO_2 in the emission have been measured by the h-NDIR. The sample gas is introduced to the h-NDIR through the heated sample line from the exhaust pipe and the measured sample gas is returned to the exhaust pipe. The exhaust emission of the vehicle is simultaneously introduced to CVS system. With such a setup, the fuel consumption and total mass emission of CO and CO_2 measured by on-board measurement system can be directly compared with CVS-Bag measurement method.

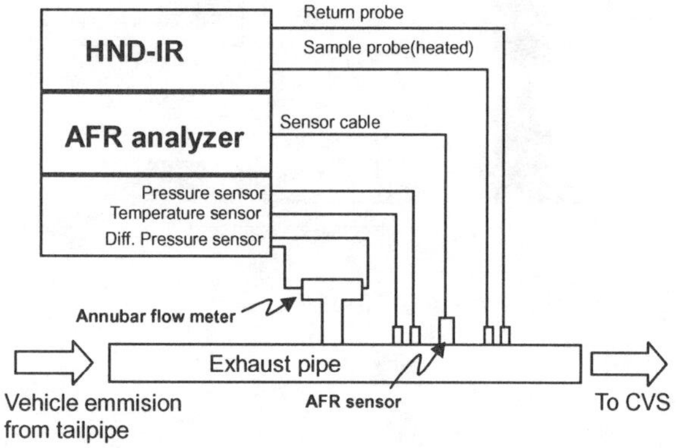

Figure 5 Configuration of on-board measurement system.

EVALUATION OF ANNUBAR FLOW METER – The Annubar® flow meter has been installed to the on-board emission measurement system to measure a vehicle exhaust flow rate. Annubar® flow meter is a differential pressure type flow meter and consists of a rod that has two separate cells. The rod is installed into exhaust pipe vertically. When the exhaust flow hits this rod, differential pressure between the upstream side and downstream side of the rod changes depending on the exhaust flow rate. One of the features of the Annubar® flow meter is a low pressure drop because of this principle. The Annubar® flow meter has been calibrated by a venturi flow meter named Smooth Approach Orifice (SAO). Figure 6 shows a calibration data of Annubar® flow meter. The parameter H is calculated from the differential pressure by equation (3).

$$H = \frac{P_{exh}}{101.3(kPa)} \times \frac{293.15(K)}{T_{exh}} \times \sqrt{\frac{h_w}{\gamma}} \qquad \text{---(3)}$$

where,

P_{exh}: Measured pressure of exhaust gas [kPa],
T_{exh}: Measured temperature of exhaust gas [K],
γ: Density of exhaust gas at standard condition at 293.15 K and 101.3 kPa [kg/m^3],
h_w: Differential pressure[kPa].

It can be seen in Fig.6 that the root of differential pressure has a good linear correlation with flow rate measured by the SAO. The actual exhaust flow rate then is calculated by the next equation.

$$Q_{exh} = K \cdot N \cdot D^2 \cdot H \qquad \text{---(4)}$$

where,

Q_{exh}: Exhaust gas flow rate[L/min]
K: Annubar coefficient
N: Unit conversion factor
D: Internal diameter of exhaust pipe[mm]

Annubar coefficient(K) is calculated by a slope of Figure 6.

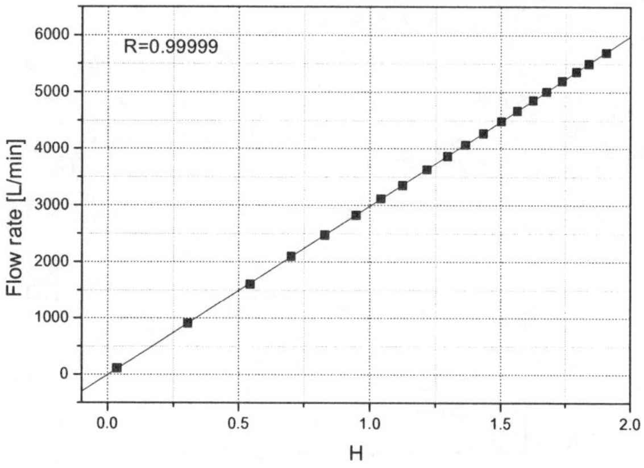

Figure 6. Calibration of Annubar flow meter using SAO.

Flow rate of vehicle exhaust measured by the Annubar® has been compared with the SAO. SAO is installed at the intake of CVS dilution air and is used to measure a dilution air flow rate continuously through the test. A vehicle exhaust flow rate can be calculated by taking a difference of dilution air flow rate and total flow rate of the CVS system. Figure 7-A shows the exhaust gas flow rate of a vehicle with a 1.6 liter gasoline engine tested on chassis dynamometer using FTP-75 cold transient phase and Fig. 7-B shows the data using 2.0 liter gasoline vehicle on the same mode.

From the Figs 7 A and B, the value of Annubar® flow meter has obvious excursion from a value calculated by SAO when engine is idling. Such difference in small engine test (A) is bigger than the larger engine (B). And in the small engine test, the values of Annubar® flow rate on high idle region are correlated with SAO values well, but on low idle region, Annubar® reading is higher than SAO. It is considered that the flow pulsation in idling has significant effect on the differential pressure of Annubar® flow meter. However, except the idle region, the data measured by Annubar® has good correlation with the value calculated by SAO.

Figure 7-C shows the exhaust gas flow rate of vehicle with IDI 2.0 liter diesel engine with inter-cooler turbo charger tested on chassis dynamometer using the same driving mode. The exhaust gas flow rate was measured by the Annubar® flow meter and intake air flow rate was also measured by Karman vortex flow meter. The vortex flow meter had been evaluated for on-board emission measurement system elsewhere.[(6),(7),(8)] The exhaust gas flow rate was calculated by the intake air flow rate and the AFR. The values measured by the Annubar® flow meter has also good correlation with the value calculated by vortex flow meter except for the idle region. The observed delay time (Td) from the vortex flow meter to the Annubar® is 1.0 second and then to the SAO is 1.1 second.

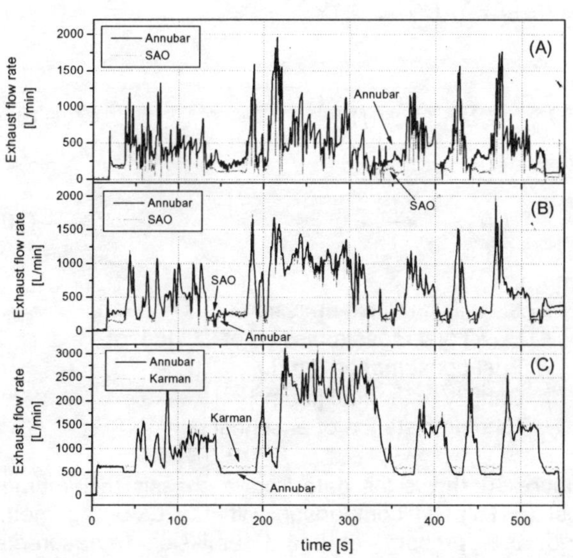

Figure 7. Vehicle exhaust gas flow rate measured by Annubar flow meter.(A);1.6 liter gasoline engine, (B); 2.0 liter gasoline engine, (C); 2.0 liter IDI diesel engine with inter-cooler and turbo charger.

TEST RUNS WITH GASOLINE VEHICLE ON A CHASSIS DYNAMOMETER

An on-board emission measurement system with the h-NDIR has been evaluated by several tests run on a chassis dynamometer. The test vehicle is a 1.6 liter gasoline passenger car, and test mode is full FTP75. Figure 8 shows the test results of phase 1. Plots are made for a speed of the vehicle, AFR measured by AFR sensor, CO and CO_2 concentration measured by the h-NDIR, and exhaust flow rate measured by the Annubar® flow meter. In these results, a sharp response of CO concentration has been observed at cold start and acceleration region. Moreover, a decrease of CO_2 concentration by fuel cut is observed with a sharp increase in AFR at deceleration region. It can be confirmed that the h-NDIR has a sufficient response to perform real-time emission analysis. The CO_2 concentration at cold start was higher than the concentration after the vehicle is getting warmed up. This can be explained by a change in volume of exhaust stream by a water condensation inside an exhaust system. At the same time, the exhaust gas flow rate is decreased for the same amount. Therefore, the resulting mass emissions of CO_2 and CO do not have significant influence from the water condensation in exhaust system.

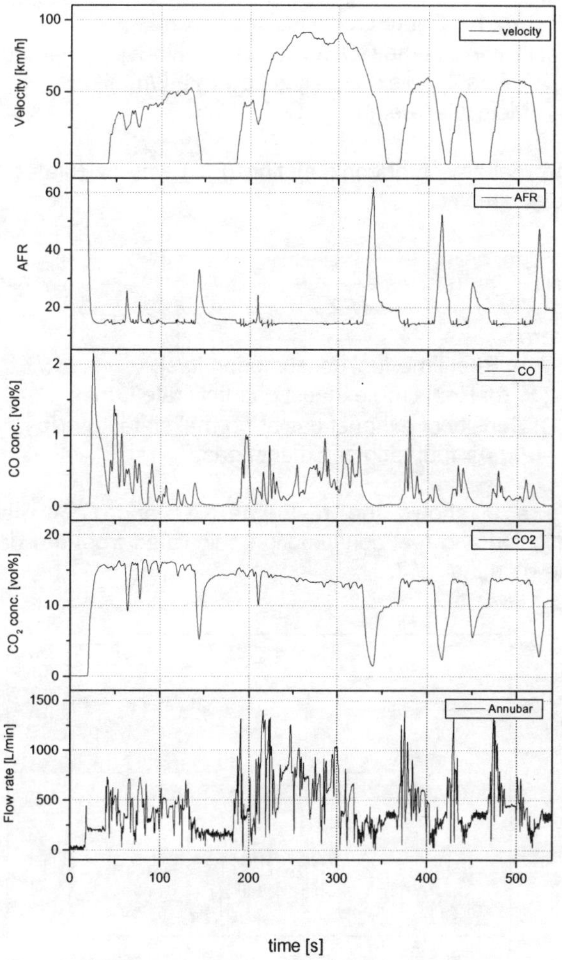

Figure 8. Emission from a LEV equipped a 1.6 liter gasoline engine in FTP75 phase 1.

CALCULATION OF FUEL ECONOMY AND MASS EMISSION OF CO AND CO2

The real time mass emission of CO and CO_2, and real time fuel consumption have been calculated from measured values of AFR, CO and CO_2 concentration and exhaust gas flow rate. The real time CO_2 and CO mass emissions,$CO_{2mass}(t)$, $CO_{mass}(t)$, are calculated by the equations (5) and (6)

$$CO_{2mass}(t) = CO_{2conc}(t) \times 10^{-2} \times Q_{exh}(t) \times \frac{W_{CO2}}{22.4} \times \frac{1}{60} \quad \text{---(5)}$$

where,

$CO_{2mass}(t)$; Real time CO_2 mass per second [g/s],
$CO_{2conc}(t)$; Real time CO_2 concentration [vol%],
$Q_{exh}(t)$; Real time exhaust gas flow rate[L/min],
W_{CO2}; Molecular weights of CO_2 [g/mol].

$$CO_{mass}(t) = CO_{conc}(t) \times 10^{-2} \times Q_{exh}(t) \times \frac{W_{CO}}{22.4} \times \frac{1}{60} \quad \text{---(6)}$$

where,
$CO_{mass}(t)$; Real time CO mass per second [g/s],
$CO_{conc}(t)$; Real time CO concentration [vol%],
$Q_{exh}(t)$; Real time exhaust gas flow rate[L/min],
W_{CO}; Molecular weights of CO[g/mol].

The real time fuel consumption (F(t)) was calculated by the equation (7).

$$F(t) = Q_{exh}(t) \times \gamma \times \frac{1}{AFR + 1} \qquad \text{---(7)}$$

where,
$F(t)$; Real time fuel consumption [g/s],
$Q_{exh}(t)$; Real time exhaust gas flow rate[L/min],
γ; Density of exhaust gas at normal state[kg/m^3],
AFR; Air fuel ratio of exhaust gas.

Figure 9 shows the real time CO and CO_2 mass emission and fuel consumption calculated from the data shown in Figure 8.

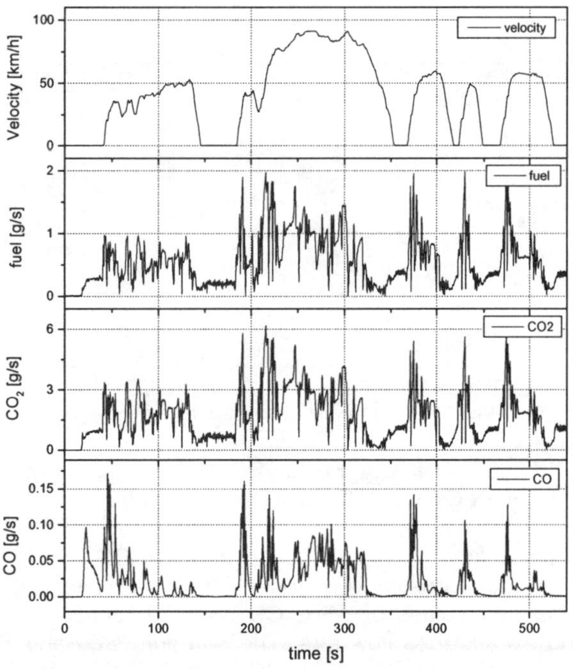

Figure 9. Real time mass emission of CO_2 and CO, and fuel consumption. LEV equipped a 1.6 liter gasoline in FTP75 phase 1.

COMPARISON OF THE TOTAL MASS EMISSION WITH CVS-BAG MEASUREMENT METHOD

The total mass emission of CO and CO_2, and fuel consumption has been compared with the CVS-Bag measurement. The total mass emission of CO and CO_2, and fuel consumption were calculated by the following equations.

$$CO_{2\,mass} = \Sigma\ CO_{2\,mass}(t) \times \frac{1}{l} \qquad \text{---(8)}$$

$$CO_{mass} = \Sigma\ CO_{mass}(t) \times \frac{1}{l} \qquad \text{---(9)}$$

$$F = \Sigma F(t) \times \frac{1}{\rho} \times \frac{1}{l} \qquad \text{---(10)}$$

where,
CO_{2mass}; Total mass emission of CO_2 [g/km],
CO_{mass}; Total mass emission of CO [g/km],
F; Fuel consumption [km/L],
ρ; Density of the fuel [kg/m^3],
l; Running distance of each test.[km]

Figure 10 shows the data from a chassis dynamometer test (FTP75) in comparison with the CVS-bag method and the present on-board emission measurement system. Differences between the two methods were within 7% for fuel consumption, within 5% for CO mass emission, within 6% for CO_2 mass emission.

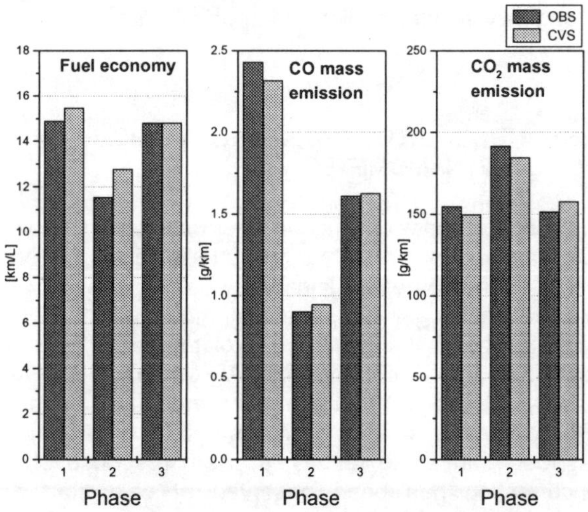

Figure 10. The comparison of on-board emission measurement system with CVS-Bag measurement method. (LEV equipped 1.6 liter gasoline engine, FTP75)

TEST RUNS WITH DIESEL VEHICLE ON A CHASSIS DYNAMOMETER

The on-board emission measurement system has been also evaluated by diesel vehicle on a chassis dynamometer. The test vehicle is a 2.0 liter diesel passenger car, and test mode is FTP75. The on-board emission measurement system for diesel vehicles equipped with the same analyzers as for gasoline vehicles, a NOx analyzer and an opacity meter. NOx analyzer has a ZrO2 type sensor installed to exhaust pipe directly. An opacity meter is a full flow type opacity meter and also installed to exhaust pipe directly to measure smoke in emission. This type of opacity meter can measure exhaust stream in-situ as shown in Figure 11.

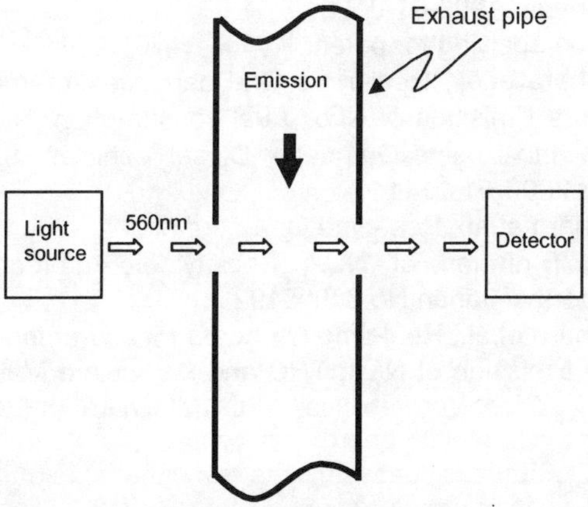

Figure 11. Configuration of the full flow type opacity meter.

Figure 12 shows the test results of FTP75 phase 1. Plots are made for a speed of the vehicle, AFR measured by AFR sensor, CO and CO_2 concentration measured by the h-NDIR, NOx concetration measured by NOx sensor, Opacity[%] measured by full flow type opacity meter and exhaust flow rate measured by the Annubar® flow meter. In these results, a sharp response of NOx concentration has been observed at acceleration region. Moreover, a decrease of CO_2 concentration, NOx concentration and opacity by fuel cut is observed with a sharp increase in AFR at deceleration region. It can be confirmed that the h-NDIR and NOx sensor have a sufficient response to perform real-time analysis also in diesel emission. Moreover, the full flow type opacity meter has a sufficient response to analyze a real time smoke in diesel emission. In this work, due to an unavailability of CVS system at the same time of data collection, mass-based correlation has not been able to be discussed.

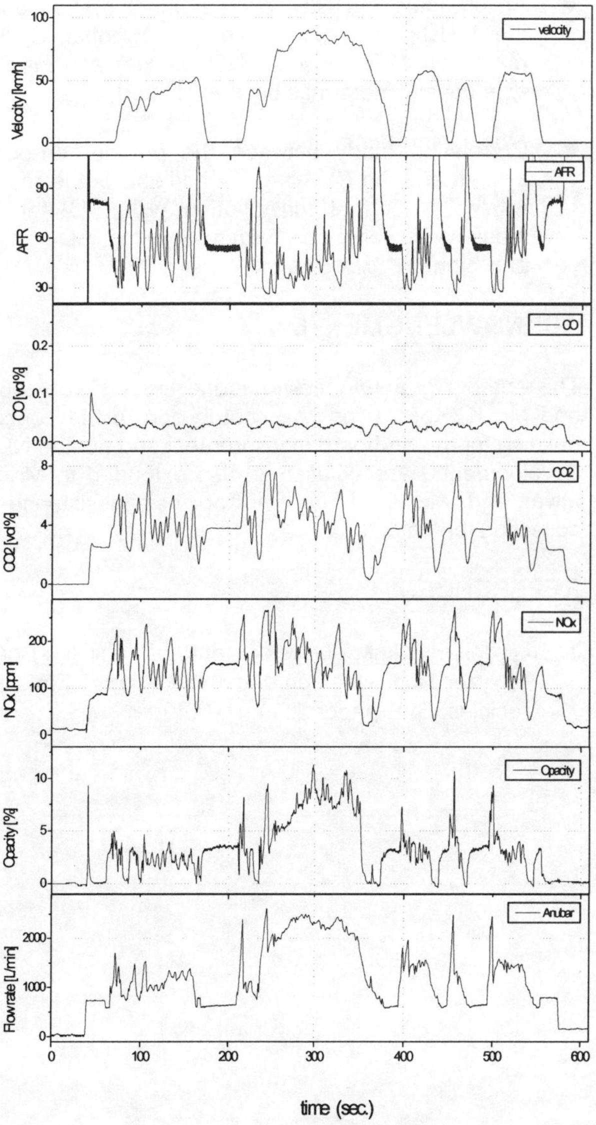

Figure 12. Emission from a 2.0 liter diesel vehicle in FTP75 phase 1.

CONCLUSION

A heated NDIR (h-NDIR) has been developed and evaluated for its potential for on-board emission measurement.

- The h-NDIR is capable of measuring wet-based CO and CO_2 by correcting H_2O interference using H_2O concentration measured by H_2O detector installed in the same optics.

- The zero point H_2O interference of CO and CO_2 detector are less than 100ppm at 0 to 12vol%H_2O. The span points error of CO detector is less than 2% and that of CO_2 detector is less than 1% at 0 to 12% of H_2O.

- An on-board emission measurement system using the h-NDIR combined with the Annubar ® flow meter, an AFR sensor, a NOx sensor and full flow type opacity meter has been evaluated.

- Correlation study between the present on-board system and the CVS-bag method shows the results within 7% for fuel consumption, within 5% for CO mass emission, and within 6% for CO_2 mass emission.

ACKNOWLEDGMENTS

The authors gratefully acknowledge Dr. T. Tsukamoto, and Mr. K. Kojima for their contribution to the concept building for the on-board measurement and heated NDIR developments. Appreciation is also extended to Mr. M. Miyai and Mr. S. Murakami for their assistance in collecting the data.

REFERENCES

1. J. Jetter et.al., Development of an On-Board Analyzer for Use on Advanced Low Emission Vehicles, SAE paper 2000-01-1140
2. T. J. Truex et.al., Measurement of Ambient Roadway and Vehicle Exhaust Emission – An Assessment of Instrument Capability and Initial On-Road Test Results with an Advanced Low Emission Vehicle, SAE paper 2000-01-1142
3. M. Lenner et.al., Influence Of Roof-Rack, Trailer Etc On Automobile Fuel Consumption and Exhaust Emissions, Measured On-The-Road, SAE paper 980682
4. K. Inoue et.al., Numerical Analysis of Mass Emission Measurement Systems for Low Emission Vehicles, SAE paper 1999-01-0150
5. Now on applying for patent.
6. N. Kihara et.al., Real-Time On-Board Measurement of Mass Emission of NOx, Fuel consumption, Road Load, and Engine Output for Diesel Vehicles, SAE paper 2000-01-1141
7. N. Kihara et.al., Real-time On-board Measurement of Mass Emission of NOx, Society of Automotive Enginers of Japan No.20005192
8. N. Kihara et.al., Real-time On-board Measurement of Mass Emission of NOx, THC and Particulate Matter from Diesel vehicles, COMMODIA 2001

2002-01-0613

Development of On-Board System to Measure Running Condition and Actual NOx Emissions from Freight Vehicle

Tomio Miyazaki, Yutaka Takada and Norimasa Iida
Keio University

ABSTRACT

It is necessary to clarify the influence on NOx emissions by driver's operation for analyzing of NOx pollution mechanism at roadside. Simple on-board measurement system for vehicle running condition and actual NOx emissions was developed to evaluate local NOx pollution caused by vehicles. The measurement system was installed in a freight vehicle whose payload is two tons, and actual driving test was conducted to evaluate the effectiveness of the system.

The tests on chassis dynamometer including steady state and transient mode indicated enough accuracy to evaluate NOx emissions from a vehicle at local roadsides.

INTRODUCTION

In major cities in Japan, roadside air pollutions caused by vehicles are serious problem. People living near the roadside have filed lawsuits over the prevention of air pollution. The Government is considering an effective scheme within the automobile sector for achieving the emission reduction target and dissemination target of low emission vehicles. However, the measures tend to be for the reduction of whole amount of exhaust emissions. As there is serious air pollution in local roadsides, the mechanism of roadside air pollution caused by exhaust emissions from vehicles have to be considered. As NOx play one of the main role for roadside air pollution, they should be focused on. In order to analyze NOx pollution at roadside, it is necessary to clarify mass of actual NOx emissions per running distance from vehicles, geography and weather etc. on actual road. In this study, measuring method for NOx emission from vehicles on actual road was focused on as a primary factor for NOx pollution at roadside.

Real-time on-board measurement of NOx and fuel consumption was held in a small vehicle [1][2]. As diesel trucks play a main role for NOx pollution in urban cities, evaluation of NOx emission from freight trucks is desirable. Some on-board measurements of NOx from a diesel truck were conducted by CVS system so far [3][4][5][6][7][8]. But the system was rather big for on-board NOx measurement. (See Appendix 1 and 2).

In this study, simple on-board measurement system for road infrastructure, traffic conditions, vehicle running conditions and actual NOx emissions was set to a medium duty freight vehicle. Zirconia NOx sensor was installed in the vehicle to make the system simple. Emissions from vehicles depend on the running condition. The mechanism is very tough to solve because running condition is affected by various factors such as traffic condition, road infrastructure, and driver's operation etc. So video tape recording was also mounted on a tested vehicle. Hereafter this system is described as "On-board measurement system".

Exhaust emission tests on chassis dynamometer and actual on-road tests were conducted by using this system. The tests on chassis dynamometer including steady state and transient mode indicated enough accuracy to evaluate NOx and CO_2 emissions from a vehicle. The system was demonstrated on road and it was shown that information to analyze the mechanism of roadside NOx pollution could be obtained.

EXPERIMENTAL CONDITIONS

EXPERIMENTAL APPARATUS – The tested vehicles are shown in Figure 1 and 2. The vehicle 1 is the freight vehicle that passed Japanese 1998-exhaust emission regulation. Naturally aspirated OHC direct injection diesel engine with EGR system is mounted on the vehicle. The vehicle 2 is the mini-bus with turbo diesel engine. The main specifications of both vehicles are listed in Table 1. Items evaluated in this study are shown in Table 2.

Appendix 3, 4, 5 and 6 describe photos of Karman vortices air flow meter in air cleaner box, of excess air ratio sensor and of NOx sensor, simply installed in the tested vehicles. Appendix 7 shows the outline of On-board measurement system, measured data and sensors used.

Figure 1 Tested vehicle : Vehicle 1

Figure 2 Tested vehicle : Vehicle 2

EXPERIMENTAL METHOD –In this study, NOx mass emission was calculated by using zirconia NOx sensor and Karman vortices sensor. Table 3 shows important equations to obtain NOx mass emissions and fuel consumption. As shown in equation (2), NOx mass emission based on time was calculated by multiplying NOx concentration by an exhaust gas flow and NO_2 molecular weight. An exhaust gas flow was a summation of an intake air flow and fuel consumption from equation (4). CO_2 emission based on time was calculated from fuel consumption under the premise of H/C ratio of a fuel, 1.875. As shown equation (1), NOx mass emission based on distance was obtained from NOx mass emission based on driving time divided by a vehicle speed. Engine power was estimated from multiplying a vehicle speed by a running resistance that can be calculated from rolling resistance, air flow resistance, climbing resistance and acceleration resistance. EGR ratio was defined as the reduction of intake air flow.

Table 1 Main specifications of tested vehicles

	Vehicle 1	Vehicle 2
Vehicle type	Van type truck	Mini-bus
Length mm	4990	4850
Width mm	1880	2000
Height mm	2940	2525
Vehicle weight kg	2510	2250
Max. payload kg	2000	750
GVW kg	4675	3000
Passenger	3	10
Coefficient of drag	0.63	0.54
Frontal projected area m^2	5.4	4.9
Tire radius mm	320	324
Transmission	5MT	5MT
Final ratio	4.777	3.91
Engine type	OHC DI diesel	OHC DI turbo diesel
Number of cyl.	4	4
Swept volume cm^3	4334	2499
Bore×Stroke mm	112.0×110.0	93.0×92.0
Compression ratio	18.5	18.0
Max. output kW/rpm	81/3100	77/3800
Max. torque N• m/rpm	275/1500	222/2200
Fuel injection system	Electronic control	Mechanical control
EGR system	With	W/O

Table 2 Items considered in this study

Circumstances concerning road infrastructure

- Traffic condition
- Traffic light position : $X_{T.LIGHT}$ m
- Width of road : $X_{W.ROAD}$ m
- Altitude : h m
- Mean vehicle speed : V km/h

Driver's operation

- Driver's motion
- Gear position
- Vehicle position : X_{VHC} m
- Braking : Br
- Accelerator stroke : Ac mm

Running condition

- Vehicle speed : v km/h
- Vehicle acceleration : a m/s^2
- Engine speed : Ne rpm
- Fuel consumption rate : B_t g/s
- Engine power : Pe kW
- EGR ratio : EGR %
- Pay load : W_L kg
- Soak condition

Exhaust emissions

- NOx concentration : C_{NOx} ppm
- NOx mass emissions based on time : $E_{NOx,t}$ g/s
- NOx mass emissions based on distance : $E_{NOx,x}$ g/m
- CO_2 mass emissions based on time : $E_{CO2,t}$ g/s
- CO_2 mass emissions based on distance : $E_{CO2,x}$ g/m

Delay time for a measurement between the intake air flow and the NOx measuring point was 0.4 second at maximum in a stand-still of a vehicle. Sampling time was 0.5 second in this study, considering the response of NOx sensor shown in Table 4. This means that NOx mass emission of every four meters on road can be evaluated when a vehicle runs at a speed of 30 km/h.

Table 3 Main equations to calculate NOx mass emission

NOx mass emissions based on distance

$$E_{NOx,x} = E_{NOx,t} \times \frac{1}{v} \text{ ------------------------------ } (1)$$

NOx mass emissions based on time

$$E_{NOx,t} = \frac{Q_{EX,273K,1atm}}{2.24 \times 10^{2}} \times \frac{C_{NOx}}{10^{6}} \times M_{NO2} \text{ ---------------- } (2)$$

Exhaust air flow

$$Q_{EX,273K,1atm} = Q_{IN,273K,1atm} + \frac{B_t \times 6.85}{M_{FUEL}} \times 2.24 \times 10^{2} \text{ ---- } (3)$$

Fuel consumption rate

$$B_t = \frac{Q_{IN,dry,273K,1atm} \times D_{AIR}}{A/F_{stoich}} \times \frac{1}{LMD} \text{ ----------------- } (4)$$

NOx mass emissions based on distance $E_{NOx,x}$: g/m

NOx mass emissions based on time $E_{NOx,t}$: g/s

Vehicle speed v : m/s

NOx concentration C_{NOx} : ppm

Exhaust air flow at 273K $Q_{EX,273K,1atm}$: m^3/s

Intake air flow at 273K $Q_{IN,273K,1atm}$: m^3/s

NO_2 molecular weight M_{NOx} : g/mol

Fuel consumption B_t : g/s

Molecular weight of Fuel M_{FUEL} : g/mol

Intake dry air flow at 273K $Q_{IN,dry,273K,1atm}$: m^3/s

Air density D_{AIR} : g/m^3

Stoichiometric ratio A/F_{stoich} : –

Excess air ratio LMD : –

Table 4 Measurement accuracy and response time

	Intake air flow Q_{IN} m^3/s	Excess air ratio LMD	NOx concentration C_{NOx} ppm
Measurement accuracy	±2.9%	±3.3%	100ppm:±23% 500ppm:±5.5% 1000ppm:±2.9%
Response time	0.04s	0.05s	0.5s

The table also shows the measuring accuracy of important three sensors, which are those for an intake air flow, an excess air ratio and a NOx concentration. Accuracy within ±3 % is guaranteed for an air flow and an excess air flow, and accuracy within ±5.5% for NOx concentration is guaranteed in the condition more than a concentration of 500ppm. It was found that a fuel consumption calculated from an intake air flow and an excess air ratio could be estimated within accuracy of ±5%, according to an experiment comparing the actual fuel consumption and a calculated fuel consumption.

EXPERIMENTAL RESULT

EXPERIMENTAL CONDITIONS

Two major tests by vehicle 1 were conducted to evaluate the effectiveness of the On-board measurement system.

The first was a comparison of exhaust emissions between CVS system on chassis dynamometer and the On-board measurement system at a steady state and a transient test mode. The steady state test was Japanese diesel 13 mode test procedure (D13 test mode) and the transient test was Japanese 10-15 mode test procedure (10-15 mode).(See Appendix 8 and 9)

The second was a demonstration with two test vehicles in an actual traffic. In order to cover various traffic conditions, the tests were carried out in varying running conditions such as running time slot, mean vehicle speed, driver, vehicle soak time, and payload. A test route was set in Beijing and near Keio University, Yokohama, which includes an intersection, a hill, a turning left, a tuning right, a wide road, a narrow road and a traffic jam etc. The distance of the route near Keio University was about 17.5 km. The outline of the experimental conditions near Keio University was indicated in Table 5 as the results of its test are discussed in this paper mainly.

COMPARISON WITH CVS SYSTEM

Figure 3 shows the correlation of NOx and CO_2 between CVS system and the On-board measurement system at each operating point of D13 mode test, which was carried out for one time with Vehicle 1. NOx of the On-board measurement system tends to be just a little lower than that of CVS system but CO_2 of the on-board measurement system relatively demonstrated a good correlation. Table 6 shows the comparison of NOx and CO_2 between the On-board measurement system and CVS system at D13 mode test by Vehicle 1. In the table, the NOx emission of the On-board measurement system at D13 mode was lower by 9.1% from that of CVS system. On the other hand, the CO_2 emission of the On-board measurement system was higher by 6.8%.

Table 5 Outline of the experimental conditions

Test route	Area	Kouhoku-ku, Yokohama
	Characteristics of route	Distance 17.5 km
		Left turn 10
		Right turn 11
		Intersection with traffic light 12
		Intersection without traffic light 3
		Overpass 1
		Relief 27 m
		Road width 4~20 m
Test hours		9:00, 11:00, 14:00, 15:00 Starting time of test
Range of mean vehicle speed		11.8~19.3km/h
Driver		Y.T Driving career 30 years T.M Driving career 5 years
Engine condition		Cold start Hot start
Load		No load (0kg) Half load (1000kg) Full load (2,000kg)

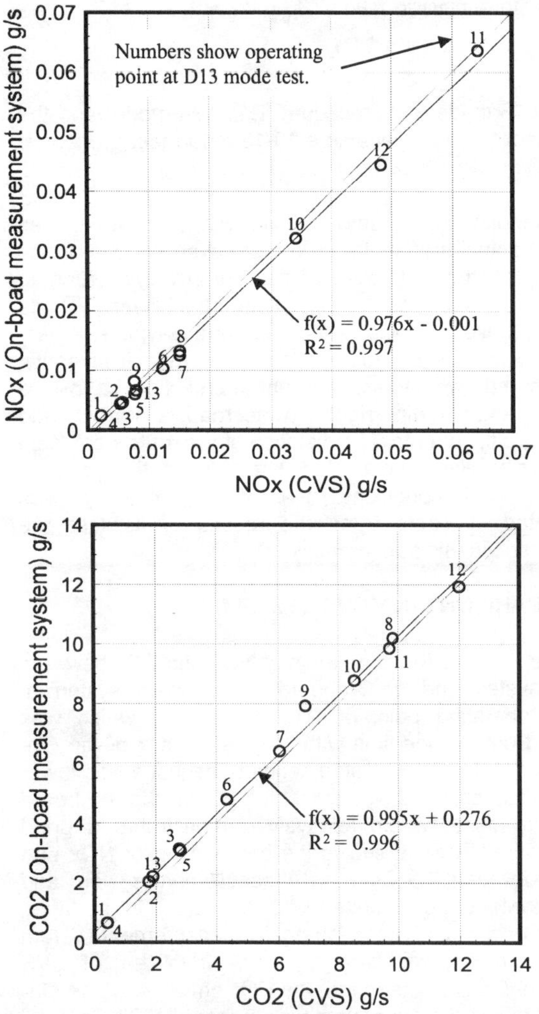

Figure 3 Correlation between CVS and on-board system at D13 mode test(Vehicle 1)

Table 6 Comparison of emissions between CVS and on-board system at D13 mode test(Vehicle 1)

	CVS system	On-board system	Emission regulation value	(CVS−On-board)/CVS
NOx g/kWh	3.669	3.33±0.20	5.8	0.091
CO_2 g/kWh	1053.0	1125±70	—	-0.068

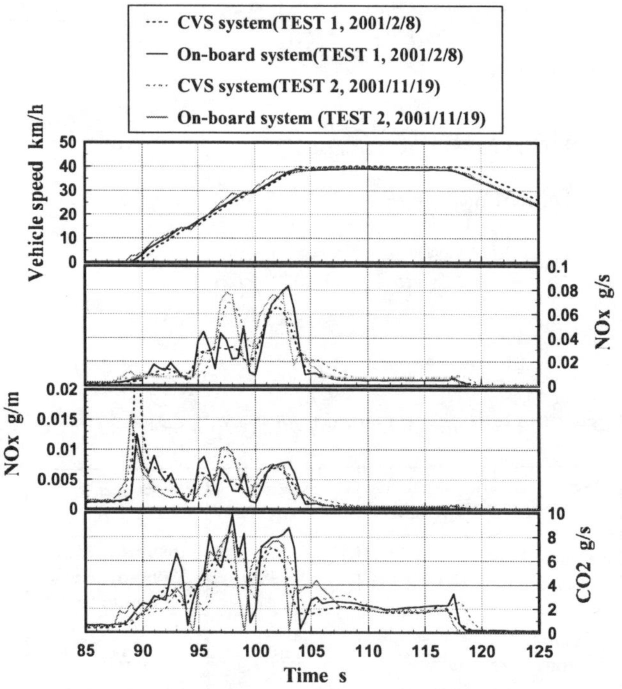

Figure 4 Comparison of transient response between CVS and on-board system at D13 mode test (Vehicle 1)

A good transient response of the On-board measurement system was demonstrated at 10-15 mode as indicated in Figure 4 which describes the history of vehicle speed, NOx and CO_2 from 80seconds to 140 seconds after starting of the test. Table 7 shows the results of 10-15mode tests, which were held for two times. A driver operated test 1, though a robot drove test 2. NOx of the On-board measurement system indicated difference of 1.4% and 8.0% with the CVS system. On the other hand CO_2 of the On-board measurement system in test 1 was measured lower by 28.1% than that of the CVS system, though it had good correlation with the CVS system in test 2. The reason why CO_2 of the On-board measurement system was higher in test 2 might derive from the variation of a driver's operation. The variation resulted in the operation of a driver will be studied in the future.

Table 7 Comparison of emissions between CVS and on-board system at 10-15 mode test (Vehicle 1)

	TEST 1 (In. air temp. 25degC, Humidity 45%)			TEST 2 (In. air temp. 25degC, Humidity 49.7%)		
	CVS system	On-board system	(CVS − On-board) / CVS	CVS system	On-board system	(CVS − On-board) / CVS
Mean vehicle speed km/h	22.6	21.9	0.031	22.2	22.3	-0.005
NOx g/km	1.423	1.404	0.014	1.359	1.250	0.080
CO_2 g/km	292.9	375.3	-0.281	314.0	304.6	0.030

RESULTS ON ACTUAL RUNNING TESTS

Actual NOx and CO_2 on road – Figure 5 shows NOx and CO_2 from Vehicle 1 on actual road in the test route near Keio University. The test conditions such as a running date and time, an engine soak condition and a payload were varied in the tests. In the tested conditions, maximum NOx emission per route was 2.5 times more than minimum one. Maximum CO_2 emission per route was 1.4 times more than minimum one. At engine hot conditions NOx was emitted 1.6 times compared to the no payload condition when the payload is full. CO_2 was emitted 1.2 times at the above-mentioned condition. Noteworthy results were obtained at engine cold conditions. NOx at the engine cold condition was higher 1.3 times than that at the engine hot condition. Figure 6 describes the history of the engine coolant temperature and NOx emission in each soak condition versus the vehicle running distance. During the period until which the engine coolant temperature reaches to some level, NOx emission level of a cold condition was about 2 times higher than that of a hot condition. As the intake air flow measured by the On-board measurement system showed that EGR system was not operating during the cold engine condition. So it was found that high NOx emission at the engine cold conditions was caused by no EGR operation.

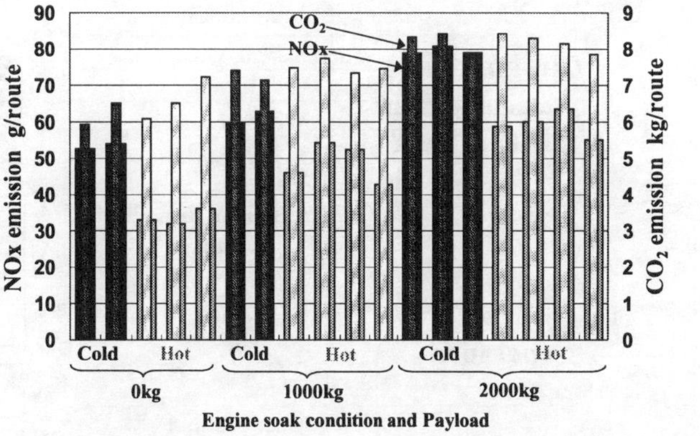

Figure 5 Effect of engine soak condition and payload on NOx and CO_2 emissions (Vehicle 1)

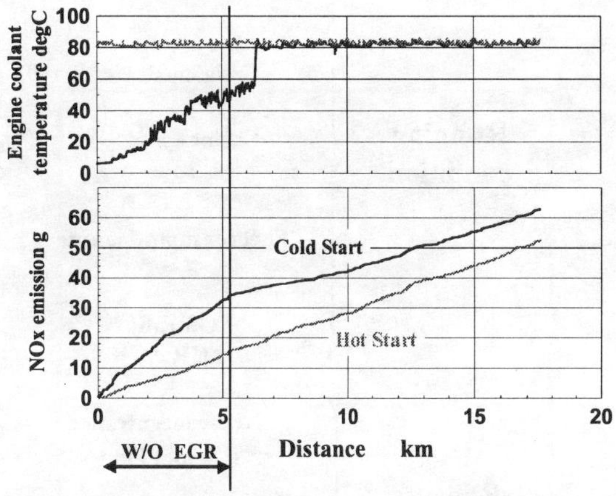

Figure 6 Influence of engine soak conditions on NOx

Figure 7 is the relationship between NOx and CO_2 on actual road in Vehicle 1 and 2. Data from Vehicle 1 were ones tested in the route near Keio University and data from Vehicle 2 were ones tested in Beijing. In Vehicle 1 in which EGR system was installed, there seemed to be two tendencies on the relationship between NOx and CO_2. It was found that there was clear difference of NOx characteristics in vehicles between with EGR system and with no EGR system.

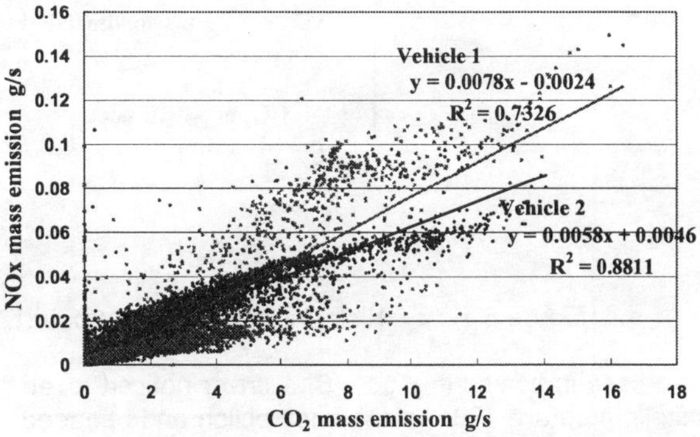

Figure 7 Relationship between NOx and CO_2 emissions on actual road in Vehicle 1 and 2

Confined roadside area NOx pollution – In order to clarify confined roadside NOx pollution, NOx emitted at an intersection was analyzed. Detailed data was measured in the case that the vehicle stopped just before the intersection according to the traffic light. Figure 8 shows the road infrastructure, driver's action, vehicle behavior and emissions around an intersection. Circles, which represent the emitted NOx around an intersection, were described at top of Figure 8. The radius of the circle represents the mass of NOx emitted. The interval between centers of the circle shows the distance that the vehicle advances by 0.5 second.

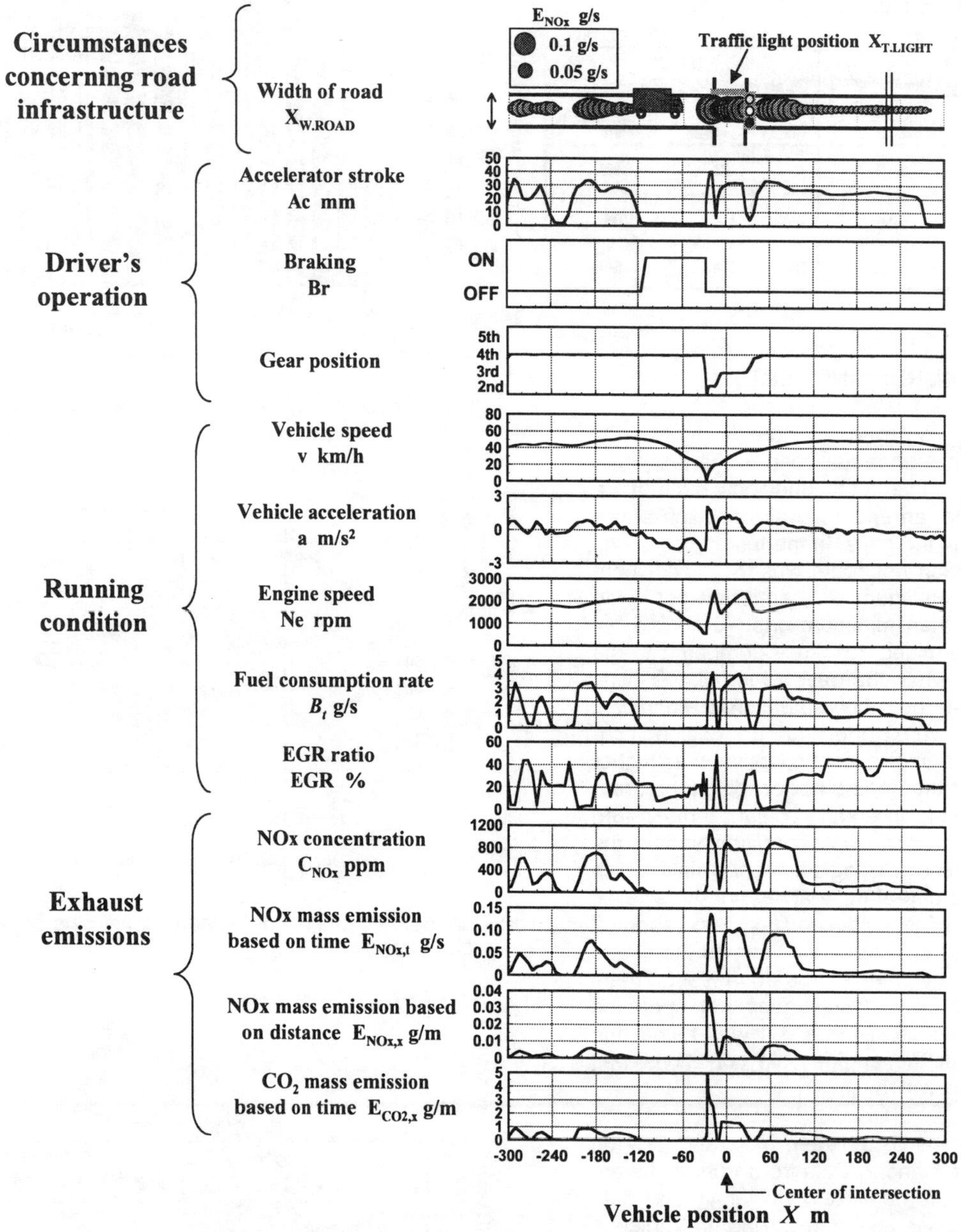

Figure 8 Histories of road infrastructure, driver's operation, running conditions and emissions

In the case indicated in Figure 8, a driver noticed a red traffic light before 120m of the intersection and released accelerator pedal. Ant then he put his leg on the braking pedal at the point of 110m to the traffic light. The vehicle decelerated gradually and stopped at the point of 30m before the center of the intersection. After 75 second stand-still, it started moreover. Regarding the distance-based NOx, it decreased to nearly zero during the vehicle deceleration. At the stand-still, it indicated maximum value. The distance-based NOx at the vehicle stand-still is defined as an accumulation for the time during the stand-still. After that, high NOx concentration was observed when the vehicle was accelerated. Although the maximum NOx concentration during accelerations was not so different, distance-based maximum NOx were quite different according to the position of the gear. That is to say, NOx is more emitted to roadside when the vehicle speed is lower and engine is operated at higher load, though the engine emit equal NOx per time during accelerations. This intensive NOx emission might be a cause of pollution at confined roadside area. As the result, it was found that it was possible to evaluate pollution at a confined roadside by this On-board measurement system.

CONCLUSION

On-board measurement system for vehicle was developed to evaluate the road infrastructure, driver's operation, vehicle running conditions, NOx mass emission and CO2 mass emission. The system was installed in a freight vehicle whose payload is two tons. Exhaust emission tests including steady state and transient mode on chassis dynamometer, and actual on-road tests were conducted by using this system. Conclusions are as follows.

- The NOx emission of the On-board measurement system at steady state mode test (D13 mode) was different by 9.1% from that of CVS system. The CO_2 emission of the On-board measurement system was correlated by the difference of 6.8%. The NOx emission of the On-board measurement system at transient mode test (10-15 mode) was different by 8.0 to 14.0% from that of CVS system. On the other hand, the CO_2 emission of the On-board measurement system was rather different in two tests. As the results, NOx distribution at roadside from a running freight vehicle could be easily evaluated by this on-board measurement system with enough accuracy.
- It was found that NOx at the engine cold condition was higher than that at the engine hot condition. The reason resulted from no EGR operation at engine cold condition.

It was demonstrated that On-board measurement system could be used to evaluate the road infrastructure, driver's operation, vehicle running conditions, NOx and CO_2 emissions at a confined roadside.

ACKNOWLEDGMENTS

This research was conducted under the support of the Organization for the Promotion of Low Emission Vehicles (LEVO). The cooperation from LEVO for this study is gratefully acknowledged

REFERENCES

1. Nobutaka Kihara, Tokihiro Tsukamoto, Koichi Matsumoto, Kozo Ishida, Masao Kon, Takao Murase, "Real-time On-board Measurement of Mass Emission of NOx, Fuel Consumption, Road Load, and Engine Output for Diesel Vehicles", SAE Technical Paper, No.2000-01-1141, p.1-9
2. Nobutaka Kihara, Tokihiro Tsukamoto, "Real-time On-board Measurement of Mass Emission of NOx, THC and Particulate Matters from for Diesel Vehicles", COMODIA, No.1-06, 2001
3. Hisashi YOKOTA, Ken NAKAMURA, Saburo FUKUOKA, Masanao FUNESHIMA, "NOx Emission Characteristics from Diesel Truck by On Board Type Measurement system (on Arterial Road)", J. Japan Soc. Air Pollut. 24 (4), 276~286, 1989
4. Joseph M. Norbeck, J. Wayne Miller, William A. Welch, Matthew Smith, Kent Johnson, David Pankratz, "Develop On-Road System for Emissions Measurement from Heavy-Duty Trucks", South Coast Air Quality Management District Contract 20906,2001
5. Koichi YOSHIDA, Noriyuki KOIKE, Yujiro TSUKAMOTO, Kazuyuki NARUSAWA, "Method for Estimation of NOx Emission from Diesel Heavy Duty Vehicles", Traffic Safety and Nuisance Research Institute, Ministry of Transport, Report, No.91, 1982
6. Staab, J.; Schurmann, D.; "Measurement of Automobile Exhaust Emissions under Realistic Road Conditions", SAE paper 871986, 1986
7. D.Bruce Harris, Foy G. King, Edward Brown, "Development of On-road Emission Factors for Heavy-duty Diesel Vehicles Using a Continuous Sampling System", Technical Report Data, No.600/A-98/125, p.1-19, 1999
8. Joseph M. Norbeck, J. Wayne Miller, William A. Welch, Matthew Smith, Kent Johnson, David Pankratz, "Develop On-Road System for Emissions Measurement from Heavy-Duty Trucks", South Coast Air Quality Management District Contract 20906,2001

APPENDIX 1 Current on board type measurement system of NOx emission [3]

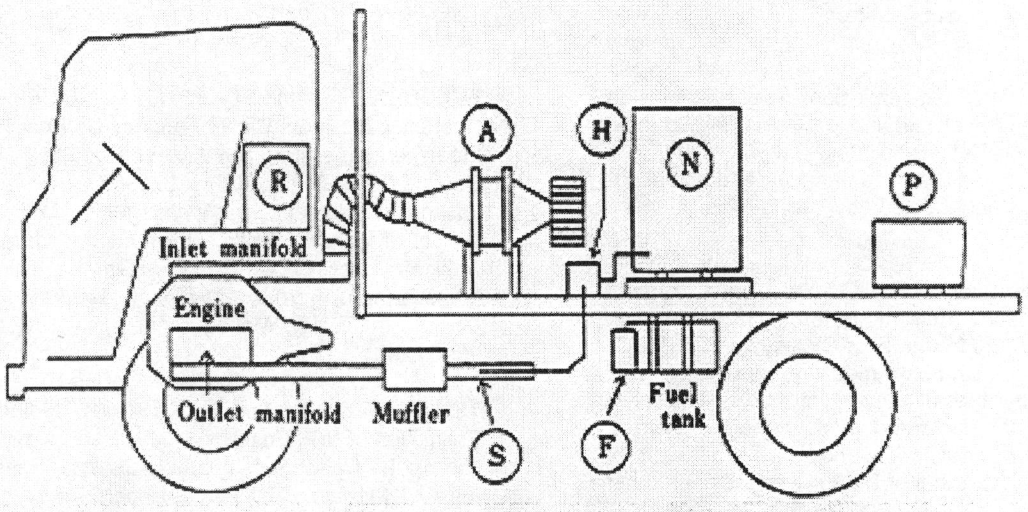

A: Air flow meter, F: Fuel Flow meter, H: Heated filter, N: NO_x analyzer, P: Engine generator, R: Data recorder, S: NO_x sampling point.

APPENDIX 2 Current on board type measurement system of NOx emission [8]

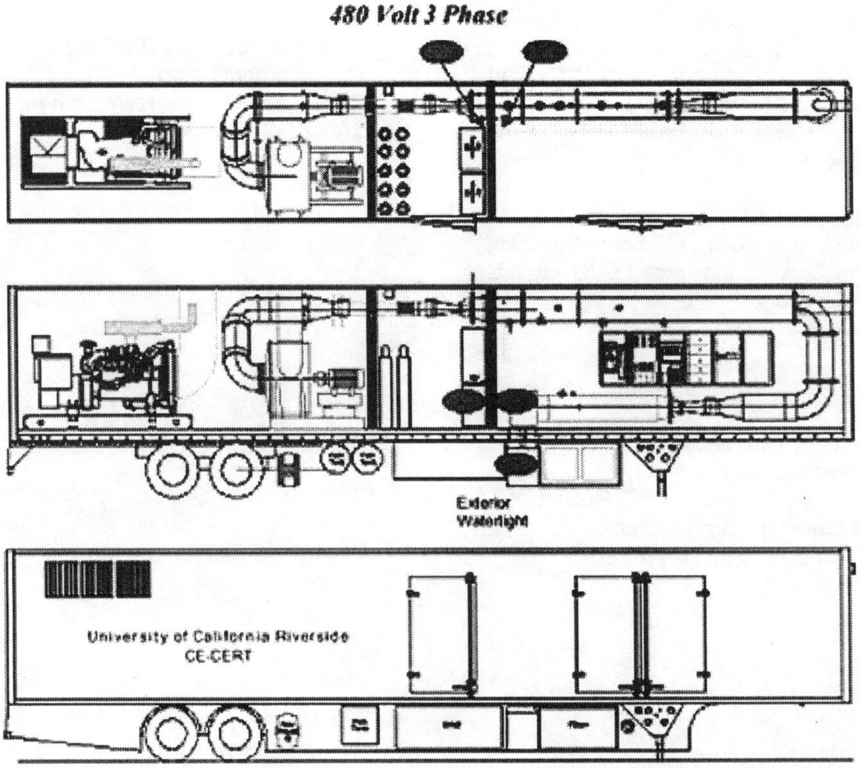

APPENDIX 3 Layout of air cleaner box (Vehicle 1)

APPENDIX 4 Karman vortices air flow meter (Vehicle 1)

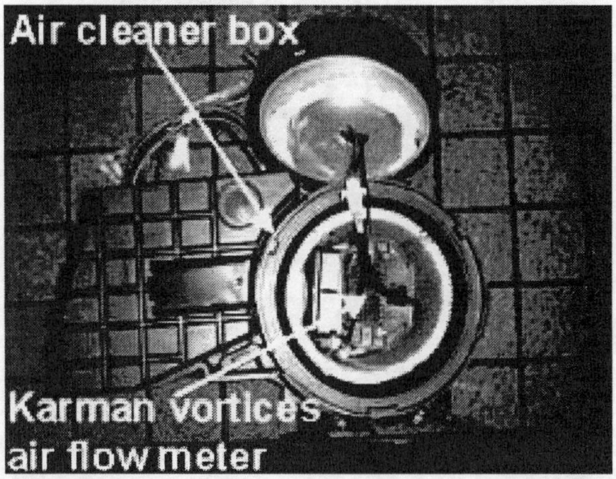

APPENDIX 5 Karman vortices air flow meter in air cleaner box (Vehicle 2)

APPENDIX 6 Excess air ratio sensor and NOx sensor at exhaust pipe (Vehicle 2)

APPENDIX 7 Outline of On-board measurement system

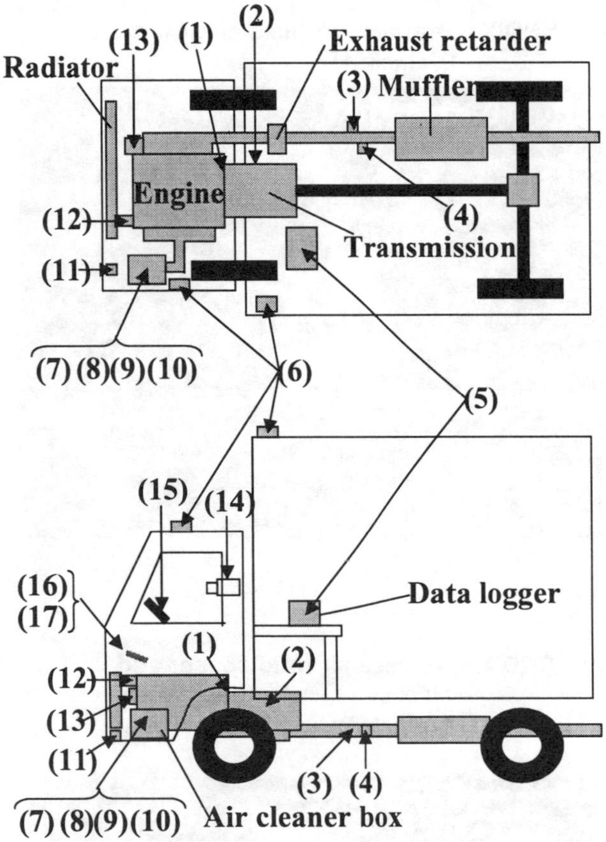

(1)Engine speed : *Ne*(Magnet pick-up)
(2)Vehicle speed : *v*(Rotary encoder)
(3)NOx concentration : C_{NOx}
(Zirconia NOx sensor)
(4)Excess air ratio : *LMD*(O_2 sensor)
(5)Atmospheric pressure : P_{ATM}
(Strain gauge pressure sensor)
(6)Vehicle position : *X* (GPS)
(7)Intake air flow : Q_{IN}
(Karman vortices air flow meter)
(8)Intake air pressure : P_{IN}
(Semiconductor sensor)
(9)Intake air temperature : T_{IN}
(Platinum resistor)
(10)Intake air humidity : H_{IN}
(Polymer humidity sensor)
(11)Atmospheric temperature : T_{ATM}
(Platinum resistor)
(12)Cooling water temperature : T_{WAT}
(K thermocouple)
(13)Air conditioner ON/OFF : Acon
(Signal of operation)
(14)Traffic condition and road structure
(Digital video camera)
(15)Steering wheel angle : St
(Digital video camera)
(16)Accelerator stroke : Ac
(Variable resistance)
(17)Brake ON/OFF : Br (ON/OFF Switch)

APPENDIX 8 Japanese D13 test mode

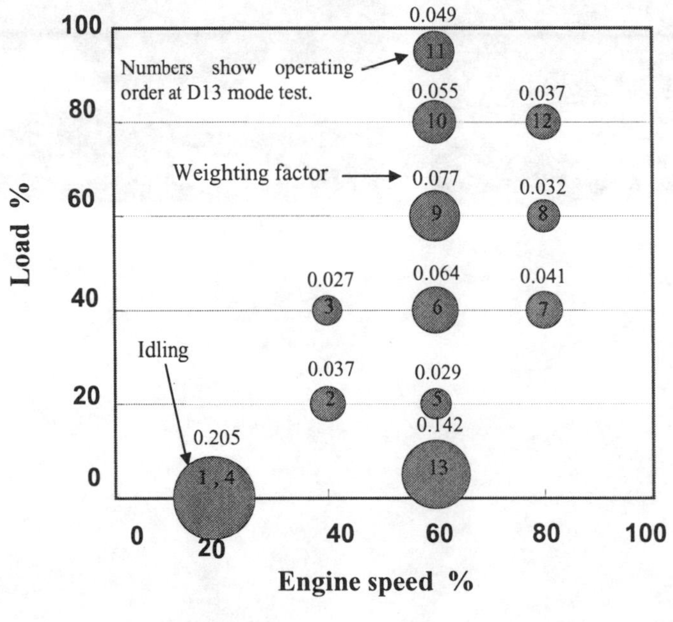

APPENDIX 9 Japanese 10-15 test mode

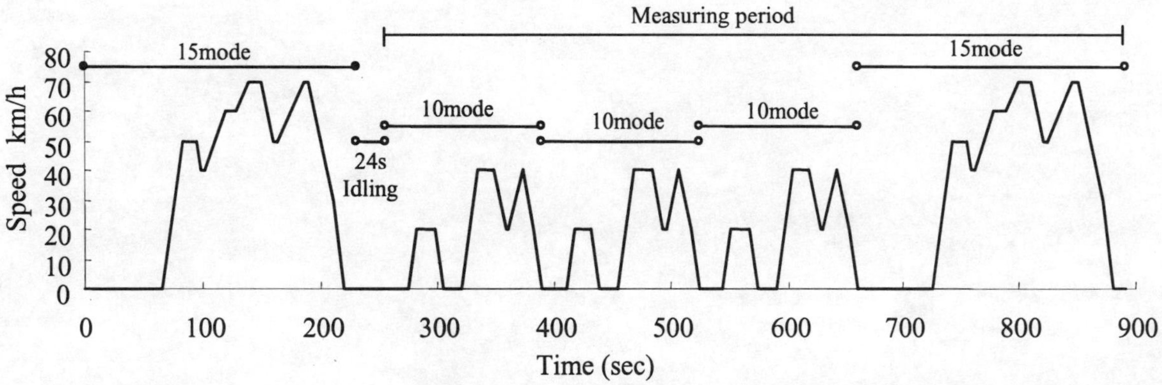

2002-01-0614

Inference of Torque and Power from Heavy-Duty Diesel Engines for On-Road Emissions Monitoring

Gregory J. Thompson, Nigel N. Clark, Mridul Gautam, Daniel K. Carder and Donald W. Lyons
West Virginia University

ABSTRACT

Increased concerns about the emissions produced from mobile sources have placed an emphasis on the in-use monitoring of on- and off-road vehicles. Measuring the emissions emitted from an in-use vehicle during its operation provides for a rich dataset that is generally too expensive and too time consuming to reproduce in a laboratory setting. Many portable systems have been developed and implemented in the past to acquire in-use emissions data for spark ignited and compression ignited engines. However, the majority of these systems only measured the concentration levels of the exhaust constituents and or reported the results in time-specific (g/s) and or distance-specific (g/km) mass units through knowledge of the exhaust flow. For heavy-duty engines, it is desirable to report the in-use emission levels in brake-specific mass units (g/kW-hr) since that is how the emission levels are reported from engine dynamometer certification testing. Only a few portable systems have attempted to report the emissions in brake-specific mass units and have met with limited success. A method to infer the load and power is presented for modern, electronically controlled, compression ignition engines. The associated errors with this method are detailed along with a comparison for three different engines. The results show that the errors are on the order of 5 to 10% but can be as low as 1% when the cumulative energy is averaged for a transient cycle for a well-maintained engine operating through the FTP cycle.

INTRODUCTION

In-use monitoring of the emissions generated by on- and off-road vehicles is becoming more prominent. Actual in-use emissions data provides for a rich dataset that is generally too expensive and too time consuming to obtain in an engine laboratory setting. Many portable systems have been developed and implemented in the past to acquire in-use emissions data with an on-road emissions measurement system (OREMS), but only a few systems have attempted to obtain brake-specific mass emissions from heavy-duty (compression ignition) engines operated in trucks, busses, and off-road equipment.

The two major components needed to measure brake-specific mass emissions are the mass of each constituent emitted over a test period and the work delivered to the engine output shaft over that same period. The mass of each constituent emitted in an OREMS environment is a daunting task and requires the measurement of both concentration of species and total exhaust mass flow. These measurements have been discussed elsewhere [1-4]. The measurement of engine power, or the product of engine speed and torque, can be obtained through the electronic control unit (ECU) that is used to control modern heavy-duty diesel engines: this is the topic explored in this paper.

On-road engine speed and torque measurements will differ from in-cell laboratory measurements due to the fundamental differences between the two types of tests. In-cell engine tests use a dynamometer to control and measure the engine speed and load on the engine to a high and verifiable accuracy. On-road tests will rely upon ECU broadcast load information and engine speed measurements provided by on-board sensors. The broadcast load is invariably computed by the ECU from engine fueling and speed information, using a model or algorithm proprietary to the manufacturer. Although in-line techniques (shaft collars) are available for measuring the torque directly, these methods fail to account for accessory work and are installed only with difficulty. The best method, therefore, to estimate output shaft power is via ECU broadcast. However, a disadvantage of relying upon the ECU data for shaft power estimation is that only electronically controlled vehicles with required signal output can be evaluated.

The methods used to arrive at engine output power derived from ECU parameters are discussed with the potential errors involved. Comparisons between manufacturer-supplied data for an average production engine and laboratory-derived data for a specific engine are examined.

LITERATURE REVIEW

ECU COMMUNICATION PROTOCOLS

Currently, there are three standards that are used for ECU communication in the United States for heavy-duty engine/vehicle applications, namely, SAE Standards J1587, J1922, and J1939 [5-7]. Generally, a protocol adaptor (hardware) is required to communicate between the ECU and a data-logging computer via a standard interface (for example, serial port, ISA card, PC104, etc.) [8]. There are two different types of packets of information that are broadcasted through the data link, termed public and private.

Although the SAE standards have provisions for a plethora of engine and vehicle information, not all of the information is broadcasted through public packets. It is desirable to use only publicly broadcast packets in inferring engine power, thus alleviating the task of implementing an engine company's proprietary hardware into an OREMS. The various packets of information are broadcast at different rates. For SAE J1587, engine speed is broadcast at 10 Hz with a 0.25 rpm resolution, engine percent load is broadcast at 10 Hz with a 0.5% resolution, and output torque is broadcast at 1 Hz with a 20 lb-ft resolution. However, output torque is generally not a publicly broadcasted parameter but the instantaneous torque can be inferred using the engine speed percent load as discussed below. The intent of broadcasting the percent load is usually for control of automatic transmission shift management and so is not required by application to be either very precise or truly linear in its representation.

ENGINE CERTIFICATION

Existing certification testing for heavy-duty, on-road engines in the United States involves testing the engine on an engine dynamometer through a prescribed test cycle. The procedures for performing this test are described in Title 40 Code of Federal Regulations Part 86 [9]. This testing is performed in a controlled laboratory environment where an engine dynamometer is used to load or motor the engine through the specified transient cycle. The exhaust stream is routed to a full-scale dilution tunnel where the raw exhaust is mixed with dilution air to simulate the real-world interaction of a vehicle's exhaust expelling into the atmosphere. In this way the measurement of vehicle exhaust flow is avoided, because only the flow of the diluted stream in the tunnel must be known. For off highway engines, a similar type of laboratory setup may be used except steady state testing is performed [10].

CONSENT DECREES

In 1998, following legal action over in-use oxides of nitrogen emissions levels, six settling heavy-duty diesel engine manufacturing companies (S-HDDE) entered into individual agreements (Consent Decrees) with the United States government [11]. The agreements state that, in addition to the standard engine-based Federal Test Procedure (FTP), engines will be tested according to the Euro III test procedure, which incorporates the steady state test and emission weighting protocols identified as the "ESC Test" in Annex III to the Proposal adopted by the Commission of the European Union on December 3, 1997. The engine manufacturers agreed also that their engines shall also be tested to demonstrate that they do not exceed prescribed emissions limits in a "Not To Exceed" (NTE) zone, Smoke or Alternative Opacity limits, and Transient Load Response limits. Engines must meet these limits when new and during in-use operations throughout the useful life of the engine, which is currently defined as 430,000 miles for engines in heavy-duty truck use.

As part of these Consent Decrees, each engine manufacturer is to obtain brake-specific mass emissions data on in-use pre and post Consent Decree engines. The development of a Mobile Emissions Measurement System (MEMS) is detailed in three reports and a publication that fulfill the on-road requirements of the Consent Decrees [1-4].

NTE ZONES

Engine speed and torque are required independently for the NTE zone determination and cannot be described using engine shaft power alone. Figure 1 illustrates the NTE area with the associated boundaries. The NTE zone is defined in the Consent Decrees and is bounded by engine speeds 15% above the ESC Speed,

$$\eta_{15\%ESCSpeed} = \eta_{lo} + 0.15(\eta_{hi} - \eta_{lo}), \qquad (1)$$

engine loads greater than 30% of maximum, and engine power greater than 30% of maximum. In Equation 1, n_{lo} is defined as the lowest engine speed at which 50% of the maximum power occurs, while n_{hi}, is defined as the highest engine speed where 70% of the maximum power occurs. The Consent Decrees require that exhaust emissions be reported for engines operating within the NTE zone averaged for a period of at least 30 consecutive seconds.

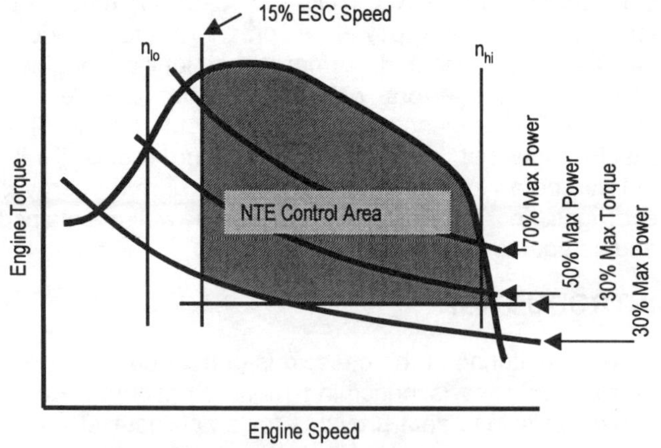

Figure 1 Example NTE area definition.

IN-FIELD EMISSION MEASUREMENT SYSTEMS

In-field emissions measurement systems have been developed for, and employed in, inspection and maintenance (I/M) programs and in various research activities, including emissions inventories and human exposure studies. A review of the work performed for portable and mobile emissions measurement systems over the last 20 years has shown that prior attempts to measure in-use emissions were limited in their success, only a few systems have attempted to measure engine power. At present heavy-duty I/M programs use only a curb-side snap idle smoke opacity test as described in the SAE J1667 Standard [12], although one county in Arizona uses a loaded smoke test. The SAE J1667 Standard provides a procedure for assessing smoke emissions from in-use vehicles powered by heavy-duty diesel engines and is intended to identify gross emitters. No gaseous species are measured in current United States heavy-duty I/M testing.

Most systems developed measured only concentration levels or included a method to measure exhaust flow to obtain time-specific and or distance-specific mass emissions. The concept of interfacing gaseous emissions analyzers with the ECU data was detailed in a 1992 patent [13]. These systems include a portable bag collection system developed by Caterpillar to quantify fuel specific NOx emission levels from in-use diesel engines [14]. Michigan Technological University (MTU) researchers developed an Emissions Measurement Apparatus (EMA) system and reported results from underground mining equipment tests [15]. A 1989 gasoline fueled passenger vehicle was instrumented and driven through city and highway routes to obtain real-world emissions data [16]. The emissions results from three different instrumented gasoline-fueled passenger vehicles are detailed in several reports [17-20]. The impetus of the study was to compare on-board measurements to remote measurement techniques. An on-board emissions measurement system for I/M was developed for natural gas-powered passenger vans at the University of Pittsburgh [21]. Vito, The Flemish Institute for Technological Research, performed on-board emission measurements with a system called VOEM (Vito's On-the-road Emission and energy Measurement system). Tests were performed on both gasoline cars and diesel buses. The emissions-assisted maintenance procedure (EAMP) for diesel-powered mining equipment was developed by the University of Minnesota [22]. The EAMP system was designed to be far more portable than the prior systems developed by Southwest Research Institute and MTU, but still very capable of detecting engine faults. A recent study by the Northeast States for Coordinated Air Use Management (NESCAUM) evaluated in-use emissions from diesel-powered off-road construction vehicles and explored the effects of various emissions control devices [23]. Ford Motor Company and WPI-Microprocessor Systems, Inc. are developing a Portable Real-Time Emission Vehicular Integrated Engineering Workstation (PREVIEW) that will sample water-laden exhaust [24]. Horiba, Ltd. and NGK Insulators, Ltd have recently presented an on-board NOx emissions measurement system for diesel vehicles [25]. Preliminary work on an FTIR-based system was recently presented by Honda R&D Americas, Ltd., Honda R&D Co., Ltd, and Nicolet Instrument Corp. for measuring real-world emissions from light duty gasoline vehicles [26].

Systems developed to measure or infer the engine load, along with the mass emissions, include work performed by Southwest Research Institute from 1978 to 1983 to develop a system to test diesel engines in a mine for an I/M program [27]. The transportable system consisted of a portable engine dynamometer, laboratory-grade emissions instruments, volumetric fuel flow meter, and a laminar air meter. A second system was developed by Southwest Research Institute to measure exhaust emissions from diesel buses and to compare the data against the U.S. Environmental Protection Agency's (US-EPA's) database of transient engine emissions [28]. The cycles ranged from idle, no-load testing to loading the engine against the transmission through prescribed accelerator pedal positions. The prescribed test procedure could only be performed on vehicles with automatic transmissions. A 1992 SAE paper and a 1997 report describe the on-board testing of U.S. Coast Guard Cutters to assess the emissions as part of the 1990 Clean Air Act for non-road air pollution [29,30]. Although the system was recognized as being too bulky and lacking portability, it demonstrated that emissions tests could be performed on-board a ship. The monitoring system incorporated air and fuel flow measurements and provided for inference of engine-out torque via driveshaft mounted strain gauges, via fuel measurement, and from an on-board power meter. The Office of Mobile Sources at the US-EPA is currently developing a mobile measurement system, termed ROVER, for light-duty gasoline vehicles and for use on heavy-duty vehicles [31].

EXPERIMENTAL PROCEDURE

Five different heavy-duty diesel engines were investigated in this work and include a 1995 Mack E7-400, a 1999 Cummins ISM-370, a 1999 Navistar T444E, a 1998 Navistar T444E, and a 2000 DDC Series 60. Each engine was tested in the engine dynamometer test cell described below over a two year research period. Although not all of the data from these engines are presented, there were no differences in the end result from one engine, or engine manufacturer, to another.

An engine dynamometer test cell at West Virginia University, capable of performing heavy-duty transient and steady state testing, was used to perform the work [32]. Although dilute emissions can be measured with the full-scale dilution tunnel in this laboratory, emissions data are not reported here. A 550 hp General Electric model 42G263AK DC dynamometer was used to absorb or motor the engine under test. An in-house developed QuickBasic code was used to control the engine and measure the necessary data. An in-house developed Visual Basic program was used for the post processing of the engine laboratory's data following the requirements of 40 CFR Part 86 [9]. It is noted that the reported laboratory torque is the dynamic torque and not the static torque measured from the dynamometer's load cell: in other words, the dynamometer acceleration was taken into account.

All ECU data were collected with a DPA II+ protocol adaptor [8]. Only the SAE J1708/1587 protocols were investigated. An in-house developed Visual Basic data collection program, incorporating ActiveX controls [8], was used to collect the data from the protocol adaptor to a Windows-based PC. Generically, the data collection program "dumped" the raw data from the protocol adaptor to disk for post processing. A second Visual Basic program was used for the off-line post processing of the raw data into engineering unit data.

INFERENCE OF ENGINE POWER

Engine speed and torque must be measured by an OREMS to report emissions data in brake-specific units. Engine torque derived from the ECU broadcast parameter is based upon fueling commands and assumed engine efficiency by the manufacturer. The ECU-derived torque approach is obtained from in-field measurement of the ECU data and from manufacturer's supplied data. The manufacturer-supplied data are for a typical engine for that engine series and it is known that engine-to-engine power variations may exceed 1%. The ECU broadcast speed and torque is reliable and can be employed directly for an OREMS measurement. The inference of power from broadcast of ECU engine speed and percent load can be accomplished with accompanying manufacturer's lug torque curve and a curbside no-load test. The lug curve is the maximum torque generated by the engine at a given engine speed as defined in 40CFR86.1332-90 [9]. The curbside no-load test is performed with the transmission disengaged and the engine operated from curb idle to governed speed to determine the required fueling (percent load) to overcome the internal loading of the engine. It is recognized that accessory loadings (which are not included during certification testing) are associated with the brake-specific mass emissions in an OREMS and must be identified during on-road testing. The addition of the accessory loadings will be significant at low to medium loadings since these loadings will influence the denominator in the brake-specific mass emissions calculation.

ECU ENGINE SPEED

Engine speed from the ECU is obtained from the protocol adaptor data stream and translated into engineering units with the relevant SAE J Standard conversion. Since electronically controlled engines rely upon engine speed (insomuch that redundant engine speed sensors may be implemented), the measurement (hence broadcast) of engine speed is very reliable. Generally, the engine speed measurement is simple and typically consists of a magnetic pickup used to detect the teeth on the starter gear. The "tooth count rate" ranges from 1000 to 4000 teeth per second for a typical on-road heavy-duty engine.

A comparison between measured laboratory and ECU broadcast engine speed is shown in Figure 2 for a heavy-duty diesel engine operating through the FTP cycle. As shown in this figure, the percent difference varies from –6.2% (during a steep deceleration) to 13.6% (during an aggressive acceleration). It should be noted that these differences might be attributed to slight time alignment errors between the two different data acquisition computers and the 5 Hz data used to generate the chart. The average absolute percent difference between ECU broadcast and measured laboratory engine speed, over the FTP cycle, is 0.55% for the test shown in Figure 2. FTP test data from a 1995 Mack E7-400, 1999 Cummins ISM 370, and a 2000 DDC Series 60 550 hp shows that the average absolute percent difference ranges from 0.4 to 0.7% for multiple (at least three) repeat runs. This suggests that ECU broadcast engine speed is a very reliable and accurate measurement.

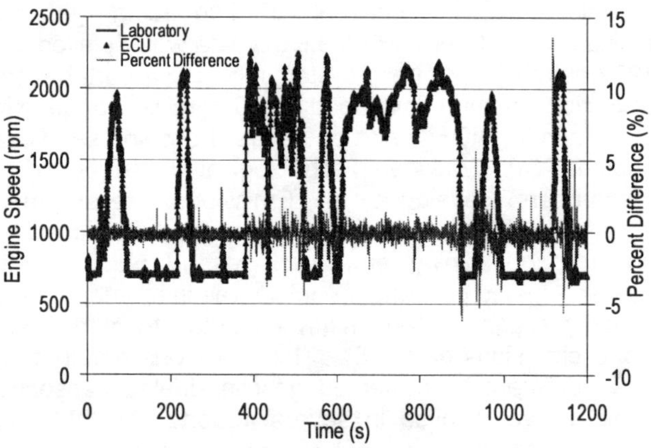

Figure 2 Comparison between measured laboratory and ECU broadcast engine speed for a heavy-duty diesel engine exercised through the FTP cycle.

ECU-DERIVED ENGINE TORQUE

The ECU-derived torque is obtained through a percent load broadcast value, an engine speed broadcast value, a curbside no-load idle curve, and a lug curve. The lug curve may be obtained from the certification testing lug curve where the maximum engine output torque is measured at a rate of 8 rpm/s or may be obtained from steady-state maximum torque data. The percent load generally varies from 0 (no fueling, friction load) to 100 (maximum torque) at a given engine speed, although there are provisions for the percent load to exceed 100%. It is recognized that torque, and hence percent load, will vary with engine operating temperature and will vary with the fuel's properties (viscosity, density, heating value, and cetane rating). Modern engines employ late injection timing, and a slow burn rate or long injection delay may reduce the indicated work on the piston. It is also recognized that the ECU-derived torque is engine total output torque and not merely the flywheel torque that is used for US-EPA certification tests. The reader must realize that auxiliary loads, including fan, alternator, and compressor loads may exceed 30 hp in some heavy-duty truck applications. To arrive at an accurate estimate of engine output torque at a given engine speed, the lug curve (100% load) must be combined with either the friction torque (zero fueling) curve or the zero flywheel (zero output shaft load) percent load curve.

The current approach involves measuring the no-load percent load (ECU^{rpm}_{noload}) through the speed domain at the curb and employing the lug curve (T^{rpm}_{max}) obtained through laboratory testing or from manufacturer-supplied data. The resulting engine torque ($T^{rpm}(t)$) at a given engine speed, percent load ($ECU^{rpm}_{\%}$) and time is obtained by

$$T^{rpm}(t)=\left(\frac{ECU^{rpm}_{\%}-ECU^{rpm}_{noload}}{ECU^{rpm}_{\%max}-ECU^{rpm}_{noload}}\right)*T^{rpm}_{max}. \qquad (2)$$

It is stressed that the above equation is a function of engine speed, as indicated by the superscript "rpm," for each of the parameters. Equation 2 assumes that the internal friction load is a function of speed only. However, friction load is also dependent upon the absolute engine load because bearing clearances change with rod load and piston ring and skirt friction change with in-cylinder pressure. Hence, the above relationship will overestimate the actual load.

A composite chart of the shaft torque measured with a dynamometer and the ECU broadcast percent load for an electronically controlled engine is shown in Figure 3. The no-load percent load (P1) varies as a function engine speed, while the output torque (T1) is constant. For the lug curve, the percent load (P2) is constant at 100% up to approximately 2670 rpm while the torque (T2) varies throughout the speed domain. At an engine speed above 2670 rpm, the percent load drops sharply from 100% down to the no-load percent load level. The lower NTE torque limit is also illustrated in this figure for the percent load (P3) and torque (T3).

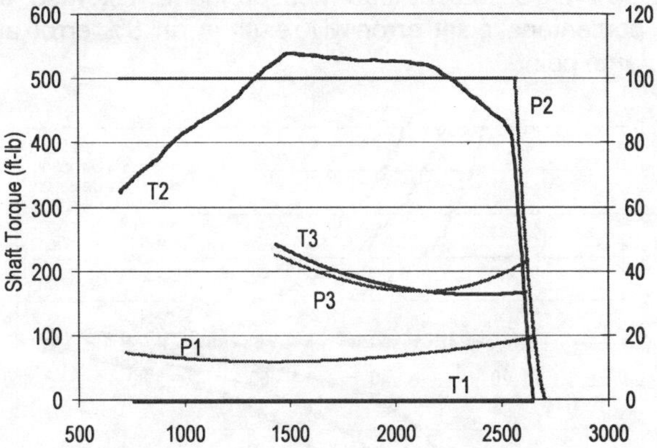

Figure 3 Shaft torque and ECU percent load variation for a modern electronically controlled engine.

As shown in Figure 3 and Equation 2, the error in estimating the torque at a given time ($T^{rpm}(t)$) will depend upon the percent load signal from the ECU ($ECU^{rpm}_{\%}$), the lug curve torque (T^{rpm}_{max}), the associated lug curve percent load ($ECU^{rpm}_{\%max}$), and the curb idle no load percent load (ECU^{rpm}_{noload}). It is stressed that the torque is not only critical in the determination of brake-specific mass emissions but also in the binary determination of whether the engine is or is not in the NTE zone.

Figure 4 illustrates the error in torque determination resulting from errors in the percent load ($ECU^{rpm}_{\%}$). The family of curves shows the error in the estimated torque due to the deviation of the actual percent load above or below the broadcasted value. Shown in this figure are the no load percent load (ECU^{rpm}_{noload}) (assumed to be at 14% for illustrative purposes), the error associated with a ±1%, and the error associated with a ±2% deviation. At low load conditions the error is greatest and approaches infinity at the curb idle location (14% in this illustration) and asymptotically approaches a minima at the 100% load conditions. For example, at a nominal 33% load, an error of one percentage point (±1) in the measured ECU load reading (32% or 34%) will result in a 5.3% error (low or high) in the torque estimation ($T^{rpm}(t)$).

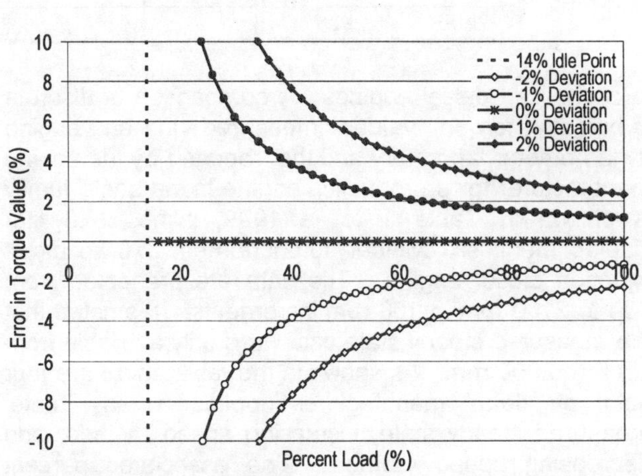

Figure 4 Error in the inferred torque due to an error in measured percent load.

An error in the lug curve torque (T^{rpm}_{max}) will result in a proportional error in the inferred torque. Errors in the lug curve torque can be from many sources. For a well-maintained engine, the lug curve torque should be the same over the useful life of the engine. However, for a poorly maintained engine, the lug curve torque could vary significantly from a well-maintained engine. A change in the oil viscosity can also influence the lug curve. It can be expected that laboratory-to-laboratory variations in measuring the lug curve would be on the order of 1 to 2%, the allowable error in calibrating the dynamometer load cell. This is the least error that should be expected in inferring power via ECU information.

The procedure used to measure the maximum torque can also influence the maximum torque value. Per 40 CFR Part 86 requirements, an increasing engine speed of 8 rpm/s from minimum to maximum speed is to be used to calculate the maximum torque curve for generating the test cycle. However, a decreasing engine speed procedure may be used for non-governed or non-

Table 1 Lug curve comparison between laboratory measured and manufacturer supplied data. Speeds increased or decreased at a rate of 8 rpm/s.

Engine Speed (rpm)	Manufacturer Reported (ft-lb)	Steady State (ft-lb)	Percent Difference (%)	Increasing Speed (ft-lb)	Percent Difference (%)	Decreasing Speed (ft-lb)	Percent Difference (%)
700	840	-	-	656	-21.9	811	-3.5
800	871	-	-	811	-6.9	820	-5.8
900	996	-	-	931	-6.5	974	-2.2
1000	1160	-	-	1074	-7.5	1122	-3.3
1100	1285	-	-	1196	-6.9	1235	-3.9
1200	1350	1236	-8.4	1247	-7.7	1282	-5.0
1300	1350	1247	-7.6	1244	-7.9	1280	-5.2
1400	1350	1249	-7.5	1241	-8.1	1279	-5.3
1500	1307	1235	-5.5	1210	-7.4	1245	-4.7
1600	1264	1192	-5.7	1163	-8.0	1212	-4.2
1700	1189	1133	-4.7	1084	-8.8	1135	-4.6
1800	1123	1092	-2.8	1021	-9.1	1071	-4.6
1900	1057	1091	3.2	985	-6.8	1043	-1.3
2000	991	1070	8.0	925	-6.7	990	-0.1
2100	925	-	-	845	-8.6	876	-5.3

turbocharged diesel engines. A comparison of discrete maximum torque values measured in an engine dynamometer laboratory and that reported by the engine manufacturer for a production engine in the same family is shown in Table 1 for a 1999, inline, 10.8 liter displacement, six cylinder, turbocharged, 370 hp direct injection diesel engine. The data are presented from 700 to 2100 rpm in 100-rpm increments. It is noted that the measured steady state data were only available from 1200 to 2000 rpm. As shown in the table, there are four sets of data: manufacturer-supplied steady state, measured steady state, increasing speed method, and decreasing speed method. The associated percent difference between the measured data and the manufacturer-supplied data is presented for each engine speed data point. For example, at 1800 rpm, the manufacturer reported a torque of 1123 ft-lb. A steady state value of 1092 ft-lb was measured (-2.8% difference), an increasing speed method measurement shows a value of 1021 ft-lb (-9.1% difference), and a decreasing speed method measurement shows a value of 1071 ft-lb (-4.6% difference).

The manufacturer-supplied data can be, on average, 3.5% different for the steady state procedure, over 8% different for the increasing engine speed procedure, and 4% higher for the decreasing engine speed procedure as shown in the table. It should be noted that in-use engines descend in engine speed as much as they ascend. If the engine speed is descending, the actual torque may be higher, resulting in the brake-specific mass emissions to be reported lower than actually determined. For the data presented, the manufacturer-supplied lug curve and an increasing speed lug curve are compared for the transient tests. In practice, a manufacturer-supplied lug curve will be used.

Figure 5 illustrates the error in the torque when the no-load (ECU^{rpm}_{noload}) percent load in Equation 2 is assumed to be 14%. The family of curves shows the error in the estimated torque value as a function of the actual no-load percent load deviation above or below the assumed/measured value. At low load conditions the error is greatest and approaches infinity. As the percent load approaches the 100% load conditions, the error asymptotically approaches zero. For example, an error of one percentage point (±1) in the no-load ECU load reading will result in a 4% error in the torque estimation at a 33% ECU percent load measurement. Likewise, a two percentage point error will result in an 8% error at the same point.

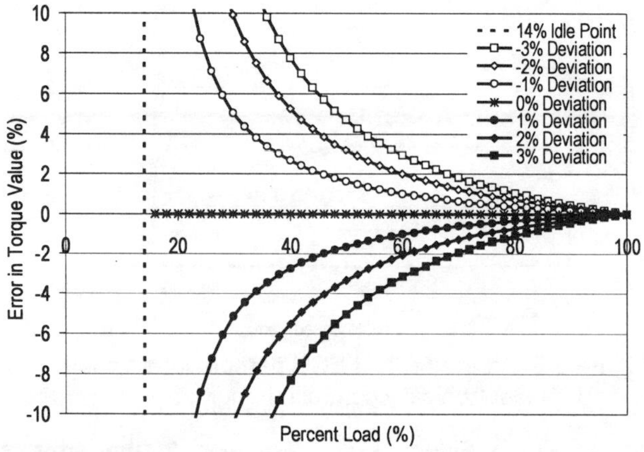

Figure 5 Error in torque due to error in no-load ECU load reading.

DERIVED POWER

The resulting inferred power at time t is the product of the engine speed and derived torque

$$P(t) = rpm(t) * T^{rpm}(t). \quad (3)$$

As shown above, the engine speed can be obtained from the ECU broadcast to a high degree of confidence. However, the derived torque is dependent upon the estimated load algorithm in the ECU, curb idle no load percent load, lug curve percent load, and the lug curve torque.

The overall error associated with inference of load from the ECU broadcast percent load and engine speed for an engine exercised through the FTP cycle is illustrated in Figures 6-8. The manufacturer supplied data and the measured data as labeled in the figures are from the lug curve listed in Table 1. As shown in these figures, there is a discrepancy between the inferred and measured power. Generally, the power inferred from the ECU is greater than the laboratory reported power. This is also borne out in Figure 7 where the 30 second integrated power is higher for the ECU derived values compared to the laboratory values.

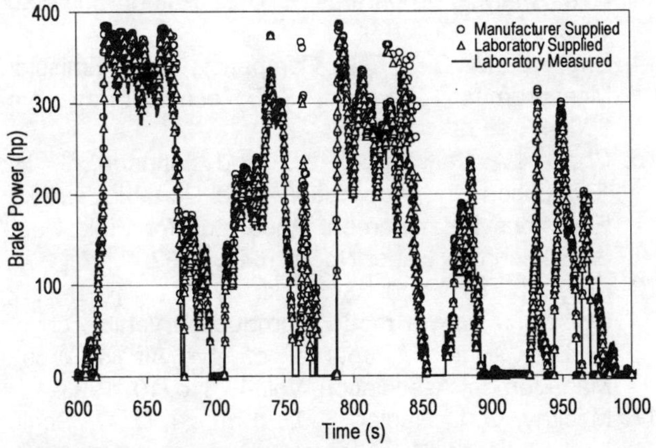

Figure 6 Instantaneous brake power comparisons between laboratory and ECU inferred data for a modern diesel engine exercised through the FTP cycle from 600 to 1000 seconds.

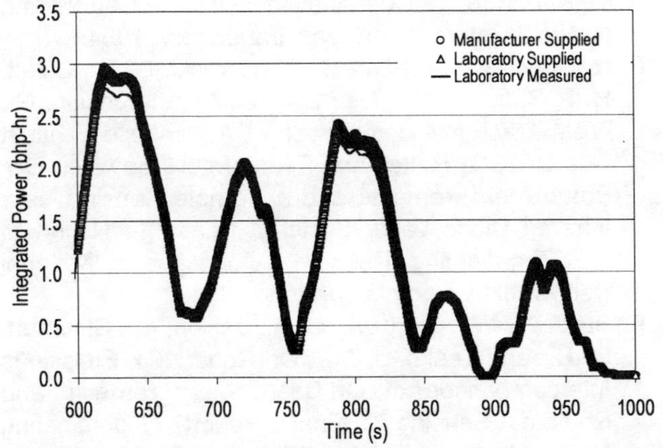

Figure 7 Integrated 30 second brake power windows between laboratory and ECU inferred data for a modern diesel engine exercised through the FTP cycle from 600 to 1000 seconds.

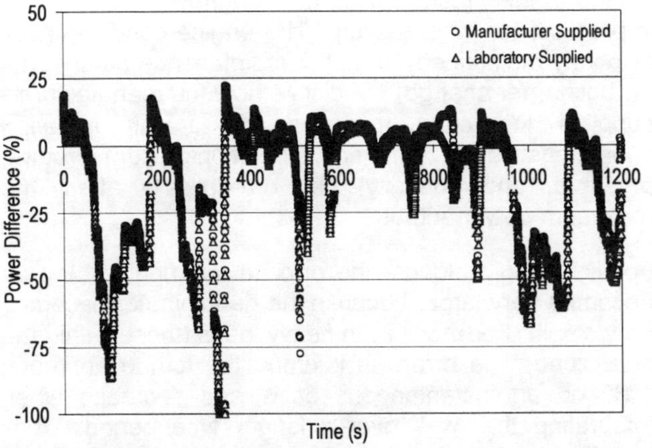

Figure 8 Integrated 30 second brake power windows percent difference between laboratory and ECU inferred data for a modern diesel engine exercised through the FTP cycle.

As illustrated in the above figures, errors will become significantly greater outside the NTE zone. Within the NTE zone, the error in the inferred power, via the ECU percent load broadcast, is of the order of 10% for a 30 second window. Merely using a larger window can reduce this error. For example, in the limiting case of integrating the power over the entire 1200 seconds, there is a 1% difference between ECU and laboratory integrated brake power when using the measured lug curve, and a 5% difference when using the lug curve supplied by the manufacturer.

A second heavy-duty diesel engine was compared on the engine dynamometer using the FTP cycle and an oscillating cycle developed for an academic research project. The oscillating cycle consisted of 10 seconds of idle, 60 seconds at a constant speed and load, and then 270 seconds where the load was varied (0.1 Hz) from 20 to 100% of the maximum load at a constant engine speed. Using laboratory data for the lug curve, the integrated brake power over the FTP cycle for the ECU-derived data was 9.3% higher the integrated power reported from the laboratory. For the oscillating test, the ECU-derived integrated brake power data was 5.5% higher than the integrated power reported from the laboratory.

CONCLUSIONS

Inference of engine power from a modern, electronically controlled diesel engine is possible through publicly broadcasted ECU data using the described methodology. The broadcast of engine speed is very accurate and can be easily checked through an external means (such as a stroboscope) at the pulleys as a quality assurance check in the field. The accuracy of the derived power is strongly dependant upon the lug curve data and the curb idle, no load data.

It is noted that good engineering judgment must be used in implementing this approach since the errors in inferring the torque is dependent upon many factors. It is always desirable to measure the lug curve for the engine under test, but it is recognized that this is impractical for most testing. The engine condition (well or poorly maintained) and prior maintenance procedures (turbocharger change) will dictate how the manufacturer-supplied lug curve matches the actual operating characteristics. Ambient conditions (temperature, pressure, and humidity) can dramatically affect the maximum power output.

For light engine loads, the error in inferring the torque becomes very large, because the denominator becomes very small. For medium to heavy loads (those within the NTE zone), the errors in inferring the torque approach 10% on an instantaneous basis and decrease when integrating the work over a longer time period. It is possible to estimate the integrated brake work, over a long enough test cycle that is operated sufficiently in the NTE zone, to within 10% of the work calculated with a research grade engine dynamometer.

REFERENCES

1. Gautam, M., Clark, N. N., Thompson, G. J., and Lyons, D. W., "Assessment of Mobile Monitoring Technologies for Heavy-Duty Vehicle Emissions," Whitepaper Submitted to the Settling Heavy-Duty Diesel Engine Manufacturers, Department of Mechanical and Aerospace Engineering, West Virginia University, Morgantown, WV, 1999. See http://es.epa.gov/oeca/ore/aed/diesel/test.html.
2. Gautam, M., Clark, N. N., Thompson, G. J., Carder, D. K., and Lyons, D. W., "Evaluation of Mobile Monitoring Technologies for Heavy-Duty Diesel-Powered Vehicle Emissions," Phase I Final Report Submitted to the Settling Heavy-Duty Diesel Engine Manufacturers, Department of Mechanical and Aerospace Engineering, West Virginia University, Morgantown, WV, 2000. See http://es.epa.gov/oeca /ore /aed/diesel/test.html.
3. Gautam, M., Clark, N. N., Thompson, G. J., Carder, D. K., and Lyons, D. W., "Development of In-use Testing Procedures for Heavy-Duty Diesel-Powered Vehicle Emissions," Phase II Final Report Submitted to the Settling Heavy-Duty Diesel Engine Manufacturers, Department of Mechanical and Aerospace Engineering, West Virginia University, Morgantown, WV, 2000. See http://es.epa.gov/oeca /ore/aed/diesel/test.html.
4. Gautam, G., Thompson, G. J., Carder, D. K., Clark, N. N., Shade, B. C., Riddle, W. C., and Lyons, D. W., "Measurement of In-Use, On-Board Emissions from Heavy-Duty Diesel Vehicles: Mobile Emissions Measurement System," SAE Technical Paper No. 2001-01-3643, 2001.
5. "Powertrain Control Interface for Electronic Controls Used in Medium and Heavy Duty Diesel On-Highway Vehicle Applications," SAE Standard, SAE J1922, 1989.
6. "Joint SAE/TMC Electronic Data Interchange Between Microcomputer Systems in Heavy-Duty Vehicle Applications," SAE Standard, SAE J1587, 1996.
7. "Vehicle Application Layer," SAE Standard, SAE J1939/71, 1996.
8. Dearborn Group, Farmington Hills, MI, 1999.
9. Title 40 Code of Federal Regulations, Part 86, U.S. Government Printing Office, Washington, DC, 2000.
10. Title 40 Code of Federal Regulations, Part 89, U.S. Government Printing Office, Washington, DC, 2000.
11. "Notice of Filing of Consent Decree Under the Clean Air Act," Federal Register, Vol. 63, No. 212, Office of the Federal Register, National Achieves and Records Administration, Washington DC, November 3, 1998.
12. "Snap-Acceleration Smoke Test Procedure for Heavy-Duty Diesel Powered Vehicles," SAE Standard, SAE J1667, 1996.
13. Fournier, T. J., Reading, A. R., Wilson, R. L., and Kapolka, M. F., "Method of Analyzing and In-flight Gas Analysis Apparatus, " U.S. Patent 5099680, Washington, DC, 1992.
14. Englund, M. S., "Field Compatible NOx Emission Measurement Technique," SAE Technical Paper No. 820647, 1982.
15. Chan, L., Carlson, D. H., and Johnson, J. H., "Evaluation and Application of a Portable Tailpipe Emissions Measurement Apparatus for Field Use," SAE Technical Paper No. 921647, 1992.
16. Kelly, N. A. and Groblicki, P. J., "Real-world emissions from a modern production vehicle driven in Los Angeles," Journal of the Air & Waste Management Association, Vol. 43, No. 10, 1993.
17. Mackay, G. I., Nadler, S. D., Karecki, D. R., Schiff, H. I., Butler, J. W., Gierczak, C. A., and Jesion, G., "Dynamometer Intercomparison of Automobile Exhaust Gas CO/CO_2 Ratios and Temperature Between On-Board Measurements and a Remote Sensing Near Infrared Diode Laser System," Phase 1b Report to the Coordinating Research Council and National Renewable Energy Laboratory, 1994.
18. Mackay, G. I., Nadler, S. D., Karecki, D. R., Schiff, H. I., Butler, J. W., Gierczak, C. A., and Jesion, G., "Test Track Intercomparison of Automobile Exhaust Gas CO/CO_2 Ratios and Temperature Between On-Board Measurements and a Remote Sensing Near Infrared Diode Laser System," Phase 1c Report to the Coordinating Research Council and National Renewable Energy Laboratory, 1994.
19. Butler, J. W., Gierczak, C. A., Jesion, G., Stedman, D. H., and Lesko, J. M., "On-Road NOx Emissions Intercomparison of On-Board Measurements and Remote Sensing," Final Report, Coordinating Research Council, Inc., Atlanta, GA, CRC Report No. VE-11-6, 1994.
20. Gierczak, C. A., Jesion, G, Piatak, J. W., and Butler, J. W., "On-Board Vehicle Emissions Measurement Program," Final Report, Coordinating Research

Council, Inc., Atlanta, GA, CRC Report No. VE-11-1, 1994.
21. Vojtisek-Lom, M. and Cobb, Jr., J. T., "On-Road Light-Duty Vehicle Mass Emission Measurements Using a Novel Inexpensive On-Board Portable System," Proceedings of the Eighth CRC On-Road Vehicle Workshop, San Diego, CA, April 20-22, 1998.
22. Spears, M. W., "An Emissions-Assisted Maintenance Procedure for Diesel-Powered Equipment," University of Minnesota, Center for Diesel Research, Minneapolis, MN, 1997.
23. "Construction Equipment Retrofit Project," Northeast States for Coordinated Air Use Management, Boston, MA, 1998.
24. Butler, J. W., Kornisk, T. J., Reading, A. R., and Kotenko, T. L., "Dynamometer Quality Data On-board Vehicles for Real-World Emission Measurements," Proceedings of the Ninth CRC On-Road Vehicle Workshop, April 19-21, San Diego, CA, 1999.
25. Kihara, N., Tsukamoto, T., Matsumoto, K., Ishida, K., Kon, M., and Murase, T., "Real-time On-Board Measurement of Mass Emission of NOx, Fuel Consumption, Road Load, and Engine Output for Diesel Vehicles," SAE Technical Paper No. 2000-01-1141, 2000.
26. Jetter, J., Maeshiro, S., Hatcho, S., and Klebba, R., "Development of an On-Board Analyzer for Use on Advanced Low Emission Vehicles," SAE Technical Paper No. 2000-01-1140, 2000
27. Branstetter, R., Burrahm, R., and Dietzmann, H., "Relationship of Underground Diesel Engine Maintenance to Emissions," Final Report for 1978 to 1983 to the U.S. Bureau of Mines, Department of the Interior Contract H0292009, 1983.
28. Human, D. M. and Ullman, T. L., "Development of an I/M Short Emissions Test for Buses," SAE Technical Paper No. 920727, 1992.
29. Bentz, A. P. and Weaver, E., "Marine Diesel Exhaust Emissions Measured by Portable Instruments," SAE Technical Paper No. 941784, 1994.
30. Bentz, A. P., "Final Summary Report on Project 3310, Marine Diesel Exhaust Emissions (Alternative Fuels)," United States Department of Transportation United States Coast Guard Systems, Report No. CG-D-08-98, 1997.
31. Breton, L. A. G., "Real-time On-Road Vehicle Exhaust Gas Modular Flowmeter and Emissions Reporting System, " U.S. Patent 6148656, Washington, DC, 2000.
32. Thompson, G. J., Atkinson, C. M., Clark, N. N., Long, T. W., and Hanzevack, E., "Neural Network Modelling of the Emissions and Performance of a Heavy-Duty Diesel Engine," *Proc Instn Mech Engrs*, Vol. 214, Part D, No. D04499, 2000.

2002-01-0615

Investigation of 'Sweep' Mapping Approach on Engine Testbed

M. C. Ward, C. J. Brace and N. D. Vaughan
University of Bath

R. Ceen
CP Engineering

T. Hale and G. Kennedy
Cosworth Technology

ABSTRACT

Steady state mapping is fundamental to optimizing IC engine operation. Engine variables are set, a predefined settling time elapses, and then engine data are logged. This is an accurate but time consuming approach to engine testing. In contrast the sweep method seeks to speed up data capture by continuously moving the engine through its operating envelope without dwelling. This is facilitated by the enhanced capability of modern test rig control systems. The purpose of this work is to compare the accuracy and repeatability of the sweep approach under experimental conditions, with that of steady state testing.

Limiting factors for the accuracy of the sweep approach fall into two categories. Firstly on the instrumentation side - transducers have a characteristic settling time. Secondly on the engine side - thermal and mechanical inertias will mean that instantaneous measurements of engine parameters differ from the steady state values. These errors can be reduced to satisfactory levels by modelling the engine and instrumentation responses. Some data i.e. air, fuel and emissions pose significant problems, with the response time of the equipment limiting the maximum sweep speed.

Despite requiring intensive data processing and test bed sophistication, sweep mapping shows the potential to give accuracy comparable to steady state testing, but with much reduced mapping times.

INTRODUCTION

In order to comply with increasingly stringent emissions standards and meet drivability requirements, modern engines are equipped with an increasing number of subsystems and controlling elements. The result has been to greatly increase the calibration effort required to find the parameter settings that offer the best global compromise across the entire engine map. This process is traditionally centered on steady state testing. During steady state engine mapping a dwell period is required at each test point to allow instrumentation, and engine parameters to settle to a steady state value. This is an accurate but a very time intensive procedure. Steady state mapping is assumed as the gold standard by which the acceptability of the engine's calibration. In practice, however most engine operation is transient in nature and so steady state mapping is being re-evaluated, as transient calibration is being more widely understood.

Aimed at reducing the time required for the calibration procedure, a current trend within engine testing is the application of statistical tools, these include design of experiments (DoE), Bayesian and stochastic methods. These seek to minimize the data captured to that required for a model fit, and originally stem from arenas of work where data is scarce e.g. pharmaceutical and biological industries. It could be argued however, that an engine test cell is a data rich environment and hence the idea behind the sweep approach is to maximize the data captured, allowing more detailed models to be developed within an acceptable period of time.

The sweep approach as first described in [1] ramps the engine between two operating points, while data is logged continuously. Of course not dwelling at individual operating points will give rise to measurement errors, which would not occur during a normal steady state test. As a starting point this paper concentrates on the application of engine torque as the variable, with the engine speed held constant. The benefit of testing at a constant engine speed is to remove the effects of mechanical inertias, which would occur if the engine were accelerating or decelerating see [1],[8].

The experimentation outlined in this paper considers two phases of engine testing:

Phase 1: Application of step changes in engine torque to model the response of the instrumentation and engine, in order to access the accuracy of the technique.

Phase 2: Application of ramps in engine torque. The results from the ramps were post processed using the analysis from phase 1 compared to the corresponding steady state results.

Although the experimentation was carried out on a diesel engine it is just as applicable to a gasoline.

TEST RIG AND EXPERIMENTAL PROCEDURE

The experimentation carried out in this paper is intended as an alternative to steady state testing. The aim therefore, was to achieve satisfactory results from a standard steady state cell, and its instrumentation, without moving the testing onto a transient test cell setup, for which the capital and running costs would be considerably higher.

The mechanical setup is shown schematically in **Figure 1**. The test cell is based upon a Ford 1.8 Lynx – common rail Diesel engine equipped with a VGT. For the purpose of the testing the EGR system was disabled, as it's transient behavior would add variables, which were not to be considered here. The engine is loaded by a Froude eddy current dynamometer, capable of absorbing up to 110 kW.

Control of the test cell, dynamometer and data acquisition is carried out by a CP Cadet V12 [2,3] host system shown in **Figure 2.** The system features an open code format affording the user control over the functionality of the system. The ECU parameters can be altered via a ASAP 3 link from a laptop running the Kleinknecht Gredi program, this also allows monitoring of the ECU variables. The exhaust emissions are analyzed using an AVL CEB emissions tower. This tower allows monitoring of pre-catalyst exhaust CO, CO_2, NOx, uHC, O_2 and EGR rate via inlet manifold CO_2 measurement. Fuelling to the engine is measured using a gravimetric fuel balance, which gives an averaged value from multiple readings (10 in this case) during a steady state test, it can also give an instantaneous reading.

Although the dynamometer is designed for steady state work, it was found that its response was more than satisfactory for the maneuvers carried out in the procedures discussed here. Looking at the torque trace in **Figure 3** there is a drop off in torque immediately after the negative step, which does not occur after the positive step. This is due to inertial effects caused by the sudden unloading of the dynamometer to a low torque level. This is not an issue on the ramp maneuvers as the rate of change in torque is more gradual.

The approach of sampling data over the whole of a ramp, as opposed to only at discrete sampling points leads to a data volume with a higher order of magnitude. It is essential therefore that the testbed data storage structure is well defined and efficient. Data is streamed directly from the rig into a Microsoft ACCESS database based on the ASAM [4] structure. From there it is accessible to MATLAB [5] using the Mathworks database toolbox. The advantage of using MATLAB as the data processing tool is the ease of implementation of mathematical methods.

For the sake of brevity this paper presents results and analysis from one of each typical transducer type:

- Engine speed (Dynamometer speed encoder)
- Torque (Dynamometer load cell)
- Inlet manifold pressure (Piezo-resistive sensor)
- Exhaust manifold temperature (K-type thermocouple)
- NOx (Chemiluminescence detector)
- Fuel flow (Gravimetric balance)
- Air Flow (Vortex shedding flow meter)

EFFECT ON HARDWARE CHANGES

Analysis of the engine and transducer responses will obviously specific to the type of engine and instrumentation employed. Therefore it is acknowledged that re-characterization of the responses will be necessary is the hardware setup is changed. This is achieved via application of the described ramp and step inputs. Once the responses of the tranducers are known suitable ramp rates can be chosen, by comparing data from a ramp with steady state data.

With the help of modern host systems this process can be automated to be carried out quickly at the beginning of a new hardware test.

SOURCES OF ERROR

ENGINE SIDE ERRORS

Estimation of port air mass flow is vital for the characterization of engine behavior. Its value at a point during a ramp will differ from the steady state value for the following reasons:

When the engine is operated under steady state conditions, interaction of pressure waves in the intake and exhaust manifolds will lead to the setup of a standing wave after a number of engine cycles, as described in [6]. When moving between operating conditions however, there is insufficient time available for the standing wave to develop. This leads to a deviation between the instantaneous air mass flow reading, and the value, which would be achieved from steady state mapping.

Similarly during a ramp, manifold temperature readings will not give the same value as a steady state test. Previous estimates of inlet manifold temperature error during a ramp compared with steady state have been up to 20% [6]. Another source of temperature variations is backflow of high temperature exhaust gas into the intake manifold, thus causing temperature fluctuations. This backflow however, is likely only to occur during maneuvers much more severe than those carried out here.

Non steady state mapping will provoke ECU strategy reactions designed to optimize engine response to transient maneuvers. For example, control of the boost pressure using the VGT vane position. This is a problem if the strategy is controlling a parameter, which is not a variable in the experiment being undertaken. Care must be taken to ensure this variable remains constant throughout its calibrated map. Alternatively a low ramp rate could be used that the ECU strategy does not consider a transient maneuver.

INSTRUMENTATION SIDE ERRORS

Instrumentation type can be grouped according to differing speed of response.

Pressure transducers and small thermocouples have short time constants and so will respond quickly, without significant delay between an engine event and the time that the transducer sees a change. A description of the dynamics of pressure transducers and thermocouples can be found in [9].

The readings from the emissions analyzer have a slower and composite type response. The heated line connecting the engine and the emissions analyzer introduces a transport delay into the emissions response. Secondly there will be a series of lags caused by the analyzer element behavior and also the thermal settling time of the exhaust temperature.

Fuel flow measurement is problematic. The test cell is instrumented with a gravimetric fuel balance, designed to give an average mass flow over a time span, for good steady state accuracy. It however is not ideal for giving instantaneous readings due to its principle of operation.

Airflow is measured using a Yokogawa vortex shedding flow meter, which is a steady state device. Eddy currents in the flow induce vibration of a wafer, proportional to the flow rate. The correlation between the ramp and steady state results was better than expected.

Strictly speaking the A/D conversion of the signals by the host system will occur over a finite timescale but this can be considered as negliable.

DISSCUSSION OF RESULTS

In order to determine the response of the transducers, step changes in engine torque from 20 up to 60 Nm and 20 up to 100 Nm were applied at a constant engine speed of 2000 rpm. For the ramp testing positive and negative ramps of between 20 and 100 Nm at rates of 80 Nm/min and 53 Nm/min were used. The settling times for the instrumentation components, derived from the experimentation can be found in **Table 1**.

NO_X ANALYSER RESPONSES

As the emissions analyzer samples at a constant flow rate, the assumption can be made that the transport delay of the exhaust gas between the engine and the analyzer will remain constant for all engine operating conditions. In contrast, the settling time of the emissions readings is significantly influenced by the operating point. This is a function of all engine parameters e.g. speed, injection timing etc.

Figure 3 shows the responses of NOx reading to a step change in engine torque. The reading exhibits a 3 phase response before reaching a steady state value, **Figure 4**.

Phase 1: A pure delay of 3 seconds while the exhaust gas 'plug' travels down the heated line to the emissions tower.

Phase 2: An initial analyzer defined settling period, characterized by an under damped second order transfer function.

Phase 3: Finally there is a secondary slower settling period while the engine temperature stabilizes, taking the form of a first order lag.

The emissions response to the positive and the negative step appears symmetrical. The modeling of the emissions response during phase 2 can be achieved by applying a second order term. The modeling of the phase 3 response requires a first order lag. The forms of these terms can be found in **Appendix 1** and described further in [7]. The corrected values of NOx compared with the steady state results are plotted on the hysteresis plot in **Figure 5.**

Figure 6 shows the NOx in comparison with the demanded torque ramp. The NOx profile tracks the torque satisfactory and the delay time can clearly be seen. During the steady state periods between the ramps the slower dynamic is due to the thermal settling of the engine at the new operating point. **Figure 5** shows the hysteresis of the NOx response during the 80 and 53 Nm/min ramps. The steady state values each side of the ramp return to the same value, but the loops are fairly 'open' showing a significant delay. The loops are symmetrical for both ramp rates, falling onto a centerline. This implies that average of the results on the positive and negative ramps can make a good correction to the corresponding steady state results.

NO_X DATA RECONSTRUCTION

Two methods were compared for the reconstruction of the NOx data.

Method1: The emissions values obtained from the ramp are corrected by assigning each value a time in advance of that which it was logged at. This time advance is the composite of the transport delay and the predicted settling time as described above. The settling times are derived through the analysis given in appendix 1.

Table 2 lists the corrected NOx results against the steady state results; these are plotted on **Figure 5**. It can be seen that the corrected results are a good approximation to the steady state values.

Method 2: An alternative approach is simply to average the values measured on the positive and negative ramps. The results can be seen in

Table 3. As before it can be seen that the corrected results are a good approximation to the steady state results, although slightly worse than method 1.

It can be seen for both these methods that at the start of the ramp the errors between the corrected values and the steady state results are greater than midway along it. This occurred for all the engine parameters examined here, and is because at the beginning of the ramp, mechanical and thermal inertias in the engine give rise to initial lags in it's operating parameters. These parameters then settle as a pseudo steady state condition is reached on the ramp itself. Additionally the slower ramp provides a set of corrected values closer to the steady state results than the faster ramp; which is to be expected.

FUEL AND AIRFLOW RESPONSE

Figure 8 shows the instantaneous fuel flow readings resulting from the step in engine torque. The reading seems to follow the torque steps without a lag or delay time. It can be seen that the instantaneous reading continually fluctuates, even on the steady state parts of the trace. **Figure 9** shows plot of fuel hysteresis for the 80 and 53 Nm/min ramps. Both fuelling loops return to a common starting point after the positive and negative ramps. They do not fall on a centerline with the faster ramp producing higher fuelling reading than the slower one. This implies that an average of the values from the positive and negative results cannot produce an accurate correction to steady state.

Figure 11 shows the airflow response to a 20 to 60 Nm torque step. It can be seen that there are fluctuations in the airflow reading directly after the positive and negative steps, which is caused by the vibrations in the transducer wafer as it adjusted to the increased flow rate. It was first thought that these fluctuations in airflow were a result of the VGT strategy, but the VGT position and speed track the torque step accurately see **Figure 7**. The airflow compared with the demanded torque ramp profile is plotted in **Figure 12**. It can be seen that the instantaneous airflow reading shows large fluctuations, symmetrical on both the positive and negative ramps. This suggests that an averaged correction could yield a good approximation to the corresponding steady state values. Which is backed up by the hysteresis plot **Figure 13**.

FUEL AND AIRFLOW DATA RECONSTRUCTION

The value of instantaneous fuel flow presents the most problematic reading. It was hoped prior to the experimentation that it would be possible to deduce decay rates, and overshoot in the response to a step input. Unfortunately this was not possible. For further work with the sweep approach a faster fuel meter is required, this is likely to be a rate meter such as a Coriolis effect device. **Table 7** shows an attempt at correcting the values of fuel flow. **Figure 9** shows the corrected values plotted against the steady state results.

Table 6 shows corrected results for the airflow averaged over both the positive and negative ramps. **Figure 13** plots the corrected results against the steady state results. A good correlation to the corresponding steady state readings can be seen. As with the other results the corrections are better at the slower ramp rate. The correlation is not good at the beginning of the ramps, but improves thereafter as a pseudo steady state condition is achieved. These results could be improved by replacing the vortex shedding flow meter with a hot wire mass flow meter which has a faster response an generally better accuracy.

TEMPERATURE & PRESSURE RESPONSES

The inlet manifold pressure response to a torque step is shown in **Figure 14**. It has the shortest settling time compared with the other measurements, **Table 1**. The response appears symmetrical on both the positive and negative ramps, which is borne out by the hysteresis plot in **Figure 16.**

Figure 17 shows the exhaust manifold temperature. The exhaust temperature has a 2-phase response: i.e. a fast first order lag which a function of the thermocouple response followed by a second slower order lag, occurring while the engine temperatures stabilize.

TEMPERATURE/ PRESSURE RECONSTRUCTION

The errors in these values are as a result of thermal and mechanical inertias in the engine during a ramp. It was found that averaging the values on the positive and negative ramps cancels the errors and gives a good approximation the corresponding steady state values. The results are listed in

Table 4 and **Table 5. Figure 16** and **Figure** 19 and show the corrected values plotted against the steady state values over the ramp. It can be seen that the ramp to steady state error is lower at the lower ramp rate. This is expected, as the engine side thermal and inertial lags will be lower at lower rates of change.

REPEATABILITY AND ROBUSTNESS

Table 8 shows the results of 5 repeats of the same ramp. For the modeling and subsequent correction stages investigated here to be considered satisfactory the results must be repeatable. It can be seen from the standard deviation that the results scatter is low. The standard deviation is higher for the emissions and fuel flow readings than it is for the temperature and pressure transducer readings.

The experimentation described in this paper is within the boundaries of normal engine operation, with the only variable being that of engine torque demand. For ramps in critical ECU controlled parameters such as injection timing, VGT position etc. there is a potential for engine damage. With traditional steady state testing a test bed operator has time to identify and resolve problems. Non steady state testing partly removes the human operator, and requires additional safety measures. These safety measures can potentially be coded into the system in the form of expert knowledge, thus taking the place of the human operator.

CONCLUSIONS AND OUTLOOK

The sweep approach moves the focus away from the engine and onto the mapping of the surrounding instrumentation. The experimentation detailed in this paper shows the potential of non-steady state testing to produce results comparable in accuracy to those from steady state testing.

At the start of the ramp the errors between the corrected values and the steady state results are greater than midway along it. This is because initially as the demanded torque is applied, mechanical and thermal inertias in the engine exhibit an initial lag, which then settles when a pseudo steady state condition is reached on the ramp itself. This means that data from the beginning of the ramp could be discarded. The time taken for this pseudo steady state condition is an important quantity, which could be assessed online for ramps in different engine parameters.

The next stage of work is to extend the approach to ramps in ECU engine parameters. This will require coding of additional safety measures in to the host system to reduce the potential for engine damage.

The results could be improved through the use of different transducers, in particular for the measurement of mass flows. An accurate or corrected reading for fuel flow was not possible as the fuel balance is only intended for steady state use. It is proposed that for further work a mass flow rate type device be used, giving better dynamic response. In the case of the air flow measurement better results could be achieved through the use of a hot wire mass flow meter, which has a faster response and a better accuracy than the vortex shedding flow meter used here. In the case of the emissions faster analyzers could easily be used to improve the accuracy of the results, albeit with a cost penalty.

If new hardware is added to the test bed, whether (engine or instrumentation), a re-characterization of the responses would be required. Then data from a number of predetermined ramps would be compared to steady state points to generate a suitable ramp rate. This process can be coded into the test bed system and executed online. This would enable the automatic re-characterization of the system if new engine or instrumentation were introduced.

Another approach to non steady state testing would be to isolate the piece of instrumentation with the slowest response (emissions/ or fuel flow) and then to set up a control loop to keep this parameter constant. This method is difficult to implement in practice due to the feedback delay of the controlled variable. The control could be improved using feed forward of prior knowledge into the loop to compensate for the measurement delay.

ACKNOWLEDGMENTS

This work has been undertaken with support from Cosworth Technology and CP-Engineering systems. The authors are grateful for the support and assistance that has been supplied.

REFERENCES

[1] Gouldburn J. R., Brown D G., Case R., 1991, "Computer Controlled Non-Steady-State Engine Testing" Int. J. of Vehicle Design, vol. 12 no.1, 1991

[2] CP Engineering, 2000, "Cadet V12 – Run test Manual"

[3] CP Engineering, 2000, "CP data Sheet - CP128 Control and Monitoring I/O system"

[4] ASAM, 2000, "ASAM-ODS Ver 4.0" 2000, ASAM Group

[5] Mathworks, "MATLAB online help", Mathworks

[6] Chevalier A., Mueller M., Hendricks E., "On the Validity of Mean Value Engine Models during Transient Operation" 2000, SAE Paper 2000-01-1261

[7] Nise N S., 1992, "Control Systems Engineering" Benjamin Cummings

[8] Bosch, 1996, "Automotive Handbook" 4th Edition Bosch

[9] Tektronix, "Engine Analysis Measurements", 1970, Measurements Concepts Series – Tektronix

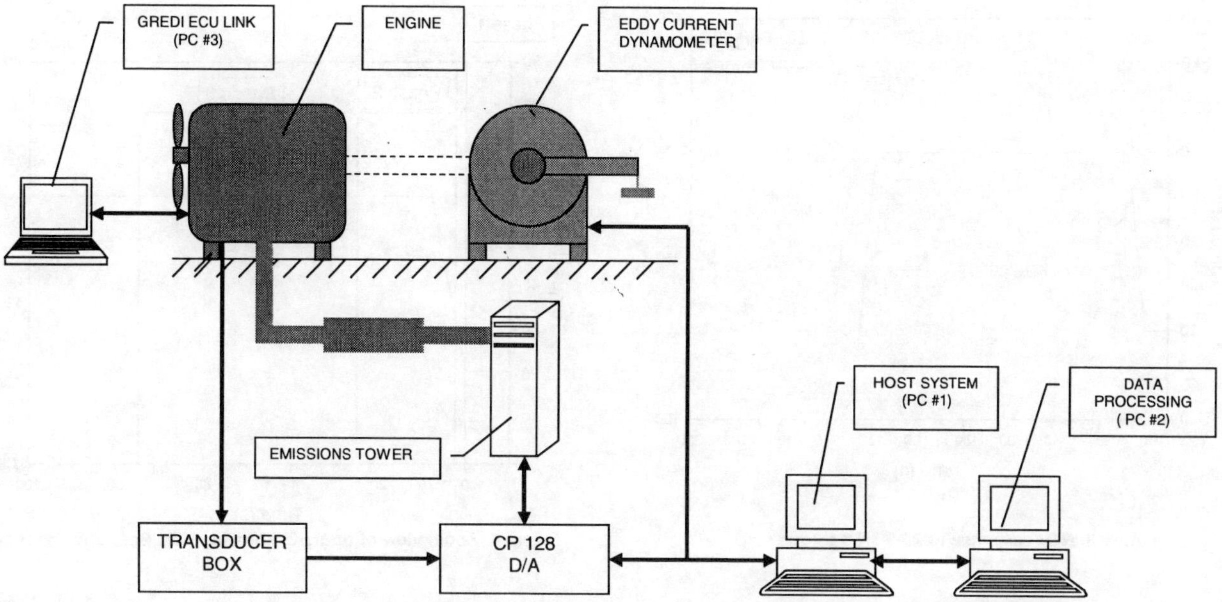

Figure 1: Test Cell Schematic

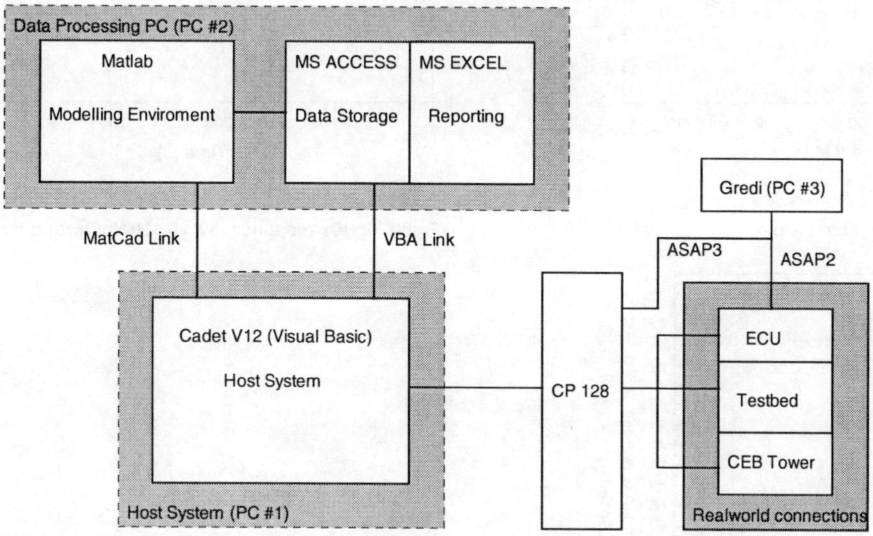

Figure 2: Host system software environment

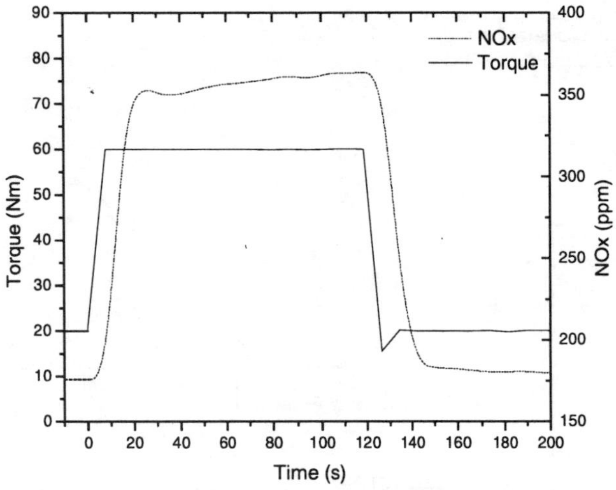

Figure 3: NOx response to 20-60 Nm torque step

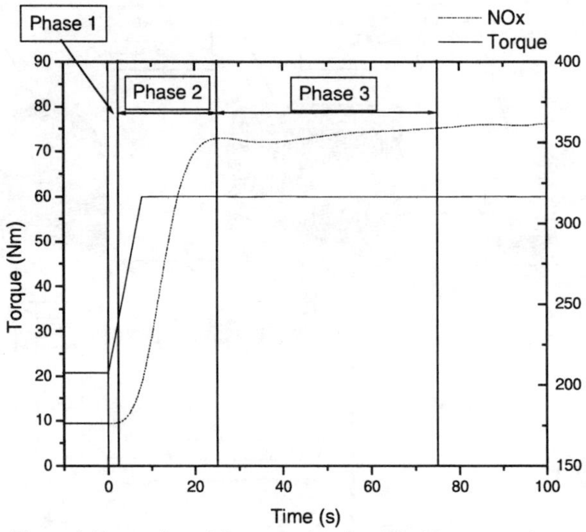

Figure 4: Zoom view of figure 3, showing NOx Response phases

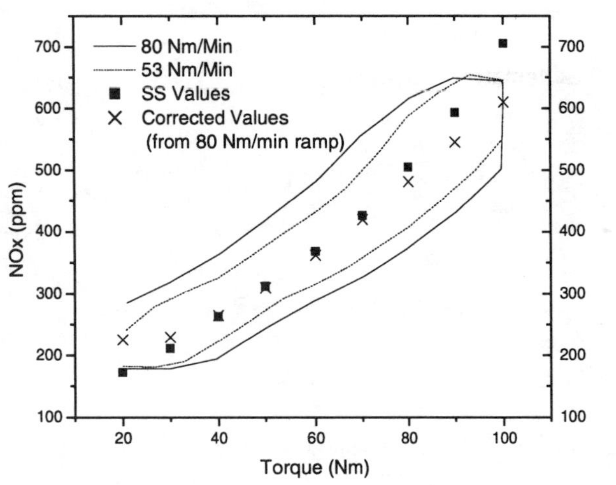

Figure 5: NOx hysteresis plot

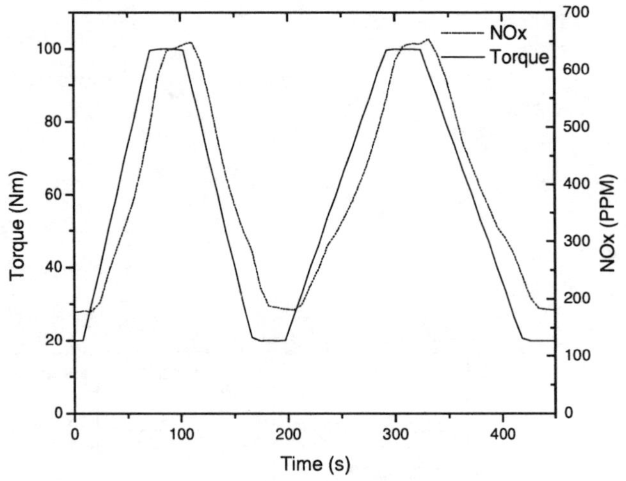

Figure 6: NOx response to a 20-100 Nm torque ramp

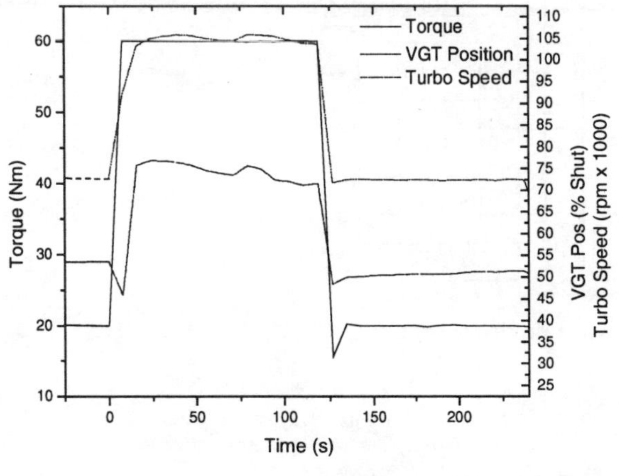

Figure 7: Turbocharger strategy response to 20-60 Nm torque step

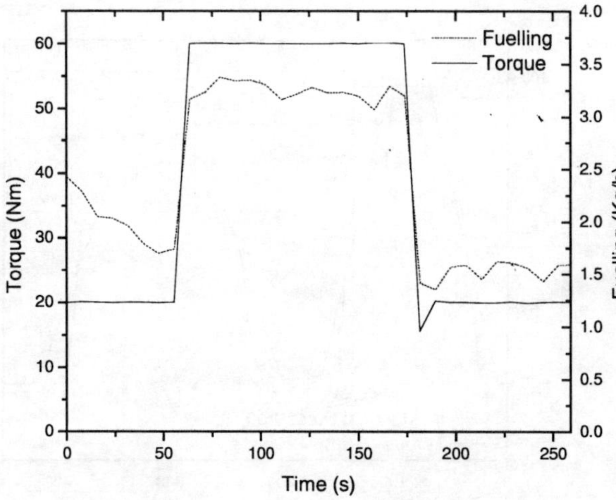

Figure 8: Fuelling response to a 20–60 Nm torque step

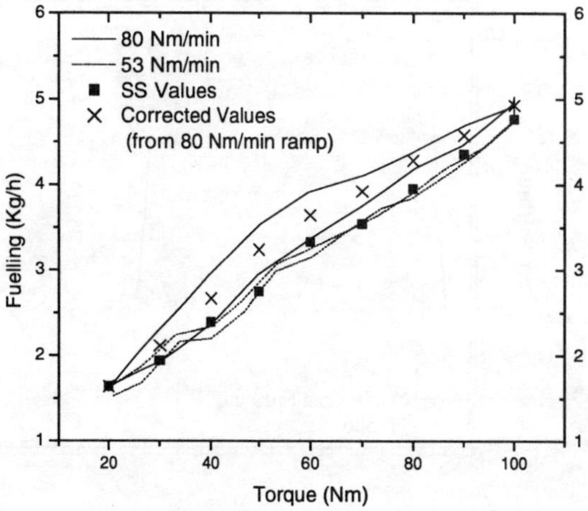

Figure 9: Hysteresis plot of fuelling

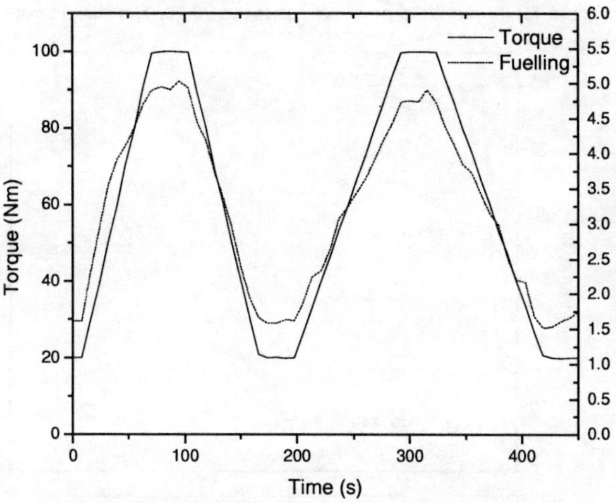

Figure 10: Fuelling response to a 20-100 Nm torque ramp

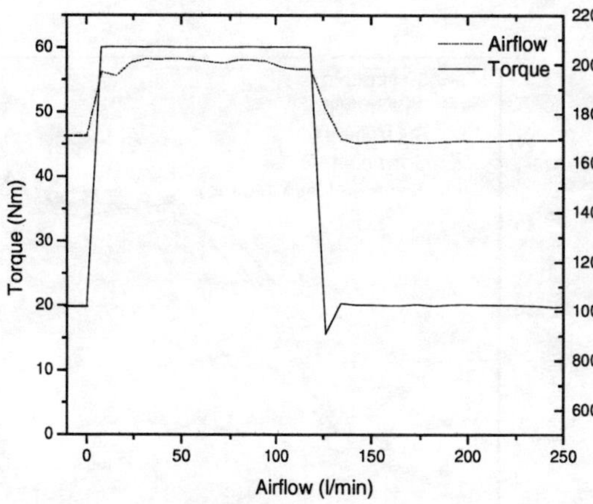

Figure 11: Airflow response to 20-60 Nm torque step

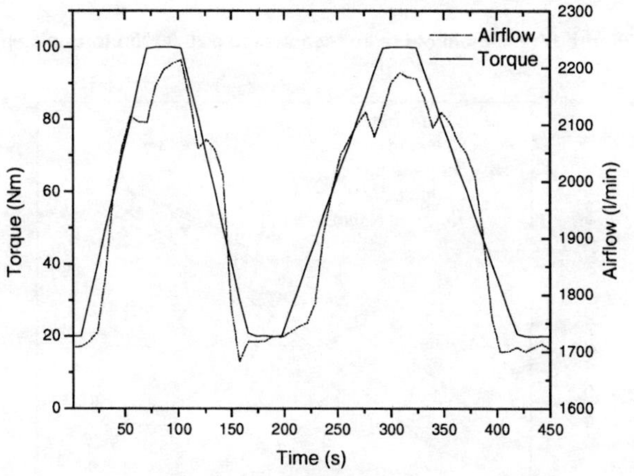

Figure 12: Airflow response 20 - 100 Nm torque ramps

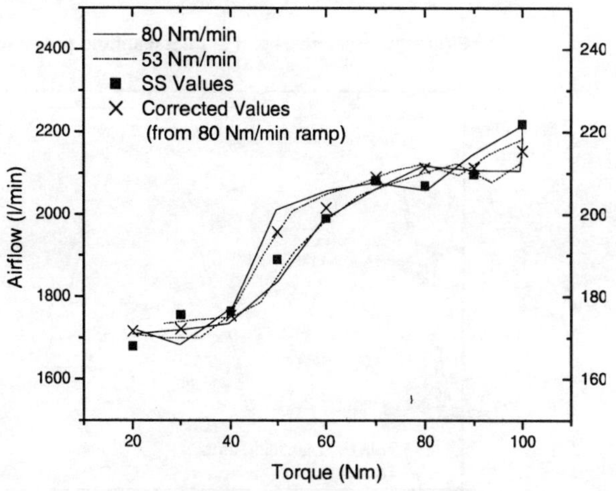

Figure 13: Hysteresis plot of airflow

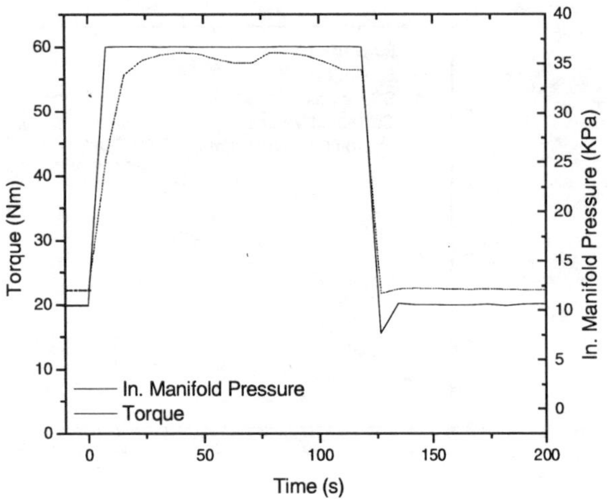

Figure 14: In. manifold pressure response to 20-60 Nm torque step

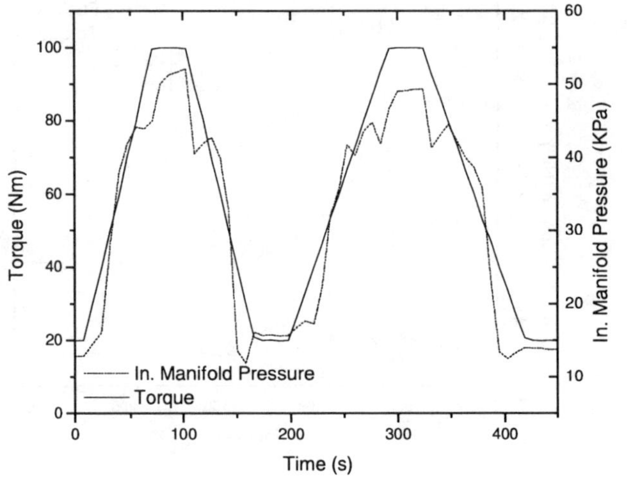

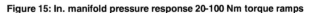

Figure 15: In. manifold pressure response 20-100 Nm torque ramps

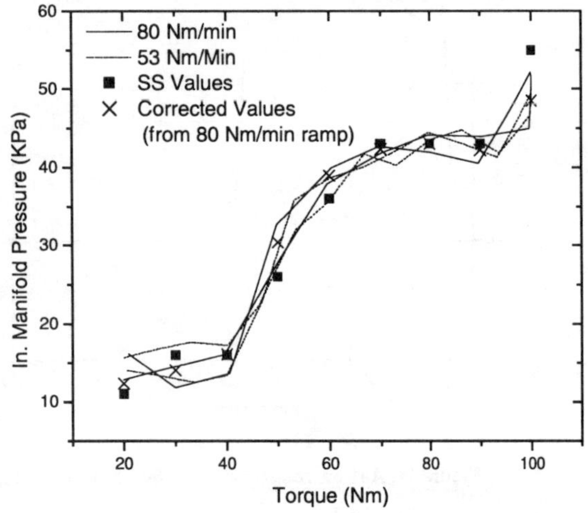

Figure 16: Hysteresis plot of Inlet manifold pressure

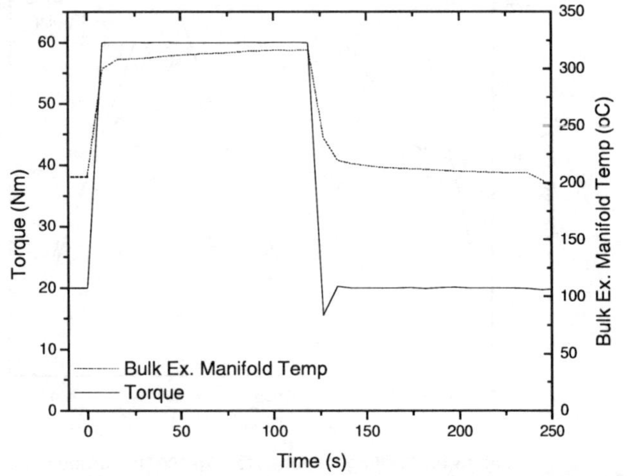

Figure 17: Ex. manifold temp response to a 20-60 Nm torque step

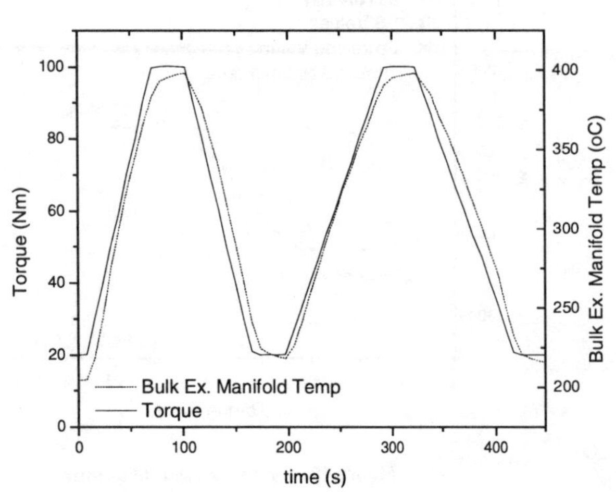

Figure 18: Ex. manifold temp response 20-100 Nm torque ramps

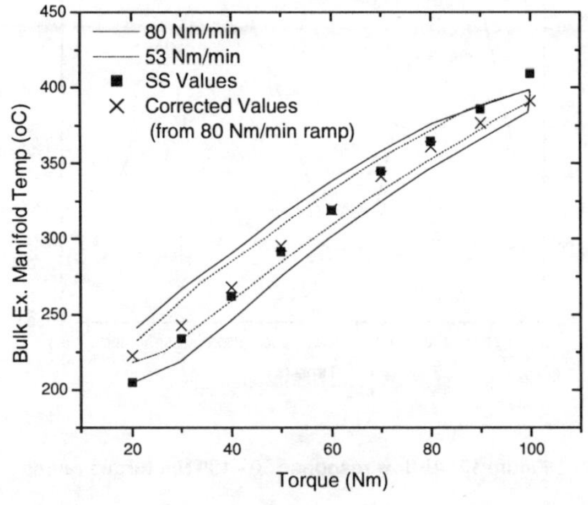

Figure 19: Hysteresis plot of exhaust manifold temp

Transducer	Delay (s)	Order	Settling Time (s)
NOx	3	2° 1°	20 - Phase 2 88 - Phase 3
Pressure	/	1°	38
Temp	/	1°	88
Airflow	/	1°	47
Fuelling	/	/	/

Table 1: Summary of responses

Torque (Nm)	NOx (ppm)			
	Ramp	Averaged	SS	% Error
20	178	225	175	-28.7
30	177	229	211	-8.6
40	193	264	263	-0.6
50	244	309	312	0.7
60	371	481	504	4.5
70	431	545	593	8
80	501	610	705	12
90	178	225	175	-28.7
100	177	229	211	-8.6

Table 2: Corrected NOx values, lag method for the 80 Nm/min ramp

Torque	SS Values	NOx (ppm) 53 Nm/min		NOx (ppm) 80 Nm/min	
		Corrected	% Error	Corrected	% Error
20	172	211	-23.0	231	-34.5
30	211	247	-17.3	247	-17.4
40	263	274	-4.3	279	-6.1
50	312	343	-10	330	-5.8
60	368	373	-1.4	384	-4.5
70	426	451	-5.9	441	-3.5
80	504	497	1.3	493	2.0
90	593	574	3.1	540	8.8
100	705	597	15.2	572	18.7

Table 3: Corrected NOx values, averaging method

Torque	SS Values	53 Nm/min		80 Nm/min	
		Corrected	% Error	Corrected	% Error
20	11	14.8	-34.9	12.3	-12.2
30	16	15	5.7	14	12.1
40	16	15.3	4.2	16	-0.6
50	26	33.8	-30.3	30.3	-16.8
60	36	37.1	-3.	38.9	-8.1
70	43	41.2	3.9	42.3	1.5
80	43	44	-2.3	43	-0.09
90	43	41.5	3.3	42.2	1.8
100	55	48	12.6	48.5	11.7

Table 4: Corrected inlet manifold pressures

Torque	SS Values	53 Nm/min		80 Nm/min	
		Corrected	% Error	Corrected	% Error
20	204.7	237.2	-15.6	228	-11.3
30	234	278.6	-19.	250.2	-6.9
40	262	295.0	-12.6	277.9	-6
50	291.2	329.7	-13.2	307.9	-5.7
60	319	347.7	-8.9	334.5	-4.8
70	344.7	378.5	-9.8	358.8	-4.1
80	364.5	391	-7.2	380.8	-4.4
90	385.75	415.2	-7.6	399.2	-3.4
100	409	423.2	-3.4	415.9	-1.6

Table 5: Corrected ex. manifold temp values

Torque	SS Values	53 Nm/min		80 Nm/min	
		Corrected	% Error	Corrected	% Error
20	1679.4	1716.5	-2.2	1961.1	-16.7
30	1755.4	1721.7	1.9	1930.4	-9.9
40	1764.0	1753.9	0.5	1896.1	-7.4
50	1890.3	1958.3	-3.5	1956.5	-3.5
60	1989.6	2015.6	-1.3	2021.7	-1.6
70	2080.9	2089.8	-0.4	2035.5	2.1
80	2069.28	2112.1	-2	1946	5.9
90	2096.2	2111.8	-0.7	1893.6	9.6
100	2218.8	2152.9	2.9	1910.8	13.8

Table 6: Corrected airflow values

Torque	SS Values	53 Nm/min		80 Nm/min	
		Corrected	% Error	Corrected	% Error
20	1.6	1.5	4.1	1.6	-0.03
30	1.9	2.2	-13.6	2.1	-9.1
40	2.3	2.2	5	2.6	-11.6
50	2.7	3	-10.1	3.2	-17.9
60	3.3	3.1	3.9	3.6	-9.4
70	3.5	3.6	-4	3.9	-10.8
80	3.9	3.8	1.7	4.28	-8.3
90	4.3	4.4	-1.2	4.5	-5.0
100	4.76	4.7	0.5	4.9	-3.4

Table 7: Corrected fuelling values

Torque (Nm)	Nox Mean (ppm)	SD %	In. Man Press Mean (Kpa)	SD %	Ex. Man Temp Mean (oC)	SD %	Airflow Mean (l/min)	SD %	Fuelflow Mean (Kg/h)	SD %
20.0	175.0	6.5	11.0	2.0	228.0	4.3	1679.5	10.0	1.6	12.1
30.0	211.0	6.0	16.0	2.3	250.2	3.0	1755.5	8.8	1.9	9.2
40.0	263.0	4.9	16.0	2.7	278.0	2.9	1764.1	6.3	2.4	6.3
50.0	312.0	3.8	26.0	3.0	308.0	3.0	1890.3	3.0	2.8	8.0
60.0	504.0	3.2	36.0	3.3	334.5	2.5	1989.6	4.2	3.3	6.2
70.0	593.0	3.0	43.0	3.1	358.9	2.9	2080.9	5.2	3.5	3.4
80.0	705.0	4.0	43.0	3.0	380.9	3.5	2069.3	6.1	4.0	5.4
80.0	175.0	4.8	43.0	2.8	399.2	4.2	2096.3	7.8	4.4	9.1
100.0	211.0	6.3	55.0	2.5	415.9	5.0	2218.8	11.0	4.8	13.0

Table 8: Repeatability Data

APPENDIX 1: RESPONSE MODEL FORMS

FIRST ORDER RESPONSE

The general form of a second order transfer function in the frequency domain is:

$$C(s) = \frac{a}{s(s+a)}$$

In the time domain the form is:

$$C(t) = 1 - e^{-at}$$

The time constant (TC) of the is the time taken to reach 63% of the final value.

$$TC = \frac{1}{a}$$

Where 'a' is the initial slope of the response.

Knowledge of 'a' from the data allows the time response to be evaluated.

In general the settling time TS:

TS = 4 x TC

SECOND ORDER RESPONSE

The general form of a second order transfer function in the frequency domain is:

$$C(s) = \frac{\omega_n^2}{s^2 + 2\xi\omega_n s + \omega_n^2}$$

In the time domain the form is:

$$C(t) = 1 - e^{-\zeta\omega_n^2} \cos\omega_n\sqrt{1-\zeta^{2t}} + \frac{\zeta}{\sqrt{1-\zeta^2}} \sin\omega_n\sqrt{1-\zeta^2}\,t$$

The time (Tp) until the peak response can be derived by setting the differential of the above to zero.

$$t = \frac{n\pi}{\omega_n\sqrt{1-\zeta^2}}$$

The maximum value occurs at the first peak, hence n=1:

$$T_p = \frac{\pi}{\omega_n\sqrt{1-\zeta^2}}$$

Defining the settling time (Ts) to be the period after which the response is less than 2% of the steady state value yields:

$$Ts = \frac{-\ln(0.02\sqrt{1-\zeta^2})}{\omega_n\zeta}$$

A knowledge of Ts and Tp from the response data enables the ω_n, ζ and hence the time response to be derived.

2002-01-0616

A Transient Hydrostatic Dynamometer for Testing Single-Cylinder Prototypes of Multi-Cylinder Engines

John L. Lahti and John J. Moskwa
University of Wisconsin - Madison

ABSTRACT

A new dynamometer system has been developed to improve the accuracy of tests that are run with a single cylinder version of a multi-cylinder engine. The dynamometer control system calculates the inertial torque and combustion torque that would normally be generated in a multi-cylinder engine. The system then applies the torque from the missing cylinders of the engine with the dynamometer. A unique high bandwidth hydraulic system is utilized to accurately apply these torque pulses. This allows the single-cylinder engine to have the identical instantaneous speed trajectory as the multi-cylinder engine, to test the single-cylinder engine at all engine speeds including very low speed operation, and to now do transient speed and load testing. Not only will this dramatically extend the capabilities of current single-cylinder engine test systems, but may open up new areas of research due to its transient testing capabilities.

INTRODUCTION

Testing single-cylinder research engines at low speed with a conventional dynamometer has always had some significant limitations. The problem of excessive speed variation between combustion cycles was in the past addressed by adding a very large non-representative flywheel. It was also nearly impossible to run tests at low speeds due to driveline vibration resulting from resonance. Since idle speed testing is critical to engine development a new dynamometer design has been developed to address these issues. This new dynamometer will be a transient hydrostatic type based on technologies developed in the Powertrain Control Research Lab (PCRL) at the University of Wisconsin-Madison. It will minimize speed variation and be capable of low speed operation because of low rotating inertia, high driveline stiffness, and fast torque response.

The new dynamometer will also be capable of simulating operation of a multi-cylinder engine by replicating the single-cylinder torque pulses in a multi-cylinder pattern. The single-cylinder will act dynamically as if it were actually in a multi-cylinder engine, providing a much more accurate testing situation than currently exists.

This paper describes the design and control of this unique dynamometer of which patents are currently pending.

GOALS OF THE TRANSIENT TEST SYSTEM

The overall goal of the transient test system described in this paper is to provide more representative operation and a more dynamic environment than is currently possible with the current state-of-the-art, single-cylinder research engines. This includes being able to test a single-cylinder engine throughout the complete expected engine speed range of the actual multi-cylinder engine in which this cylinder will be used. Also, the authors wish to replicate the instantaneous dynamic operation and speed trajectory throughout the engine cycle that would be expected in a multi-cylinder engine. Let us first examine what is commonly current practice in the use of single-cylinder research and test engines in order to appreciate the contributions of the system that is being developed.

Single-cylinder internal combustion engines are widely used throughout the world, in research and development laboratories in industry and academia, as tools for studying details of the engine's fluid dynamics, heat transfer, thermodynamics, emissions and other characteristics. They are also used in the early stages of new engine development to work out design challenges associated with combustion chamber shape, timing, or other geometric and thermodynamic issues, or to experimentally validate computation fluid dynamic results or predictions made on computers. The advantage of using the single-cylinder engine is to allow maximum accessibility to the combustion chamber for instrumentation, and to have a hardware test system with a minimal amount of expense and investment of time when compared to what would typically be associated with building a multi-cylinder engine. However, there are significant differences between the single-cylinder engine and its multi-cylinder counterpart.

Because the internal combustion engine is typically comprised of a slider-crank arrangement to convert linear piston motion to rotational crankshaft motion, there are many dynamic balancing problems with the single-cylinder engine. When the cylinder is integrated into a multi-cylinder engine then the dynamic balancing issues can be minimized significantly by either careful choice of the cylinder configuration, balance and timing, or dynamic balance shafts can be added to the engine. The unbalanced forces and torques of each cylinder are carefully phased so that they become nearly or completely balanced by the forces and torques from other cylinders. However this is a difficult challenge in a single-cylinder engine. There have been some innovative designs of single-cylinder test systems with balance shafts to address some of the slider-crank's imbalance issues [1] but these are not yet very widely used in practice.

Another even greater challenge with the single-cylinder engine is the lack of torque that comes from additional cylinders in a multi-cylinder engine, which provide energy to the crankshaft between cylinder firing. In four-stroke engines, commonly used in most transportation applications, the engine fires and provides energy to the crankshaft once every two revolutions. Therefore, during the two crankshaft revolutions between firings, the engine loses kinetic energy and slows down considerably. This problem is especially pronounced at low engine speeds such as idle because the time between cylinder firing increases. To minimize this problem a large flywheel is typically added to the single-cylinder crankshaft, and it stores most of the combustion energy in the form of kinetic energy (*KE*). Also, dynamometers with large polar moments of inertia are typically coupled to the single-cylinder engine, further increasing the inertia.

$$KE = \frac{1}{2} \cdot J \cdot \omega^2 \tag{1}$$

As the kinetic energy in the flywheel is lost through friction, windage, and compression work between cylinder firings ($KE_2 - KE_1$), the magnitude of the polar moment of inertia (J) of the flywheel and other rotational components will strongly affect the change in engine speed ($\omega_2 - \omega_1$) that will result.

$$\omega_2 - \omega_1 = \sqrt{\frac{2}{J}} \cdot \left[\sqrt{KE_2} - \sqrt{KE_1}\right] \tag{2}$$

$$\omega_2 = \sqrt{\frac{2}{J} \cdot (KE_2 - KE_1) + \omega_1^2} \tag{3}$$

There is an additional effect from the changing magnitude of the polar moment of inertia due to the dynamics of the slider-crank configuration that changes instantaneous speed, but these effects are averaged out over an engine cycle.

It is very difficult and often not possible to study combustion and other effects at low idle speeds on a single-cylinder engine due to this problem (i.e., system rotational energy increases only once every two revolutions for the 4-stroke engine). However, this is an operating region of the engine which is of great importance because of its strong correlation with engine fuel economy, especially in spark ignition engines.

One of the most widely used platforms for the testing of single-cylinder engines is the Ricardo Hydra base [2]. This is a standardized base that contains the housing and bearings for the single-cylinder crankshaft, the lubricating oil sump, as well as various accessories for engine timing, cylinder valve actuation, flywheel and other necessities for engine operation. Because of the Hydra's wide use throughout the world, the authors also use this engine base for the transient test system being developed. Therefore, researchers and engineers that wish to use this new system will not be required to make any changes to their engine cylinder set-up as long as they are currently using the Hydra base. There is also interest in applying the transient test system described in this paper to other single-cylinder test devices.

IMPORTANCE OF TRANSIENT TESTING CAPABILITIES

In essence, the goal of this system is to make the single-cylinder engine behave dynamically as it would in a multi-cylinder engine. This is accomplished by first carefully calculating in real time all of the dynamic torques that would be applied to the single cylinder if it were in a multi-cylinder engine. These torques would be typically applied in a multi-cylinder engine by means of the crankshaft which connects the cylinders. Then an apparatus must be developed that can apply these loads quickly enough to meet the required frequency or bandwidth in order to replicate the calculated dynamic torques. In the embodiment presented in this paper and developed in PCRL, this would be a hydrostatic hydraulic system controlled by high-speed servo-valves as described later in this paper.

If these results described above can be achieved with the proposed test system, then the implications for engine research and development would be extensive and can be summarized in three major points. The first point is that the single-cylinder test engine would now behave as if it were in an actual engine, unlike systems currently in use. The speed trajectory throughout the engine cycle would be the same as in the multi-cylinder engine, and therefore the test results from the single-cylinder engine would more accurately represent what would be expected in the multi-cylinder engine. This is not currently a major concern at higher engine speeds, but becomes especially important for low engine speed operation.

The second point is that the test system would allow detailed testing to be accomplished throughout the entire engine range, and no compromises would need to be made for low speed testing.. This would be a major contribution to the area of single-cylinder engine testing and research because of the difficulties currently experienced in this low speed regime. The proposed system would even allow the single-cylinder engine to be easily tested at speeds below what would be expected in the multi-cylinder engine in order to explore potential gains in further lowering idle speeds and its effects on combustion.

The third point is that the single-cylinder engine could now be tested in transients or rapidly changing speed and load conditions. Because of the large flywheel and dynamometer polar moment of inertia on current test systems, transient engine testing is not a possibility, except for some experiments in modifying spark, fuel injection or fuel timing at constant speeds. However, the system described in this paper would allow full transient testing on a single-cylinder engine set-up, and would open up new areas of research and provide these engines with a more realistic test environment. Engineers and scientists could now explore all areas of the transient engine operation while still enjoying free access to the combustion chamber for instrumentation such as laser diagnostics or other measurements.

Because only one cylinder is actually tested and the other cylinders of the "virtual engine" in this hardware-in-the-loop (H-I-L) transient test system are based in software, engine configurations could be rapidly changed to study individual cylinder effects from various configurations or a different number of cylinders. Also, the effects of cylinder cut-out or engine re-start, which are becoming more widely accepted and commonly used in new vehicles, can be studied in greater depth with wide access to the cylinder for diagnostics and instrumentation. Cut-out and re-start can be easily simulated in the real time software. With faster computer processor speeds, the single cylinder test engine could also be effectively connected to an entire virtual powertrain, to study any effects from dynamic coupling of these systems. The powertrain models would have to run in real time, but this has already been accomplished on other transient engine dynamometers in PCRL.

DYNAMOMETER HARDWARE

The hardware for the new dynamometer has some functional similarities to that of a hydrostatic transient dynamometer that was recently developed at the University of Wisconsin-Madison. A brief review of that design will help clarify the hydrostatic dynamometer concept.

A simplified drawing of the present hydrostatic dynamometer is shown in Figure 1. The hydraulic pump/motor, which is attached to the engine, applies an absorbing torque when the pump/motor outlet pressure is greater than the inlet pressure. A motoring torque is generated when the pump/motor inlet pressure is higher than the outlet pressure.

The inlet pressure to the pump/motor is maintained at a nearly constant level by the variable displacement hydraulic power supply. Motoring or absorbing torque is created by controlling the pressure on the outlet side of the pump/motor with a servovalve. Although servovalves are usually used to change flow direction, in this application it is used to restrict flow. A servovalve was selected because it provides very fast response while allowing proportional control of the spool position.

Since the servovalve is only used to restrict flow, the valve spool is only moved to one side of the center position. The valve is attached to the system in a double flow configuration that minimizes pressure drop when the valve is open.

The original plan for the single-cylinder dynamometer was to utilize the design shown in Figure 1 and simply raise the pump/motor inlet pressure to allow sufficient motoring torque to be generated. The high inlet pressure would enable both motoring and absorbing capability with a constant supply pressure. Torque would be adjusted by controlling the pump/motor outlet pressure with a servovalve.

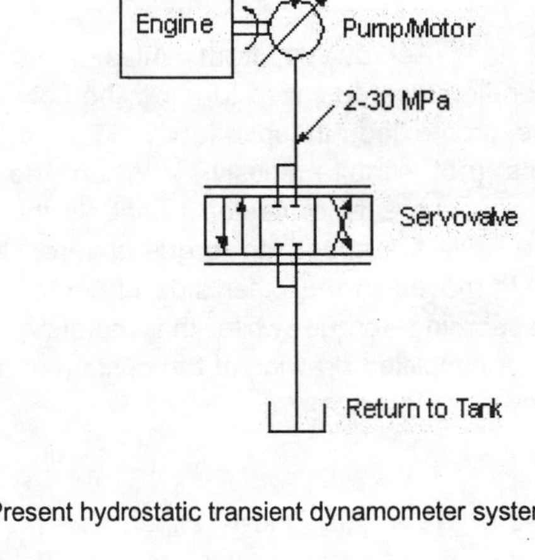

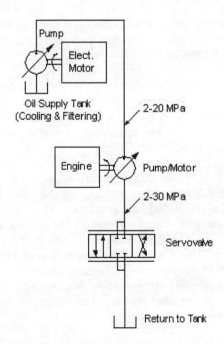

Figure 1: Present hydrostatic transient dynamometer system.

Although this design provides good dynamic response, analysis showed it to be very inefficient in this new application because of the high inlet pressure requirements. More than 100 kW would be required to deliver fluid at the flow rate and pressure required by this design. In addition, the wasted energy would heat the hydraulic fluid in the system. The hydraulic power supply for this project had neither the power nor cooling capacity to make the design shown in Figure 1 work for this application.

To improve efficiency and minimize cost the design shown in Figure 2 was developed. It allows the existing hydraulic power supply to be used while still meeting all of the design requirements. It incorporates the following new features:

- Separate pump and motor are used to apply absorbing and motoring torque. The pump circuit has an independent return loop with heat exchanger to minimize the flow requirement from the hydraulic power supply.
- The motor circuit has a return loop that recirculates flow when a motoring torque is not required. This recirculation reduces power consumption by approximately 50% since motoring torque is only required about 50% of the time.
- Accumulators were added to maintain a uniform supply and return pressure form cycle to cycle in the motor circuit.
- Restrictor valves were added to the motor and pump circuits to allow some fluid to flow back to the main tank for filtering.
- Variable displacement pumps and motors were selected for the dynamometer to allow the displacement to be reduced at high engine speeds, making it possible to optimize the efficiency of the system under all conditions.
- A custom servovalve from Atlas Fluid Controls Corporation was designed to allow the flow restriction to be controlled independently in two hydraulic circuits (not simultaneously). When the spool is moved to one side of center it controls the motoring torque while the absorbing torque is zero. When the spool is moved to the other side of center it controls the absorbing torque while the motoring torque is zero. A simplified drawing of the custom spool design is shown in Figure 3.

The servovalve is the most expensive part of the dynamometer. By using a custom servovalve it was possible to use one valve instead of two, which provided a significant saving.

Conventional engine dynamometers have a resonant point at low engine speed. When the engine is started it passes through this point. The resonance limits the low speed range of the dynamometer because driveline vibration becomes excessive. This is especially a problem with single-cylinder engines due to the low frequency of combustion events. Sometimes large flywheels are used to reduce the magnitude of the transmitted torque and to lower the natural frequency. This severely limits the transient response capability of the dynamometer. Damping is often added to the driveline by using flexible couplings that have a plastic or rubber-connecting element. Although this reduces the magnitude of the vibration, it is still difficult to operate the engine at low speed and driveline vibration may damage the flexible coupling.

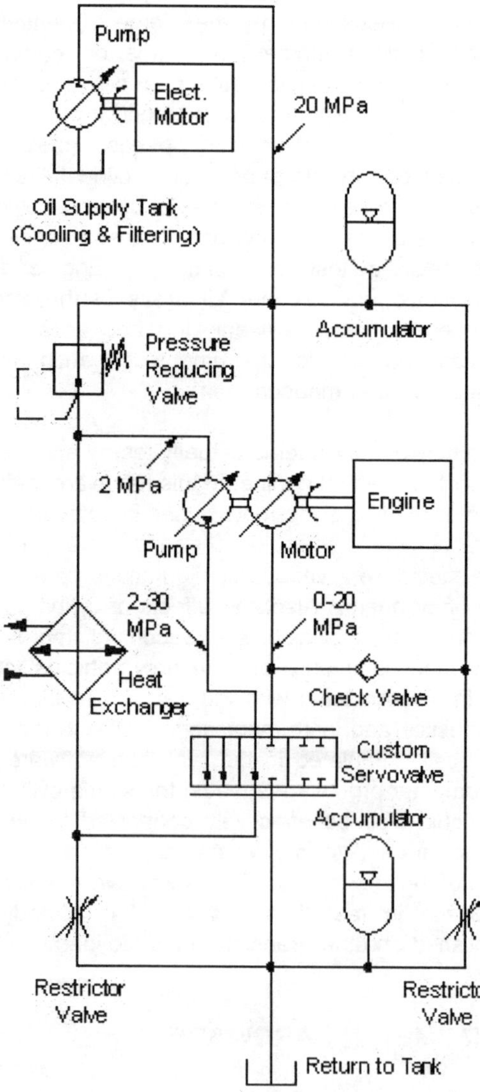

Figure 2: New single-cylinder hydrostatic transient dynamometer.

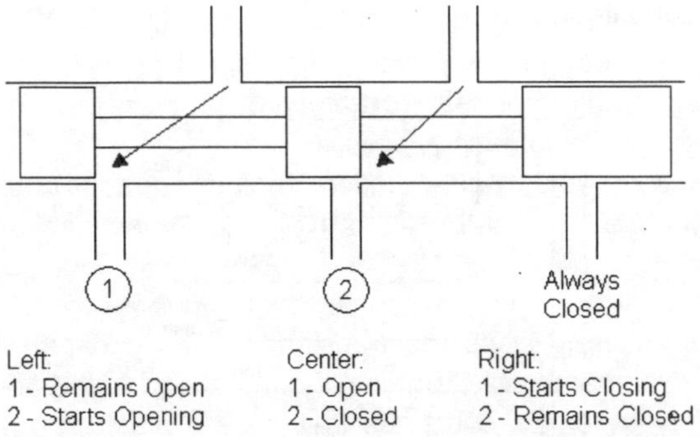

Figure 3: Custom servovalve spool design that provides flow control in two passages.

Figure 4 shows a single-cylinder engine attached to an electric dynamometer. This is how single-cylinder engine have traditionally been tested. The engine and dynamometer are each attached independently to a steel bedplate in the floor. The dynamometer is trunnion mounted with torque being measured by a load cell. The drive shaft has universal joints or flexible couplings to compensate for misalignment.

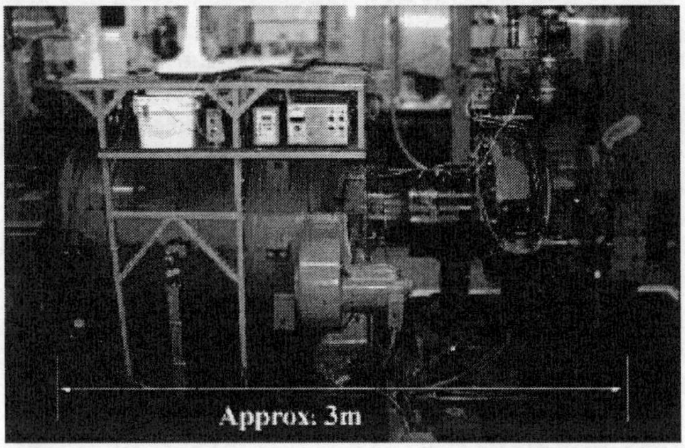

Figure 4: Conventional dynamometer set up for testing a single-cylinder research engine.

The new dynamometer configuration shown in Figure 5 is attached to the engine with a torque tube rather than being attached to the bedplate. An in-line short flange torque transducer from Lebow Products measures the dynamometer torque.

A special driveline was designed with high stiffness and low inertia. The driveline includes the following components: an adjustable flywheel to allow the inertia of the system to be sized to match that of the multi-cylinder engine that is being simulated, a flex plate to allow for slight misalignment in the driveline, and an in-line torque transducer. The distance from the crankshaft to the dynamometer shaft is less than ten centimeters. Alignment with this system is less of a problem because of the short driveline and the accurately machined torque tube that attaches the dynamometer to the engine.

In addition to low driveline inertia, the hydraulic pump and motor have significantly less inertia than a traditional electric dynamometer. The low inertia and high stiffness of this system positions the resonant point above the operating range of the engine, eliminating the low speed resonant vibration problem.

INTEGRATION HARDWARE AND SOFTWARE

Nearly all of the engine and dynamometer controls will be performed real time with a rapid-prototyping control system from dSPACE. MATLAB/Simulink software from The MathWorks will be used to generate a block diagram model of the control system. Real –Time Workshop and Stateflow Coder will then generate C code from the block diagrams. Further processing by Real-Time Interface from dSPACE prepares the code for implemented with dSPACE hardware.

All instrumentation and data acquisition will be performed on a personal computer running dSPACE Control Desk software. The PC interfaces with the dSPACE control hardware through an Ethernet connection.

Since one controller will manage all engine and dynamometer functions the control of each component will have complete access to the variables used by the other. This greatly simplifies the integration of the engine torque model into the controls for the dynamometer torque output.

To provide an additional level of safety a programmable logic controller from GE Fanuc will be used to shut down the system in the event of a malfunction. Completely redundant sets of sensors will be used with this system.

REAL TIME SIMULATION

To better understand the interactions between engine and dynamometer a model of the system was developed. The model simulates the torque output of the engine, calculates the desired dynamometer torque, and evaluates the transient response of the dynamometer/hydraulic system. The model was developed using MATLAB/Simulink.

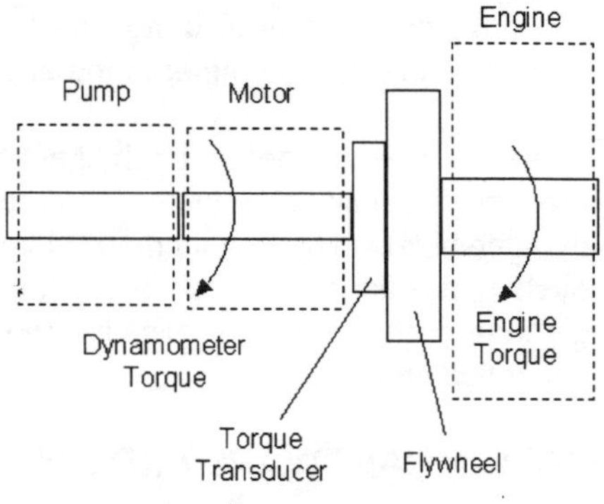

Figure 5: Free body diagram of engine crankshaft, driveline, motor shaft, and pump shaft.

Figure 5 shows the layout of the dynamometer. This figure is the basis for the system model that was developed. Since the dynamometer uses a very short stiff driveline, the engine crankshaft, flywheel, flex plate, torque transducer, motor shaft, and pump shaft were assumed to act as one rigid component with constant inertia, unlike conventional dynamometer systems. When engine torque and dynamometer torque are applied to this inertia it causes a change in speed as indicated by Equation 4.

$$J\ddot{\theta} + \frac{1}{2}\frac{\partial J}{\partial \theta}\dot{\theta}^2 = \sum Torques \qquad (4)$$

Here J represents the polar moment of inertia of the rotating components. The reciprocating mass of the piston and connecting rod were not included in the inertia term. This allowed the changing inertia term to be dropped but required the torque from the reciprocating mass to be included in the sum of the torque terms. The equation was then simplified to the form shown by Equation 5.

$$J\ddot{\theta} = T_{engine} + T_{dynamometer} \qquad (5)$$

Here $T_{dynamometer}$ represents the torque generated by the fluid pressure in the dynamometer pump or motor. T_{engine} is the torque created by the force of the connecting rod on the crankshaft. The connecting rod force has several components: combustion force, piston friction force, force due to reciprocating mass at constant engine speed, and force due to angular crankshaft acceleration. The friction force and force due to angular crankshaft acceleration were not included in this analysis because they are much less significant than the other two forces and they make the analysis and models much more complicated. The effect of the combustion force and reciprocating mass will be discussed in more detail as part of the engine torque model.

Figure 6 shows a flow chart of the overall system model. Torque from the engine and dynamometer act on the rotating components of the system to produce angular acceleration. The engine speed and position are evaluated through integration.

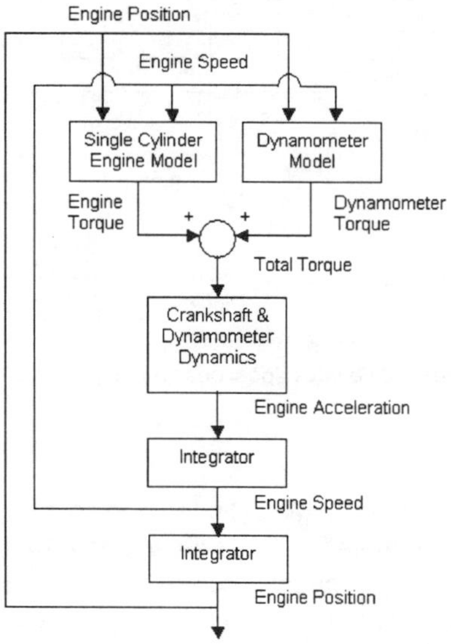

Figure 6: Block diagram model of engine and dynamometer interaction

ENGINE TORQUE MODEL

An instantaneous torque model of the engine was developed for two reasons: to simulate the operation of the single-cylinder engine, and to calculate the torque required by the dynamometer. The torque resulting from cylinder pressures as well as the torque caused by the reciprocating mass of the slider crank mechanism (piston and connecting rod) were included in the model. The intent of this analysis was component selection. As the dynamometer controls are developed a more detailed torque model will be implemented.

Cylinder pressure data were used to estimate the torque at each crank angle through one complete engine cycle. Wide-open throttle data were utilized to obtain the highest torque levels. Figure 7 shows the resulting torque for a production 4.2L V-6 engine.

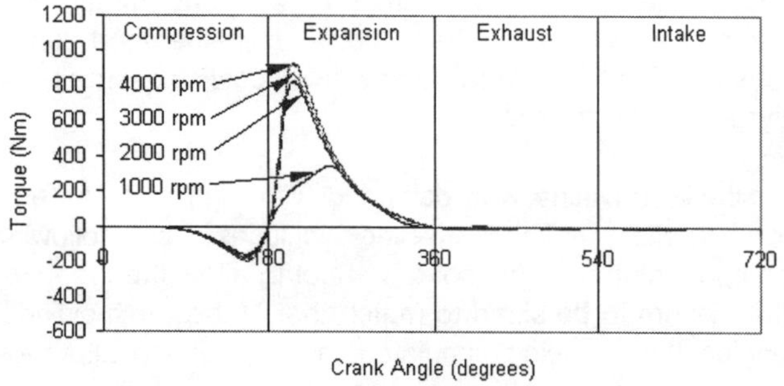

Figure 7: Crankshaft torque resulting from cylinder pressure at wide-open throttle.

To evaluate the inertia torque that is applied to the crankshaft by the reciprocation mass of the slider crank mechanism, a lumped mass model of the piston and connecting rod was used. Figure 8 shows the resulting torque from those calculations.

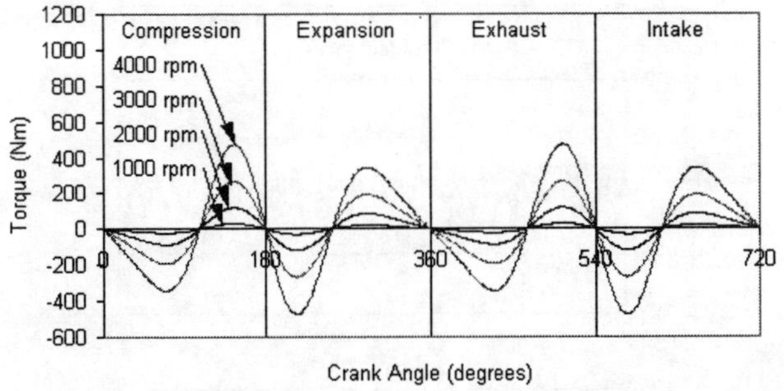

Figure 8: Torque applied to the crankshaft by the reciprocating mass of the piston and connecting rod.

Figure 9 shows the total torque resulting from cylinder pressure and reciprocating mass. This is the torque that is applied to the crankshaft rod journal by the connecting rod. Notice the change in shape of the torque trace as the engine speed increases. While the inertia torque is almost insignificant at 1000 rpm, it becomes the dominant torque at 4000 rpm.

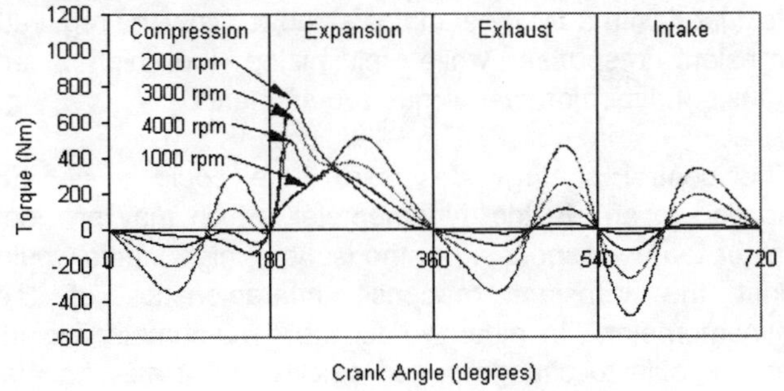

Figure 9: Total torque applied to the crankshaft by the piston and connecting rod assembly in a single-cylinder engine at wide-open throttle.

The engine model used in the simulation was based on the wide-open throttle torque trace shown in Figure 9. Using these torque values the dynamometer response was evaluated under worst-case torque conditions so that the appropriate sized components could be selected for the dynamometer. As the controls for the dynamometer are developed a more detailed model will be used to account for different conditions such as part throttle operation.

DYNAMOMETER TORQUE MODEL

The dynamometer model includes calculations for the desired dynamometer torque as well as compensation for driveline inertia and the hydraulic system response. First, the engine torque is canceled by applying a torque equal in magnitude and opposite in direction to that of the engine. Then torque ripple is added to simulate an engine with a certain number of cylinders and inertia. Finally, transfer functions for the servovalve and other hydraulic components are utilized to predict the actual dynamometer torque response. Figure 10 shows the block diagram for this model.

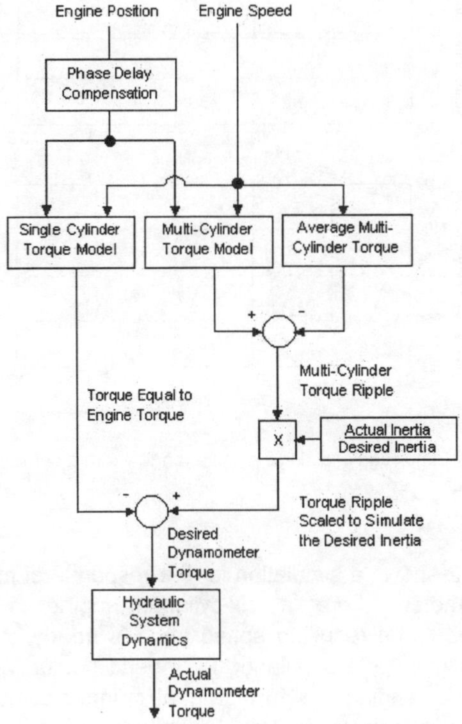

Figure 10: Block diagram model of dynamometer torque calculation

SIMULATION RESULTS

Simulations were run using the model to evaluate changes in hydraulic pump displacement, servovalve response time, inertia, and engine speed. Modifications to the models were also developed to simulate operation of a conventional dynamometer.

Figure 11 shows the engine speed variation that would occur in a conventional dynamometer with a constant load torque. The engine speed increases during the expansion (power) stroke of the engine then decreases through the remaining three engine strokes of the combustion cycle. Increasing the inertia of the system can reduce the magnitude of these speed fluctuations but the general shape of the curve will remain. Adding inertia hinders the transient response performance of the dynamometer. It also causes the resonant point of the driveline to be lowered, which could limit the operating range of the dynamometer.

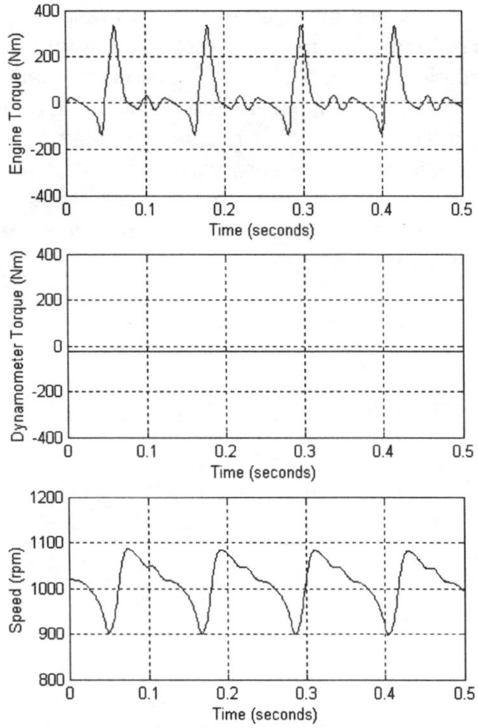

Figure 11: Simulation of a single-cylinder engine with a constant dynamometer load

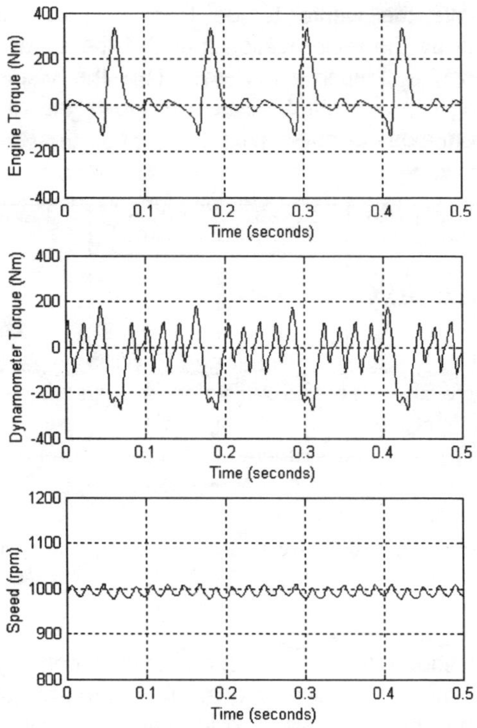

Figure 12: Simulation of a single-cylinder engine with a torque pulse replication dynamometer (simulating six-cylinders)

Figure 12 shows a simulation for the response of the new dynamometer where a six-cylinder engine is being simulated. The resulting speed trace is nearly constant with a slight ripple simulating the speed fluctuation of a six-cylinder engine due to individual cylinder combustion and inertia effects. The magnitude of the speed fluctuation can be adjusted by changing the desired inertia value used in the torque ripple calculation.

At high engine speeds the torque pulse frequency makes it difficult for the dynamometer to accurately generate the desired torque. At speeds above 3000 rpm it may not be possible to absorb each individual torque pulse due to the limited bandwidth of currently available servovalves. The bandwidth of the valve used in this simulation was 120 Hz, which is the fastest valve available for the required flow rate.

Although the dynamometer response is not quite fast enough at high engine speed to match the desired torque profile, this is not a significant problem. Each combustion event induces a certain amount of kinetic energy. As the engine speed increases the speed change due to combustion becomes smaller due to the squared relationship between engine speed and kinetic energy as shown by Equation 3. Also, at high-speed the speed variation depends less on combustion and more on reciprocating mass.

To minimize high-speed torque control error the system is designed with a flywheel that is sized to allow the required transient response while minimizing the high-speed transient dynamometer torque requirements.

The control strategy described here could easily be applied to an electric dynamometer, which may provide faster torque response, but the issue of high inertia would limit the transient response characteristics of the dynamometer. In other words, the dynamometer would not be able to change speed quickly and it may not be able to simulate the speed fluctuations of a multi-cylinder engine.

CONCLUSION

The single-cylinder dynamometer system described here has many advantages: stable low speed operation with no resonate vibration issues, the ability to generate a speed profile that matches that of a multi-cylinder engine, transient test capability, and very low cost. The system improves test data accuracy by eliminating the large speed fluctuations that typically occur with a single-cylinder research engine.

ACKNOWLEDGMENTS

The authors would like to thank General Motors for supporting work on this project and especially Charles Folkerts who worked to get final approval on the contract. They would also like to thank Frank Fronczak at the University of Wisconsin-Madison for discussions and suggestions relative to the hydraulic circuit.

REFERENCES

1. J. Allen, D. Law, G. Pitcher and P. Williams, "A New Optical Access Research Engine for Advanced Fuel Spray and Combustion Analysis Using Laser Based Diagnostics", Lotus Engineering, International Symposium on Automotive Technology and Automation, 00POW004, 2000

2. G. Cussons Ltd., "Ricardo Single Cylinder Research Engines", Technical Specifications, Manchester, England, 1987

3. R. Dorey, D. Wang, "A Hydrostatic Dynamometer for Engine Testing", Second Bath International Fluid Power Workshop, September 1989

4. H. Ghaffarzadeh, M. Guebeli, N. Vaughan, "Linear Model Identification of a Hydrostatic Dynamometer through Dynamic Simulation", Fifth Bath International Fluid Power Workshop, September 1992

5. G. Babbitt, "Transient Engine Test System for Hardware-in-the-Loop Powertrain Development", Department of Mechanical Engineering, University of Wisconsin – Madison, Ph.D. dissertation, 1999

6. S. Seaney, "Design and Construction of a High Bandwidth Dynamometer with Absorbing and Motoring Capability", Department of Mechanical Engineering, University of Wisconsin – Madison, Masters Thesis, 1994

7. G. Babbitt, J. Moskwa, "Implementation Details and Test Results for a Transient Engine Dynamometer and Hardware in the Loop Vehicle Model", IEEE CACSD-569, 1999

8. H. Mabie, C. Reinholtz, "Mechanisms and Dynamics of Machinery", Fourth Edition, John Wiley & Sons, Inc. 1987

2002-01-1300

An Investigation of SF_6 Gas for Testing Instrumental Integrity of the Emerging SULEV/PZEV Measurement Technology

Scott Porter, Richard T. Rooney and Karl Oestergaard
Horiba Instruments, Inc.

Zlatko Rauker
Ford Motor Company, Ford Research Laboratory

Satoshi Ohtsuki
Horiba, Ltd.

ABSTRACT

For the past several years, manufacturers have been developing emission measurement systems for Super Ultra Low Emission (SULEV) measurements. The Bag Mini-Diluter (BMD) with an advanced exhaust flow measurement device is designed as an alternative to the traditional method for sampling vehicle exhaust, the constant volume sampler (CVS).

Exhaust sampling instruments require system verification tests. The system verification test described and mandated for the CVS in the Code of Federal Regulations (CFR) §86.119-90(c) is a simulated test with propane. The very low concentration measurements required for SULEV regulations demand a more enhanced and accurate verification technique and procedure than the method described in the CFR.

This investigation focuses on the technique and necessary equipment for verifying system integrity of the entire emission sampling system, including the Bag Mini-Diluter and the exhaust flow measurement device in the test cell. The technique involves the precisely controlled injection of an inert check gas (SF_6) into the exhaust gas stream. This technique has significant advantages over the propane injection for the CVS check since the check gas (SF_6) is absent from the background measurement providing more accurate detection. Additionally, the check test is also run *in situ*. This test is a real verification of system integrity performed during an automotive test cycle, not just a simulation like the propane injection. Measurement of this test gas (SF_6) relies on a sensitive NDIR analyzer. In this investigation, practical operational suggestions are presented along with data to show the efficacy of the verification system.

INTRODUCTION

The traditional system verification technique as mandated by the (CFR) 86.119-90(c) employs recovery of an injected propane mass to ensure the integrity and accuracy of the emission sampling system. A small mass of Propane (C_3H_8) is typically injected into the Constant Volume Sampler (CVS) system to verify the CVS flow and to ensure system leak integrity. These techniques can use either gravimetric or constant mass flow methods for injecting the mass of propane. This mass of propane injected can then be 'recovered' using the emission sampling system by measuring the concentration collected in a CVS sample bag, as read by a hydrocarbon analyzer. The calculated recovered mass must be within ± 2% of the injected mass to consider the system acceptable.

Decreasing SULEV and Partial Zero Emissions (PZEV) levels create system measurement challenges due to the very low concentrations emitted from these vehicles. Determination and measurement of the low-level emission concentrations required development of new sampling techniques and analysis systems. Dilution ratios associated with the traditional CVS introduced ultra-low level emission concentrations in sample bags that analysis systems could not accurately and repeatably measure.

Bag mini-diluter technology was developed to obtain optimum dilution ratios that produce higher bag sample concentrations needed for more accurate and repeatable measurement. In addition, the bag mini-diluter technology eliminates ambient air concentrations in the sample bag thus eliminating the need for background concentration corrections for mass determination. Using the CVS, these corrections for ambient concentrations add variability to mass calculations since they are constantly changing.

To verify these low emission levels a precise verification method must be employed to measure the system reliability and accuracy. The accuracy of the bag mini-diluter and exhaust flow measurement device must be verified at these ultra-low concentrations.

An *in-situ* method for determining the system verification is required to completely ensure system reliability. Sulfur hexafluoride (SF_6) may be injected into the exhaust stream during an actual vehicle emission test as a trace gas. This injection technique incorporates actual test conditions while performing the verification. Conditions such as actual exhaust gas temperature, moisture, and exhaust gas constituents are experienced while performing the recovery of the SF_6. This injection method may be used to verify the entire analytical system sample stream including the following systems:

1. Bag Mini-Diluter
2. Exhaust flow measurement system
3. CVS system
4. Analysis system
5. Gas analyzer

SF_6 is advantageous to use as an injection gas due to its inert properties and its lack of existence in normal ambient air. Therefore corrections for ambient concentrations are not required while performing a system verification.

SF_6 INVESTIGATION

SF_6 ANALYZER - The Horiba Instruments, Inc. model AIA-722 was developed to measure sulfur hexaflouride (SF_6) by nondispersive infrared (NDIR) analysis. The analyzer was developed during the development of the BMD to improve the verification of the SULEV test site. The following is a brief description on the analyzer's operational principle.
Figure 1 illustrates the construction of the NDIR analyzer. The method of detection is based upon the principle of selective absorption. Infrared radiation from a single infrared light source is directed through the two separate cells, the sample cell and the reference cell. These cells are identical in length and volume. The sample cell is designed to allow the sample gas to flow through it on a continuous basis. The reference cell is filled with nitrogen (N_2). As the infrared radiation passes through the sample cell, the SF_6 present in the sample gas absorbs the radiation at its characteristic wavelength. The radiation passing through the reference cell is not absorbed at these characteristic wavelengths.

The radiation passes through the cells to the two detectors filled with equal amounts of SF_6 gas. These detectors are separated by flexible diaphragm. This diaphragm is close to a stationary metal plate. The diaphragm and the metal plate constitute a capacitor whose capacitance will vary as the diaphragm moves. The gas in the detectors is heated by the radiant energy of the infrared source. This intensity of infrared radiation at the SF_6 absorbance wavelengths will be greater in the detector for the reference cell than the detector for the sample cell. The incident radiation at the characteristic SF_6 wavelengths will be absorbed in both detectors. The temperature in the detectors will increase as the radiation is absorbed. The temperature of the detector for the reference cell will increase more than the temperature of the detector for the sample cell. This temperature differential will cause a pressure differential.

The pressure differential will cause the diaphragm to move. This movement will change the capacitance and can be measured electrically. The variation in capacitance is proportional to the absorbance of infrared energy in the sample cell. The absorbance of infrared energy is proportional to the concentration of SF_6 in the exhaust sample.

This temperature increase must be modulated to prevent damage to the membrane, improve the dynamic range and reduce response time. As shown in Figure 1, a chopper modulates the infrared radiation emitted from the light source. This allows the temperature in the detectors to equilibrate when the light is blocked and returns the membrane to its starting location before the next period of irritation.

The AIA-722 SF_6 analyzer was developed with a range of 1000ppm. At the present time, this range will be used for all ongoing development tests. The specification sheet for the AIA-722 analyzer is included in the Appendix.

<u>Interferences -</u> It is important to note that errors will occur in the NDIR type of analyzer if the exhaust gas contains

other species which will absorb radiation at the same frequencies as the gas (SF_6) will absorb in the detector. The AIA-722 SF_6 analyzer has been checked for interferences during the design process. The interferences and the concentrations tested are listed in Table 1.

Interference	Within +/-0.5%FS	H_2O(25degC Sat.)
	Within +/-0.5%FS	16vol%CO_2
	Within +/-0.5%FS	10vol%CO
	Within +/-0.5%FS	1000ppmNO
	Within +/-0.5%FS	200ppmSO_2
	Within +/-0.5%FS	50ppmC(w/ C_3H_8)
	Within +/-0.5%FS	200ppmCH_3OH
	Within +/-0.5%FS	200ppmNH_3
	Within +/-0.5%FS	200ppmC_2H_4

Table 1: SF_6 AIA-722 Interferences

Essentially, all of the listed gases and their concentrations provide only minimal interference to the SF_6 in the AIA-722 analyzer. ± 0.5% Full Scale is equivalent to 5 ppm based upon the 1000 ppm full scale of the analyzer

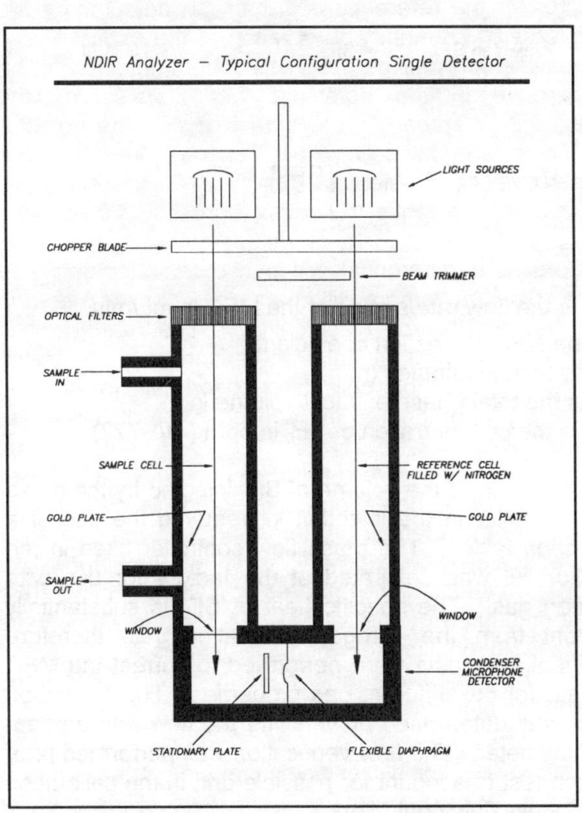

Figure 1: Construction of the NDIR analyzer

<u>SF_6 gas</u> - SF_6 has been chosen for the purpose of this investigation because it is not present in the exhaust gas and does not react with the exhaust components. SF_6 is a synthetic gas. It is used as an insulating medium in the electric power industry. Some other uses of SF_6 are as follows; as a cover gas in the magnesium industry, used for plasma etching in semiconductor manufacturing, as a reactive gas in aluminum recycling to reduce porosity, as thermal and sound insulation, and in atmospheric tracer studies as well as medical applications.

The most common use for SF_6, worldwide, is as an electrical insulator in equipment that transmits and distributes electricity. It is because of its dielectric strength and arc-quenching characteristics that the U.S. electric power industry has used SF_6. It has been used since the 1950's. SF_6 is used widely in gas-insulated substations, circuit breakers, and switchgear.

Historically, in the automotive industry, SF_6 has been used for the development of new exhaust emissions measurement systems. Because of its inert properties, it has been used as a trace gas in developing sampling systems and multi-component analytical systems. As part of this investigation, SF_6 will be looked at as a method for measuring the exhaust flow. For exhaust flow determination, the SF_6 method will be looked at as a replacement for currently used methods; the Helium and CO_2 tracer techniques.

DEVELOPMENT TESTS – Initial tests were performed to verify the design of the AIA-722 SF_6 analyzer as well as to create a set of baseline data for the development of the BMD verification. The test set-up is shown in figure 2. The sampling and analysis equipment consists of a Horiba Critical Flow Venturi (CFV) type CVS and a MEXA 7000 analyzer system respectively, along with the SF_6 injection kit. For the purposes of this test, 100% SF_6 was used.

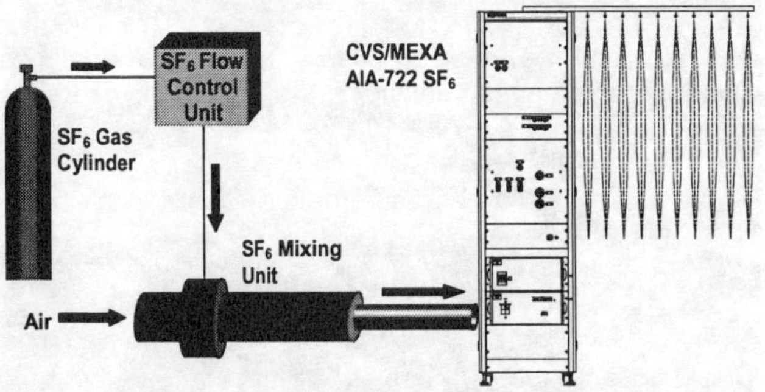

Figure 2: SF_6 Equipment Layout for Initial Tests

The SF_6 prototype injection kit consists of two components; the SF_6 flow control unit containing a variable mass flow controller (MFC), shown in Figure 3, and the SF_6 mixing unit.

Figure 3: SF_6 Flow Control Unit

The Mass Flow Controller (MFC) was used for development tests to vary the flow rate of the injected gas to obtain different concentrations and mass amounts. For a standard propane injection test, a Critical Flow Orifice (CFO) kit is typically used. With a CFO, changes in the flow rate of the injected gas are not easily attained, so the use of a mass flow controller was desirable. A mixing unit, as shown in Figure 4, was designed to provide homogeneous mixing of the SF_6 in the exhaust flow stream. The mixing unit provides an injection point for the SF_6 gas and produces turbulent flow. Attention is required regarding the pressure drop across the mixing unit. For vehicle tests, excessive back-pressure created by the mixing unit may have an adverse effect on vehicle performance and mass emissions produced by the vehicle. Initial tests were performed without the use of a vehicle. The test set-up is similar to the traditional CVS set-up for performing propane injection tests.

Figure 4: SF_6 Mixing Unit

The CVS critical flow venturi (CFV) flow rate was set at 21 m^3/min. This is the maximum flow rate of the CVS used for these tests. The tests were run at the maximum flow rate so that the concentration read by the AIA-722 analyzer was low, therefore the recovery test was done under the most extreme dilution conditions. During the tests, 100% SF_6 was injected into the flow stream. Three flow rates were used for this experiment; 500, 750, and 1000 mL/min. Also three tests at each flow rate were performed. The data from these tests is shown in Table 2.

A	B	C	D	E	F	G		
MFC Set (mL/min.)	MFC Corr. Coef. αQ	Time (s)	V_{mix} (m3)	SF6. Conc. (ppm)	SF_6 Vol MFC (mL)	SF_6 Vol CVS (mL)	Error (%) (F-G)/G*100	
500	1.0446	505	165.16	26.80	4396.03	4426.29	***-0.68***	%
500	1.0446	505	165.10	26.62	4396.03	4394.96	***0.02***	%
500	1.0446	505	165.13	26.35	4396.03	4351.18	***1.03***	%
750	1.0492	505	165.13	39.49	6623.08	6520.98	***1.57***	%
750	1.0492	505	165.16	39.48	6623.08	6520.52	***1.57***	%
750	1.0492	505	165.04	39.48	6623.08	6515.78	***1.65***	%
1000	1.0506	499	163.15	53.28	8737.49	8692.63	***0.52***	%
1000	1.0506	505	165.15	52.75	8842.55	8711.66	***1.50***	%
1000	1.0506	505	165.13	52.81	8842.55	8720.52	***1.40***	%

Table 2: Results of Initial Injection Tests

Based upon these test results, the calculations are as follows:

$$V_{injected\ SF_6} = Q_{SF_6} * \alpha * t \qquad (1)$$

and

$$V_{CVS\ SF_6} = V_{mix} * [SF_6] \qquad (2)$$

Where:

Q_{SF_6} is the flow rate setting of the MFC in mL/min
α is the MFC correction coefficient
t is the time in minutes (min)
V_{mix} is the total phase exhaust volume in m^3
$[SF_6]$ is the concentration of SF_6 in ppm (AIA-722)

Determination of the volume of SF_6 injected by the mass flow controller in the injection kit required the use of a correction factor. The mass flow controller used in the injection kit was calibrated at the factory for use with nitrogen gas. The specific heat of SF_6 is substantially different from the nitrogen calibration gas, therefore empirical flow tests were performed to correct the MFC flowrate for the different gas properties. The correction factor was determined by verifying the flow with a soap-film flowmeter. The flow verification was performed prior to each test to account for possible drift in the calibration of the mass flow controller.

The total exhaust volume determination was derived from the CVS flowrate monitored throughout the test cycle. If a BMD sampling system is used, exhaust volume must be determined by the exhaust flow measurement system.

The error listed in Table 2 is a comparison between the injected volume vs. the measured volume. When performing a standard propane injection test, ± 2 % error is used as the acceptance criteria. According to the CFR, the cause for any discrepancy greater than ± 2 % must be found and corrected. The same criteria was followed when testing SF_6 injection using the AIA-722 analyzer. As shown in the data, the results of the injection tests are within specification.

It is important to note that when examining the initial test data, the concentration of SF_6 measured by the AIA-722 is lower than optimal conditions. This concentration level was applied in these experiments to simulate extreme dilution conditions. The optimal performance of the analyzer is 80 to 90 percent of the full-scale value. Therefore, optimal values are desired near 800 to 900 ppm for SF_6 recovery concentrations.

<u>In Situ –</u> The next progression in testing was to insert a vehicle into the model. These tests were performed to provide baseline data for verification in the SULEV test site. The objective of this test was to obtain the data by running the test under actual vehicle conditions. The mixing unit was installed between the tailpipe and the CVS connection as shown below in Figure 5. The same concentrations of SF_6 were injected during the first phase of an LA-4 test cycle. The results are shown below in table 3.

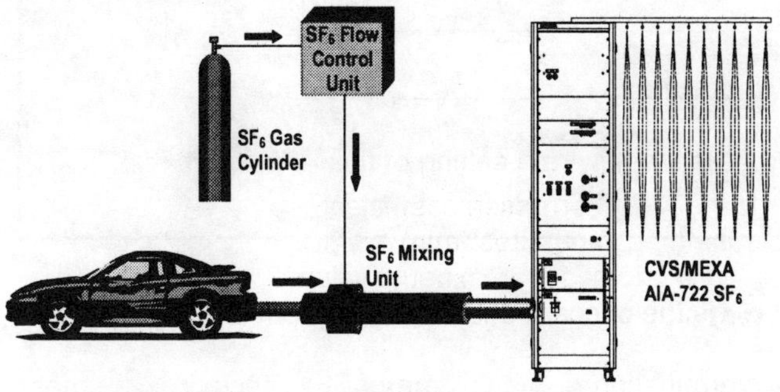

Figure 5: Equipment Layout for Injection Test with vehicle

MFC Set (mL/min.)	MFC Corr. Coef. αQ	Time (s)	V_{mix} (m^3)	SF_6 Conc.(ppm)	Vol $_{MFC}$ (mL)	Vol $_{CVS}$ (mL)	Error (%)	
500	1.0446	515	167.21	26.30	4483.1	4397.6	***1.94***	%
750	1.0492	510	165.06	41.21	6688.7	6802.1	***-1.67***	%
1000	1.0506	515	167.11	52.92	9017.7	8843.5	***1.97***	%

Table 3: Results of Vehicle Injection Tests

The same calculations were performed as in previous tests. The MFC flowrate was verified and the calculated recovery results were compared to the injected mass. The percent error was within the required tolerance for the standard propane injection test. This data provided a baseline for the next progression in this investigation. The next step is to test the BMD in conjunction with the CVS in a SULEV test site. The traditional CVS system, which is the industry standard, will be used as a standard for this investigation.

<u>*SULEV test site –*</u> This test cell was manufactured and assembled over the course of this investigation. Figure 6 shows the SULEV test site as it exists today. The equipment used is as follows:

Sample Handling System:

- Horiba CVS
- Horiba BMD
- E-Flow exhaust flow measurement device

Analytical Equipment:

- MEXA-7200SLE
- MEXA-7500DEGR
- MEXA-7100D

Host Computer:

- Vehicle Emissions Test System

This test cell was designed to provide test data using both the CVS and the BMD. The development tests run using both of the sampling systems and the resulting calculations will be furnished using the host computer. The host computer provides data for both the CVS and BMD injection tests and corrects for the bleed-off of the BMD. The results of the injection tests using both the BMD and the CVS are illustrated in Table 4.

A	B	C	D	E	F		
MFC Set	MFC Corr.	Time	CVS flow rate	BMD SF_6 Mass	CVS SF_6 Mass	BMD vs. CVS Error (%)	
mL/min	Coef. αQ	(s)	(scfm)	(grams)	(grams)	(E-F)/F*100	
2000	2061	300	175	63.966	64.034	***-0.11***	%
2000	2061	300	175	64.064	64.107	***-0.07***	%
2000	2061	300	175	63.627	63.889	***-0.41***	%
3000	3064	300	175	94.429	95.119	***-0.73***	%
3000	3064	300	175	94.203	94.918	***-0.75***	%
3000	3064	300	175	94.406	95.112	***-0.74***	%

Table 4. SF_6 BMD/CVS Test Data

EXHAUST FLOW MEASUREMENT – Since it's inception, it has been discussed that the SF_6 analyzer will be used for other verification tests, above and beyond recovery tests. Figure 7 shows the engine and exhaust system for a vehicle, and the injection points: A, B and C. Sampling points are indicated by points D and

E. There is concern that not enough information is available on SF_6 gas and its possible byproducts created during the combustion process of the engine. The purpose of the next series of tests is to check the stability of the SF_6 when it is injected into various points along the exhaust stream. The different concentration of byproducts and SF_6 between injections at points A and B will be directly attributable to SF_6 reactions during combustion. Any differences in the measured gas components between injections at points B and C will correspond to hang-up or reaction in the catalyst.

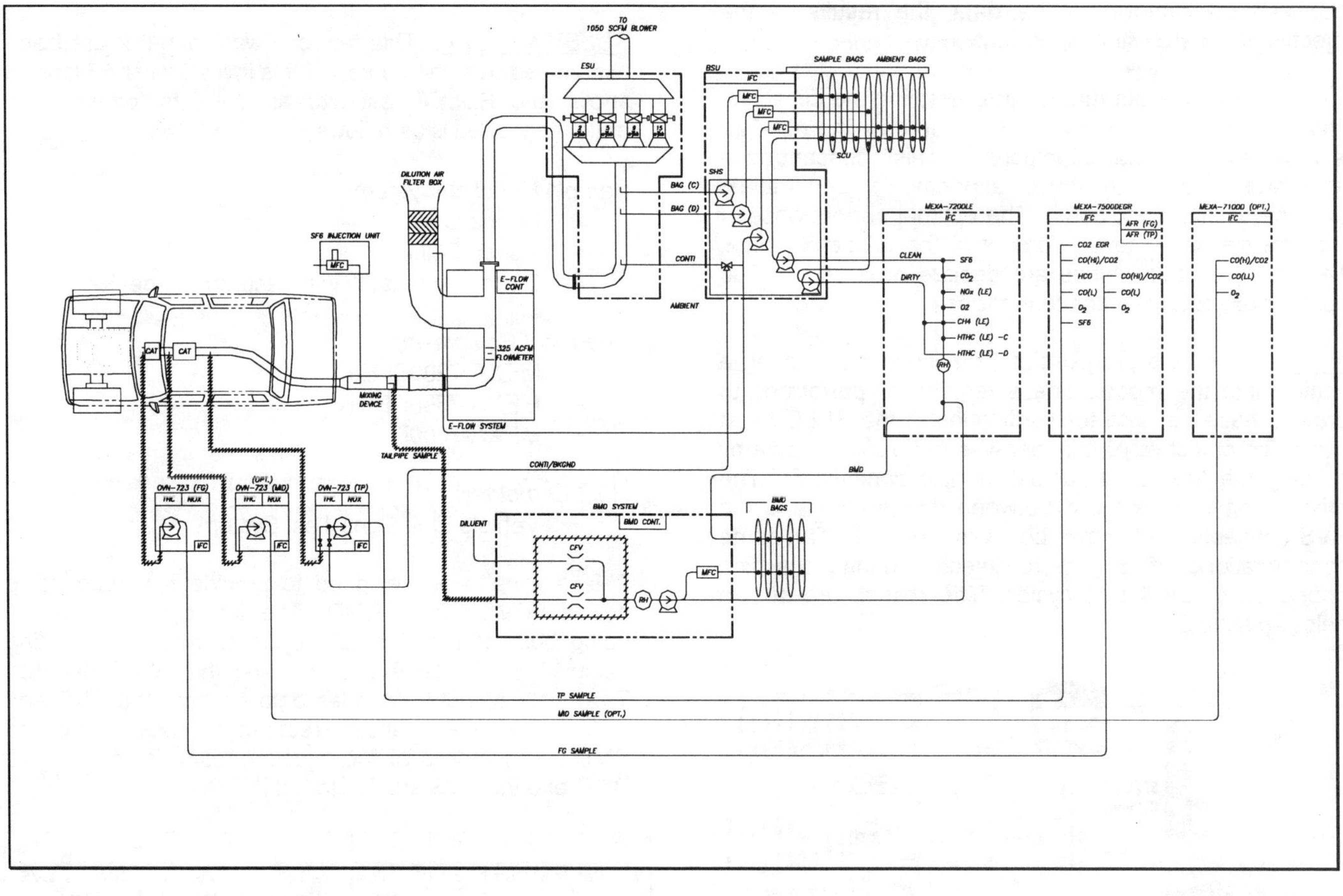

Figure 6: SULEV test site

The first set of tests is being performed while injecting a known concentration of SF_6 into the intake manifold of the vehicle. The gas is injected at point A. It is believed that the SF_6 may produce SO_2 and other byproducts during the combustion process.

<u>*Intake Manifold Problems –*</u> Initial tests involved measuring the CVS collected sample with SF_6 and SO_2 NDIR analyzers, with the vehicle at idle. Injections at point A resulted in a drastic increase in SO_2 concentrations at SF_6 flow rates as low as 750 mL/min. The SO_2 concentration during SF_6 injections was out of range of the SO_2 analyzer (300 ppm full scale) at SF_6 flow rates greater than 1000 L/min. SO_2 concentrations of 185 ppm and 144 ppm were measured for SF_6 flow rates of 1000 and 750 mL/min, respectively. SF_6 concentrations ranged from 36.7 ppm at 3000 mL/min to 11.7 ppm at 750 mL/min.

When injection point C (tailpipe) was used to inject SF_6, the SO_2 readings were all approximately 5 ppm. The SO_2 concentration was 4.6 ppm even when no SF_6 was injected. The SF_6 concentrations were measured at 721.8 and 461.9 ppm for injection rates of 3000 and 2000 mL/min, respectively. These tests possibly indicate that the breakdown of SF_6 in the engine and in the catalyst is a significant hindrance to using SF_6 in determining exhaust flow rates when injecting at the vehicle intake manifold. If SF_6 did not break down, measurement of the concentration at point E would readily allow the user to measure exhaust flow rates using this method.

To measure other byproduct components beyond SO_2, it will be necessary to use a multi-component analyzer. The MEXA 4000FT has the ability to measure a number of constituents simultaneously. NH_3 and SO_2 are two of these components, which may be measured. The MEXA 4000FT operates by measuring infrared energy absorption of the sample gas by using an interference technique and resolving the energy into its constituent frequency components by Fourier analysis (FTIR).

A second series of tests will be performed in future investigations while injecting a known concentration of SF_6 before the catalyst. The MEXA 4000FT will be used to measure the exhaust gas at point C.

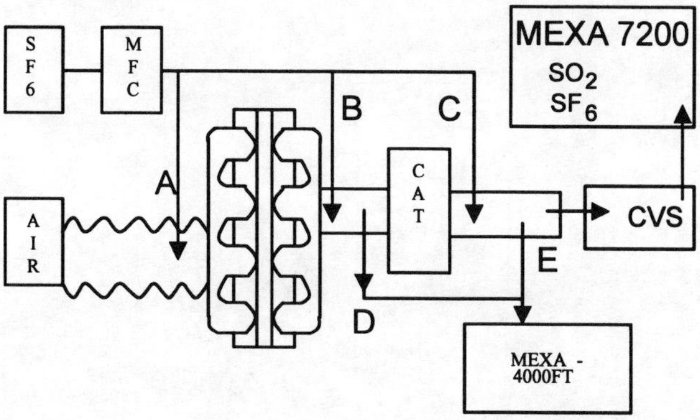

Figure 7: Exhaust flow tests

During the course of this investigation, preliminary tests were run injecting the SF_6 into the intake manifold of the vehicle. Figure 8 shows the exhaust gas spectrum output from the MEXA 4000FT. Concentrations of SO_2 and NH_3 were observed close to 100ppm, using a gasoline engine test vehicle.

The test results indicate that the SF_6 might have been combusted by the catalyst or the engine. The SF_6 seems to be detected at a constant concentration regardless of the changing concentration of SO_2. High SO_2 concentrations were still measured after shutting off the SF_6 injection gas. SO_2 produced in the engine may have been absorbed on the catalyst and released significantly after its production in the engine.

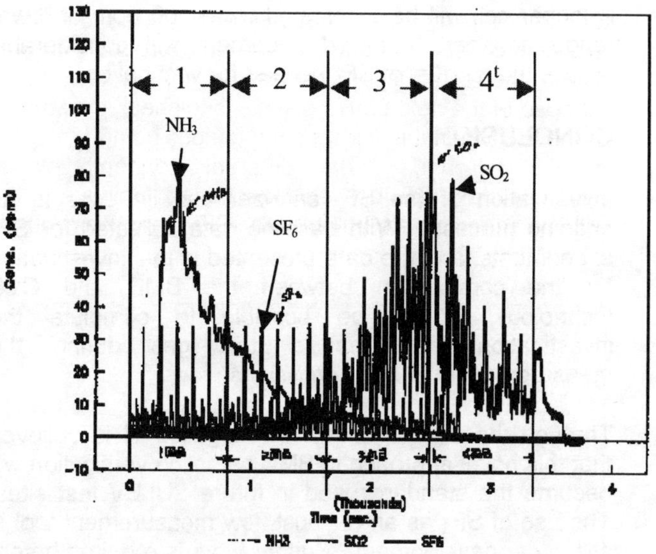

Figure 8: Exhaust spectrum for MEXA 4000FT

Tracer – The SULEV test site contains a second AIA-722 SF_6 analyzer to measure emissions from the tailpipe. The MEXA 7100D analysis system has been equipped with this analyzer for the purpose of researching the possibilities of using this technology for measuring the exhaust flow rate.

Previously, two other approaches utilizing a "Trace Method" have been investigated. These are the Helium trace method [3] and the CO_2 tracer technique [4].

In future studies, the SF_6 concentration measured at the tailpipe will be compared with the SF_6 concentration in the diluted exhaust stream to obtain the exhaust flow rate.

Lower Concentrations – SF_6 gas is an unregulated gas with practically no atmospheric background. SF_6 has very strong IR absorbance characteristics. It was because of this characteristic that it was determined to be a useful part of the verification in the SULEV test Site. At this time, experimentation is continuing with this gas for the purposes of reducing the equipment required to run and verify the integrity of a SULEV test site. It is part of an industry wide trend to reduce the physical size of the test site itself. In the future, it may not be necessary to use a CVS which consists of Exhaust Sampling Unit (ESU), a Bag Sampling Unit (BSU), a Remote Mixing Tee (RMT), and a high powered blower. The BMD technology requires considerably less space and facilities management for installation.

Further development will include investigation of an NDIR analyzer able to measure lower concentrations. Currently, the AIA-722 incorporates a 10 mm cell limiting the measurement accuracy to approximately 1000 ppm.

A longer cell will be prototyped for a 100 ppm or lower range analyzer. This development will considerably reduce the amount of SF_6 needed for verification.

CONCLUSION

Investigation of the SF_6 analyzer and its uses is an ongoing process. With baseline data collected for SF_6 experiments, and the data presented in this investigation for the comparison between the BMD and CVS technology, it will be possible to complete the investigation for uses of this gas during the measurement of SULEV emissions.

Through the course of this investigation, it is believed that this novel approach to BMD sampling verification will become the standard used in future SULEV test sites. The use of SF_6 as an exhaust flow measurement tool is still under development. Further work is required before this application can be validated.

ACKNOWLEDGMENTS

The authors would like to acknowledge the most helpful contributions of Arthur Maczei, Mike Akard, Bill Boles, and Kazuya Tsurumi of Horiba Instruments, Inc. We would also like to thank all of those involved in the testing and the commissioning of Horiba Instruments, Inc. most advanced SULEV test site. They should be commended on their teamwork and incredible effort.

REFERENCES

1. W. Herget, J Staab, H Klingenberg and W Riedel: "Progress in the Prototype Development of a New Multicomponent Exhaust Gas Sampling and Analyzing System" SAE paper 840470.
2. W. Wiers and C. Scheffler: "Carbon Dioxide (CO2) Tracer Technique for Modal Mass Exhaust Emission Measurement" SAE paper 720126
3. M. Adachi, T. Hirano, K Ishida, C. Cepeda, Y. Nagata, A. Kubo, and S. Nakamura: "Measurement of Exhaust Flow Rate: Helium Trace Method with a Mass Spectrometer"

APPENDIX

SF6 Specification Sheet

Analyzer Type	AIA-722	10mm Cell length
Range	SF_6 ; 0 – 1000ppm(Only)	
Repeatability	Within +/- 0.5% of FS for Zero & within +/-0.5% of point for span	
Response time	T90 ; within 1.0sec T98 ; within 2.0sec	
Zero Drift	Within +/-1.0%FS / 24hrs	
Span Drift	Within +/-1.0%Pt / 24hrs	
Noise	Within 5ppm	p-p for 5min(except for 10Hz vibration)
Linearity	+/-1.0%FS or +/-2.0%Pt	
Interference	Within +/-0.5%FS	H_2O(25degC Sat.)
	Within +/-0.5%FS	16vol%CO_2
	Within +/-0.5%FS	10vol%CO
	Within +/-0.5%FS	1000ppmNO
	Within +/-0.5%FS	200ppmSO_2
	Within +/-0.5%FS	50ppmC(w/ C_3H_8)
	Within +/-0.5%FS	200ppmCH_3OH
	Within +/-0.5%FS	200ppmNH_3
	Within +/-0.5%FS	200ppmC_2H_4
Temp Condition	5 through 40degC	
Drift by Temp	Zero ; within +/-1.0%FS / 10degC Span ; within +/-1.0%Pt / 10degC	Between 20 and 30degC
Humidity	85% RH or less	
Vibration	Within +/-1.0%FS	except for resonance point
Dimension	76(W) X 400(D) X 160(H)	
Weight	3kg	
Power supply	DC24V, DC+/-15V, DC5V	
AC	DC24V-60VA, DC+/-15V-2VA, DC5V-1VA	
Warming Up1	2hrs (drift ; within +/-1.0%/30min)	
Warming Up2	8hrs (drift ; within +/-1.0%/24hrs)	
Sample Pressure	25+/-2.5kPa	2.0L/min
Sample Flow rate	Within +/-1.0%Pt(Span)	2.0+/-1L/min
Back Pressure	Within +/-1.0%Pt(Span)	+/-100Pa